AGGRESSION, HOSTILITY, AND VIOLENCE

Nature or Nurture?

Edited by
TERRY MAPLE
University of California at Davis

and

DOUGLAS W. MATHESON
University of the Pacific

HOLT, RINEHART AND WINSTON, INC.

New York Chicago San Francisco Atlanta
Dallas Montreal Toronto London Sydney

PREFACE

Modern man is a believer in the power and utility of aggression, or so it seems. Perhaps it has always been so. Some use it as their primary means of acquisition. For others it is a last resort—a testimony to their implicit belief that "if all else fails, *fight* for it." To fight! Is this what man lives for? Is man by nature aggressive and cruel? Some think so. In fact, when we surveyed our students recently an impressive number of them agreed that "aggression is natural," and "we will always have wars." Despite such pessimism (and one has reason to believe that it is even more widespread among the general public), Americans spend millions of dollars each year in the fight against crime. Similarly, a large segment of the population still seems to support the idea that we can end wars by fighting them. It is a strange kind of ambivalence. But the confusion does not end with the lay public. Scientists and their literary and philosophical counterparts have long argued over the question of why men fight.

We designed this volume originally to fit a seminar course on the psychology of aggression. Not far into the task it became apparent that there was a rich literature in this field produced by many disciplines. We have included, therefore, a wide variety of work both old and new. Were there room we would have traced the problem chronologically but that was not possible. However, we have included some early papers which are fairly representative of their time. Many of the ideas in these papers are just as vital and important today as they were 20, 30 or 50 years ago when they were first published. We have provided explanatory footnotes where dated statements require some elaboration. The background covered is purposely broad. Both animal and human aggression is represented but the former always has definite implications for the latter. After all, it is *man's* peculiar problem with which we are most concerned. No one is really worried about the outcome for other animals since it is highly unlikely (as you will see) that rats, or lions, or elephants will exterminate themselves. In every case, research with animals other than man illuminates *our* problem and contributes to the development of general theories of aggression and its function in life.

Since we assume no bias in presenting this work, no theoretical or methodological framework is intended to dominate. We have chosen rather to present some of the many findings of behavioral science in general. We contend that there is no "right" or "wrong" approach to the study of aggression. Researchers have traveled in many different directions, all of which are likely to contribute important data to the research literature. We encourage the reader to draw his own conclusions after he has carefully examined each chapter.

The unifying theme which we have tried to adhere to is the so-called "nature-nurture" question. Aggression is a most popular—and fitting—focal point for this discussion as expressed in the question, "Does man fight (and kill) his own kind because he harbors an innate (built-in) predisposition for such behavior, or is it learned/cultural factors that account for its expression?" Prevailing scientific opinion has favored each of these extreme positions at various times. The European ethologists have done the most to reopen the question. One need only be familiar with the popular works of Konrad Lorenz [*On Aggression* (New York: Harcourt Brace Jovanovich, 1966)] and Robert Ardrey [*African Genesis* (London: Collins, 1961) and *The Territorial Imperative* (New York: Atheneum, 1966)] in order to recognize that the controversy today is still very much alive.

We are well aware that some readers will object to the dichotomy which we have suggested. Favoring one side or the other depends largely on the position from which one is looking. That all behavior, including aggression, reflects an interaction of hereditary and environmental elements is an axiom we are unwilling to challenge. However, as long as efforts to quantify the relative effects of heredity and environment continue, and behavioral scientists choose to argue over the extent to which human aggression, and other behaviors, is governed by internal and external variables, we believe that it is appropriate, indeed, *necessary* to provide the reader with relevant information required for an evaluation of the evidence.

In our selection of papers we have not been able to include all of those that we favor. This means that some important work has, inevitably, been excluded. To help counter this we have included some annotated references accompanying each of the four sections of the book. It is our hope that this will aid the reader in his search for further and more detailed information. The first of the four parts supplies the material for a proper understanding of the kind of theory and research that has been generated in this area. As these parts may be considered the "cause" portion of the book, so the final part may provide the "cures." This final part is very optimistic. The fact that most scientists believe there are solutions and alternatives to aggression gives one hope. We believe that man can be *tamed* if he will consent to it, for he is the only animal that may choose to tame himself and thus insure his survival.

Davis, California   T. M.
Stockton, California   D. W. M.

ACKNOWLEDGMENTS

The editors wish to thank the authors and publishers of the works herein who have been most cooperative in making their material available to us. We are grateful to Gwenda Dalman who secured many of the permissions and assisted us in many other ways. Doctors Robert L. Baird, Kenneth Beauchamp, Martin Gipson, W. Michael Kaill, and Gary Mitchell are remembered for their many conversations (both in and away from the academic scene) which helped to formulate our early interest in the subject matter. One of us (T. M.) is particularly grateful to Adelaide Petersen for her many contributions to the manuscript and continued encouragement throughout its preparation. Of course, we are happy to acknowledge the comments and criticisms of our students who struggled with our first attempts to establish a seminar course on aggression in 1968. Their interest and enthusiasm helped create this volume.

CONTENTS

Preface   iii
Acknowledgments   vii

Introduction to the Scientific Study of Aggression   1

PART 1 NATURE THEORY   **11**

1. Why War?—*Sigmund Freud*   16
2. The Predatory Transition from Ape to Man—
   Raymond Dart   27
3. Ritualized Fighting—*Konrad Lorenz*   44

PART 2 NURTURE THEORY   **59**

4. Why Do Animals Fight?—*Wallace Craig*   65
5. An Anthropological Analysis of War—
   Bronislaw Malinowski   76
6. The Frustration-Aggression Hypothesis—
   Neal E. Miller   103

PART 3 ANTECEDENT CONDITIONS:
   Recent Research   **109**

7. Some Patterns in the History of Violence—
   Frank H. Denton and Warren Phillips   114
8. The Stimulating vs. Cathartic Effects of a
   Vicarious Aggressive Activity—*Seymour Feshbach*   132
9. Weapons as Aggression Eliciting Stimuli—
   Leonard Berkowitz and Anthony Lepage   141
10. Pain as a Cause of Aggression—*Roger Ulrich*   152
11. Studies on the Basic Factors in Animal Fighting:
    VII. Inter-Species Coexistence in Mammals—
    Zing Yang Kuo   182
12. Ghetto Sentiments on Violence—*Harlan Hahn*   197

                                                                        ix

PART 4   MODIFICATION AND CONTROL:
   Alternatives to Aggression, Hostility, and Violence   **209**

13. The Reduction of Intergroup Hostility: Research
    Problems and Hypotheses—*Irving L. Janis and Daniel Katz*   213
14. Modification of Severe Disruptive and Aggressive Behavior
    Using Brief Timeout and Reinforcement Procedures—
    Darrel E. Bostow and J. B. Bailey   236
15. Treatment of Aggression: Aggression in Childhood—
    Lauretta Bender   248
16. The Moral Equivalent of War—*William James*   258
17. The Psychiatric Aspects of War and Peace—   269
    Franz Alexander   269
18. On War and Peace in Animals and Man—*N. Tinbergen*   284
19. Possible Substitutes for War—*Anthony Storr*   306
20. A Systematization of Gandhian Ethics of Conflict
    Resolution—*Arne Naess*   315
21. Three Not-So-Obvious Contributions of Psychology
    to Peace—*Ralph K. White*   338
22. The Convict as Researcher—*Hans Toch*   356

Name Index   363
Subject Index   366

Introduction to the Scientific Study of Aggression

—TERRY MAPLE

Scientists have only recently begun to investigate the phenomenon of aggression. Traditional sources of knowledge have filled the void, however, and we are able to call upon a massive literature devoted to the problems of fighting, hatred, and warfare. The poets, philosophers, theologians, and political thinkers have not been without insight over the years. However, their arguments have almost without exception been heedless of scientific principles. The most notable exception is the work of Charles Darwin. In his well-known volume *Expression of the Emotions in Man and Animals* (1872), Darwin devoted an entire chapter to the topic of hatred and anger. Moreover, he enumerated other allied emotions which we now know to be important in the continuum of responses known as *agonistic*[1] behavior. Among these (as listed by Darwin) are ill-temper, sulkiness, determination, disdain, contempt, disgust, guilt, pride, fear, and many others. Although Darwin did not concern himself with the problem of aggression *per se*, and his book did not progress much beyond description, he influenced later work in this area in at least two ways: (1) He made a strong attempt to study behavioral phenomena *objectively*, and (2) he emphasized the acquisition of complex behavioral patterns through evolutionary mechanisms. The former would have great importance for the new science of behavior, which was yet to appear. The latter provided the community of scholars with an expanded subject population. The door was therefore left open for aggression in man. As Darwin suggested, "He who admits on general grounds that the structure and habits of all animals have been gradually evolved will look at the whole subject of expression in a different light." For our purposes, this statement takes on even greater meaning if we substitute the word "aggression" in place of "expression." In this century, some of the most important contributions to the scientific literature on aggression have been provided by experimenters in the field of comparative animal behavior. It is my objective to intro-

1

duce the reader in this Introduction to some of the major problems encountered by contemporary behavioral scientists who study aggression. By discussing some of the things that these scientists do, we will be in a better position to reflect on some of the things that need to be done.

DEFINITIONS

A major concern of those who study aggression is the overabundance of terms used in the research literature. There is considerable overlap in these terms, and distinctions are often subtle and confusing. *Behavioral* definitions are generally the most precise. For example, Buss (1961) defines aggression as a "response that delivers noxious stimuli to another organism." This definition is rather narrow, and does not include any references to states of "mind" or attitude. Used in this way aggression is easy to specify and measure in an experiment. It is interesting that Darwin (1872) seems to have had a behavioral definition in mind when he said:

> Most of our emotions are so closely connected with their expression that they hardly exist if the body remains passive—the nature of the expression depending in chief part on the nature of the actions which have been habitually performed under this particular state of mind. A man, for instance, may know that his life is in the extremest peril, and may strongly desire to save it; yet, as Louis XVI said, when surrounded by a fierce mob, "Am I afraid? Feel my pulse." *So a man may intensely hate another, but until his bodily frame is affected, he cannot be said to be enraged.* [italics added]

The term *aggression* has also been used in referring to more wide-ranging phenomena. This generic sense is well expressed by Stone (1971).

> Aggression ranges far in psychoanalytic thought. It may include manifest bodily or verbal action; conscious or unconscious wishes and tensions; a specific qualitative type of psychic energy and the final broad and inclusive idea of the death instinct . . .

Carthy and Ebling (1964) have suggested that an animal "acts aggressively when it inflicts, attempts to inflict, or threatens to inflict damage on another animal." This definition agrees in substance with the one offered by Buss, but it goes a bit farther by extending aggression into the realm of *intention*. In the natural world of animals (that which has remained relatively undisturbed by human culture excesses) one observes two distinct forms of aggression, *intra*specific and *inter*specific. The former refers to fighting that occurs within a species group (i.e., cat vs. cat) while the latter denotes between-species combat (cat vs. bird), which usually occurs in the context of predation (killing for food).

Other terms that come to mind when we begin to think about aggression are "hostility," "anger," "hatred," "violence," and a host of others. For experimental purposes each of these terms must be defined precisely by the experimenter. Likewise, events of historical origin are assigned labels according to conventional usage. In anthropology distinctions are made between levels of conflict; some events may be classified as *wars* while others are termed *armed conflicts* (Lesser, 1968). The layman, however, tends to use these terms interchangeably. For the average citizen casually reading the morning newspaper, there is not much difference between aggression and violence. Our average citizen is also likely to inquire as to the meanings of hostility and anger. Are they really so different? The form of the word can be another source of misunderstanding. For example, a salesman is often described as "aggressive," but this use differs greatly from that in an account of the "aggressive" activities of a nation-state. One leads to success in business, the other to widespread death and destruction. We are therefore led to inquire into the relationship between aggressiveness and aggression. Some writers solve this problem by using the term "aggressive" to describe an attitude, whereas "aggression" is reserved for an event such as a fight or a war. In this sense, a man may be aggressive in his everyday activities, but he need not be involved in an *act* of aggression. Ghandi, therefore, could be described as an aggressive exponent of civil liberties, but as an advocate of nonviolence he did not take part in individual acts of aggression. It is worth noting that a legitimate distinction can be made between *assertiveness* and *animosity* both of which may be involved in the behavior of an "aggressive" person (see De Monchaux, 1964). Since the present discussion requires some internal consistency, I will use aggression to refer to the general subject area; that is, aggression research may be thought of as research dealing with collective or individual fighting behavior in man and animals and with all those emotional states which may accompany it.

SUBJECTS

Studies of aggression are conducted with both human and nonhuman animal subjects. Human studies employ subjects from a number of different populations, many of which are chosen because of convenience. Selecting a sample for this reason is obviously hazardous. If the research is meant to be generalized to the population at large, the sample should be selected in such a way as to minimize bias. Sampling on the basis of convenience pays no attention to this requirement. Most research on human aggression is conducted with college students, children, and mental patients as subjects. If studies are designed to evaluate a treatment or disciplinary technique (with reference to the latter two groups, for example) it is not necessary to worry about generalizing further than the sample under study. Research with college students as subjects, however, is often cited as evidence for a more general hypothesis or theory. Because of the acknowledged uniqueness of this group it should not be considered an adequate research sample on which to base general principles of behavior. This is especially true of aggression studies. In light of recent events on many college campuses it is possible that the widespread use of student subjects in aggression research will be more carefully scrutinized in the future. Of course, a study that is concerned with those variables responsible for *campus* violence is not objectionable on these grounds. If some segments of the general population express aggression differently (students, "hard-hats," military career-men, etc.) it is up to future research to determine these differences. The critical student of research would be wise to pay close attention to the choice of sample in every case.

The majority of nonhuman aggression studies use rats and mice as subjects. Fighting can be readily induced in these animals and they are easily cared for and observed. The experimenter who attempts to derive more comprehensive principles will use a greater variety of species in his experiments. This is a difficult but not impossible task. Recording equipment is often not designed for unusual species and many types of animals cannot be easily obtained for research purposes. With a little perserverence, however, these obstacles can be overcome, and it is frequently rewarding to study an unusual species. One which has been rarely studied (if ever) is likely to exhibit behavior that is unique to its own kind. Occasionally, the observation of an animal previously unexamined will lead to a restructuring of existing theories. For many years it was believed that man was the only tool-user, but recent work has shown that other animals use tools in their natural habitat (see Eibl-Eibesfeldt, 1970, Chapters 14 and 15). It is highly unlikely that chimpanzees, for example, learned how to use tools in the past ten years. How long the phenomenon existed before it was discovered no one knows. Since most of our behavioral principles are built upon a foundation supported by the slender "appendages" of rats and pigeons, it is important that we expand our work to include as many different species as possible. The efforts of Roger Ulrich (see Reading 10), N. H. Azrin, and their colleagues illustrate what can be accomplished if many species are utilized in the investigation of behavior.

METHODOLOGY

The two approaches that dominate the scientific study of aggression are *naturalistic observation* and *laboratory experimentation*. They are not mutually exclusive but for any particular problem one or the other is usually preferred. The former approach requires repeated observations of the subject in its natural habitat. The organism and its environment are thought of as a unified whole, a gestalt. This method is exemplified by the science of ethology, a branch of biology. Although ethologists have traditionally used "lower" organisms in their investigations, they have recently begun to apply their techniques to the "higher" primates (for example, Morris, 1967) including man (Eibl-Eibesfeldt, 1970; Grant, 1965). As the natural habitat of man is highly variable, ethologists have begun a concentrated effort to understand man as he behaves in the many environments in which he lives, adding to the already considerable information accumulated by anthropologists over the years.

A skilled observer of animals, the ethologist usually attempts to be as objective as possible. Interaction with the organism is avoided except when it is a stated objective of the research. Since the scientist is going into the subjects' domain, he is maximizing the chances of recording "natural" behavior under "natural" conditions; that is, behavior which would occur whether the observer were present or not. The "tools" of the naturalist are patience, a disciplined eye, and an objective pen. Unfortunately, this latter requirement has not always been adhered to, leading to misunderstandings and discrepancies in the literature (see Montagu, 1970, for a more detailed discussion of this point).

Naturalistic observation may be contrasted with the experimental approach. Experimentalists derive their notions from the mechanistic school of behavioral science, placing emphasis on the cause-effect relationship and attempting to break down behavior into its component parts. The experimental approach is really an extension of naturalistic observation, since experimentalists must also rely on careful observation. Experimental work usually employs animal subjects that are adapted to laboratory life (for example, the Norway rat) but occasionally other types of animals are experimented upon in semi-naturalistic environments constructed by the experimenter.

One difference in the two approaches is that the experimentalist tends to use more sophisticated "hardware" in his investigations. However, a more important difference is the use of *controls*. Experimental studies generally employ what is known as the *control* group. This technique allows the researcher to compare his experimental group to a similarly constituted group (control) that differs only with respect to the manipulations. If we wish to learn about the affect of alcohol on aggression, for example, we might use the following procedure: (1) Random assignment of subjects to each of two groups, ideally of the same size; (2) Ingestion of alcohol by the experimental group, of water by the control group; (3) Observation and collection of data. Our manipulations (amount of alcohol ingested) should give us some differences in behavior and these are attributable to the manipulations we have performed. Although this example is somewhat oversimplified it illustrates the procedure that is generally followed in such experiments. Experimental work attends to more segmented aspects of behavior. Such phenomena can be reliably extracted under laboratory conditions. The naturalist, on the other hand, must contend with a myriad of variables, making it difficult to pinpoint specific causes.

MEASUREMENT

Regardless of methodology, the scientist must devise an accurate way to measure his findings. Aggression takes many forms (depending, of course, on the definition of the term) and because of this variability many kinds of measurement techniques have been utilized. One may use both physiological and behavioral measures of aggression. Physiological changes in heart rate, blood sugar levels, and respiration may be correlated with aggressive behavior (including verbal behavior) or intention movements. Such measurements can be made easily enough in the laboratory. Behavioral measures may be taken in the laboratory or in the natural environment of the organism under study. A useful technique for recording behavioral measures of aggression is motion-picture filming. With films a permanent record of the behavior can be made, allowing others to verify the observations. This is a particularly important advance when the organism under study is not easily accessible for replications of the original research.

In the experimental jargon aggression is usually a *dependent variable*. It is the behavior one is interested in observing and recording; that which is affected by manipulations; that which is measured. It is measured by both direct and indirect means. For example, the number of blows struck, incidence of physical postures or expressions, or the number of shocks administered to another subject may be defined as the relevant dependent variable. Indirect means of detecting aggressive tendencies are personality inventories such as the Minnesota Multiphasic Personality Inventory (MMPI), or projective tests such as the Rorschach inkblot test and the Thematic Apperception Test (TAT). A number of different tests have been used through the years, and it is difficult to compare results obtained with one test to data extracted by other methods. Similarly, a variety of responses that may all be labeled aggressive are capable of occurring under the same conditions. Thus a frustrating event arranged by the experimenter may elicit profanity, blushing, screaming, stamping of feet, striking, or responses correlated with aggression on a written test. Which response will occur is due in large measure to the circumstances in which the experiment is arranged.

The measurement of nonhuman behavior is considerably easier than the measurement of human behavior since the nonhuman subject usually is smaller and its behavior rather stereotyped and more easily controlled. Aggression can be readily produced in nonhuman subjects as ethics do not necessarily restrict the intensity of the conflict, nor the means by which aggression may be elicited. Although the many variables that contribute to animal fighting have been investigated in the laboratory, the best accounts of nonhuman combat have been provided by ethologists. However precise the laboratory fight may be, it lacks the spontaneity and the variety of the natural encounter. On the other hand fights in the wild happen so rapidly that they are nearly impossible to study without the aid of elaborate recording equipment. Some animals fight so rarely that a camera is most important if those variables which elicited the encounter are to be discovered.

DIRECTIONS

Aggression is by no means an easy form of behavior to study. One encounters at the outset a long tradition of mythology and belief associated with "man's inhumanity to man." It is a rather emotional issue as well, surrounded by controversy. Is man evil? If so, what can we do about it? If not, why does he do so many evil things? Efforts to control aggression often meet with opposition because they invoke images of 1984. This is particularly true today in view of the popularity of armed revolutionary struggle throughout the *third world*. In our own country the National Rifle Association has mounted a powerful campaign against gun-control legislation. It is obvious that aggression can be quite useful to any organism, including man, when it comes to the acquisition of status, property, or mates. For all of these reasons it is difficult to make sense of all the information at our disposal, even more difficult to make intelligent decisions about the proper course of action.

In the face of this there is some reason to be optimistic. Many recent studies have indicated that some of the more troublesome forms of aggression can be controlled (see, for example, Reading 14) and scientists are only beginning to scratch the surface in this area. At the present time there seems to be a trend toward the construction of experimental treatment centers in order to put new techniques to tests in a natural setting. Ultimately, one can hope that the "dream" of nonviolent problem-solving will become a reality, but this is not likely without widespread training in nonviolent techniques. It is obvious that *working on the person* will not guarantee a problem-free society. Aggression is, in one sense, a reaction to environmental conditions. Even the most naturally passive animal can be induced to fight in self-defense (there are some exceptions, however, such as clams whose physical abilities are rather limited) and therefore efforts to create a peaceful environment must continue. Peace research centers and societies pledged to study social problems are a rich source of information for the student. With this acceleration of interest it is quite likely that new and provocative solutions to aggression and war will be proposed.

Since the literature on aggression is so diffuse, it will be necessary to bring investigators together in order to synthesize their findings. This has already begun on a limited scale. It would be particularly helpful if some standard definitions could be agreed upon. The organization of yearly symposiums, establishment of interdisciplinary research centers, and creation of journals (such as the newly formed *International Journal of Group Tensions*) and newsletters will do much to speed progress. There is much to be done. As you read and think about the papers that follow, more questions than answers are likely to arise. If this state of affairs creates a sense of urgency and a desire to know more and do more, then this book will have served a useful purpose.

REFERENCES

Buss, A. *The Psychology of Aggression*. New York: Wiley, 1961.

Carthy, J. D., and Ebling, F. J. (eds.) *The Natural History of Aggression*. New York: Academic, 1964.

Darwin, C. *The Expression of the Emotions in Man and Animals*. New York: Appleton, 1872.

De Monchaux, C. Hostility in small groups. In J. D. Carthy and Ebling (eds.), *The Natural History of Aggression*. New York: Academic, 1964. Pp. 83-89.

Eibl-Eibesfeldt, I. *Ethology: The Biology of Behavior*. New York: Holt, Rinehart and Winston, 1970.

Fried, M., Harris, M., and Murphy, R. (eds.) *War: The Anthropology of Armed Conflict and Aggression*. New York: The Natural History Press, 1968.

Grant, E. C. An ethological description of some schizophrenic patterns of behaviour. *Proceedings of the Leeds Symposium on Behavioral Disorders*, 1965, *12*:3-14.

Lesser, A. War and the state. In Fried, Harris and Murphy, *War: The Anthropology of Armed Conflict and Aggression*. New York: The Natural History Press, 1968. Pp. 92-96.

Montagu, M. F. A. (ed.) *Man and Aggression*. London: Oxford University Press, 1970.

Morris, D. (ed.) *Primate Ethology*. Chicago: Aldine, 1967.

Scott, J. P. *Aggression*. Chicago: University of Chicago Press, 1958.

Stone, L. Reflections on the psychoanalytic concept of aggression. *Psychoanalytic Quarterly*, 1971, *40*, 195-244.

FOOTNOTES

[1] This term has been used by J. P. Scott (1958) to describe a continuum of behaviors including fighting, escape, passivity, and defense.

PART 1   Nature Theory

Man is a predator whose natural
instinct is to kill with a weapon.
  Robert Ardrey: *African Genesis*

In this section we use the term "nature" to group together those views of behavior which, though considerably different in detail, can be collectively juxtaposed with the nurture position. This position (nurture) we will describe in greater detail in the next section of readings. If we argue by exclusion, then, we may say that nature theory encompasses those ideas which reject culture or learning as factors in the development of an aggressive drive. That is, to a nature theorist it is sufficient to say that aggressive behavior is inborn rather than acquired. In its extreme sense this argument portrays nature variables as those that are strictly internal rather than experiential. This inner biological force is selected for (through phylogenetic evolution) and passed on (through patterns of inheritance) to the offspring of the species. Frequently used to describe this inner force are the terms "innate" and "instinctive." In this regard Troland's (1928) definition, reported by Cofer and Appley (1964), is instructive:

It seems possible on the basis of contemporary discussions to formulate an exact definition of the term, instinct, which will apply satisfactorily to all the uses to which the word is put. In a general way, it is agreed that an instinct is an inherited tendency to action of a specific kind, usually set off by a limited range of stimuli, and having definite survival or biological value—in the struggle for existence. As a rule, an instinct is conceived as a purely physiological mechanism, although the term may be used loosely as if it stood for a physical

11

force having a teleological or purposive form. The most mechanistic conception of an instinct regards it as being comprised of a group of reflexes or processes of a fixed type, energizing the muscles via the outgoing nerves.

The idea that man is ruled by inner urges is not a new one, but the controversy surrounding this issue has seldom been as vigorous as in recent times. In the early 1900's, particularly, battle lines were drawn between those who favored the classical instinct position and those who had become hard-core environmentalists (not, of course, in the ecological sense as used today). On the former side were men like William James and William McDougall. Both of these men, well-known behavioral scientists of their time, believed that man harbored many instinctive tendencies. Among these, according to James, were locomotion, vocalization, imitation, rivalry, pugnacity, sympathy, hurting, fear, acquisitiveness, constructiveness, play, curiosity, sociability, secretiveness, cleanliness, modesty, love, jealousy, and parental love. So far as our subject matter is concerned, nearly all of these instincts could be seen as playing a role in the development of aggression. Similarly, McDougall postulated in 1908 a list of major instincts which are even more strikingly related to the subject of human aggression. These ten instincts were as follows: flight, repulsion, curiosity, pugnacity, self-abasement, self-assertion, reproduction, gregariousness, acquisition, and construction. Both of these lists share the problem of covering such a wide range as to be all-inclusive and, needless to say, have been subject to attack from many quarters over the years.

Freudian psychoanalytic theory gave a boost to those who backed the concept of man's behavior being inner-determined. Freud himself wrote extensively on the question of psychically determined events that were directed from within the organism. In Freud's system aggression did not have to appear in its usual form to be verified. Frequently it was not directed to humans at all, but might find expression by being displaced to inanimate objects or household pets. Similarly, it was possible to direct aggression inward, the ultimate form of which was suicide. Most important in Freud's idea was the notion of energy. Men were driven, according to Freud, to behave in certain ways and the energy behind these urges had to find its expression in some form or other. Thus the energy potential for behavior was always present at some level. If the potential were high, then a behavioral outburst was more probable than if the potential were low. With all men possessing a fixed capacity for storing this energy it was merely a matter of time before they would be compelled to act. If men, then, retain this energy for aggression it is inevitable that aggression will occur, often spontaneously and in unusual forms. Although the idea put forth by Freud and his followers is intriguing, it is very difficult to verify empirically. It is for this reason that many of Freud's ideas have fallen from favor in the academic world.

More recently, the instinct concept has been revived and revised by the European school of ethology. Much of the early work with animals was conducted in Germany during the time that our "instinct" battles were raging in this country. However, because of geographical and language difficulties much of the literature was not readily available to the academic community in the United States. During these early years, it was sufficient for the pioneers of ethology to demonstrate the incidence of instinctive or innate behavior in a wide variety of non-human organisms. Konrad Lorenz, a leading spokesman for this group, played an influential role in establishing major hypotheses extracted from experimental work. Lorenz' behavior model, like that of his student Niko Tinbergen, is similar to the Freudian energy model which we have discussed. It is a hydraulic system that accounts for behavior through the effects of action-specific energy. This energy resides in the central nervous system and is released when the *innate releasing mechanism* (IRM), acting like a lock, is opened by the keyline *releasing stimulus*. For example, a releasing stimulus could be some physical feature of a conspecific, such as the color of its breast. In the event that the particular color is present, the IRM is unlocked and the energy is allowed to flow leading to a smooth and coordinated form of instinctive behavior.

The way that ethologists generally get at the question of innate behavior patterns is to raise an animal apart from others of its own kind (in isolation). If, in the absence of stimulation from species-mates, the animal shows the particular behavior unique to its species, the behavior is said to be innate or instinctive; that is, no learning or imitation appears to be necessary for that behavior to occur. We will not go into great detail regarding the criticisms of the isolation experiment, though there are many. It is sufficient to say that the assumption that isolation from species-mates eliminates the possibility of learning is subject to some question. The key to this objection is that we must consider the *kind* of environment produced by isolation. Regardless of whether the organism under study is completely isolated from all animal life or is simply raised in a foster home by a different species, particular kinds of stimulation are still present. In no case is an animal without some kind of environmental stimulation, particularly from itself.

In reviewing this section, particular attention should be paid to the paper by Raymond Dart. Considerable public notice has been given to two recent books by Robert Ardrey (*African Genesis* and *The Territorial Imperative*) whom we have mentioned earlier. Ardrey, in turn, has given much credit to Dart for influencing his (Ardrey's) work. If man's ancestors did (as both Dart and Ardrey believe) live a carnivorous existence and hunt with weapons, what implications does this have for the *evolution* of aggression in man? Can Dart's findings be offered as conclusive proof that our basic nature is "aggressive" or could these man-apes under less harsh environmental conditions have been more "peaceful"? The reader should keep this question in mind when he reads Dart's fascinating paper.

REFERENCES

Cofer, C. N., and Appley, M. H., *Motivation: Theory and Research*. New York: Wiley, 1964.

Troland, L. T., *The Fundamentals of Human Motivation*. New York: Van Nostrand, 1928. (As quoted by Cofer and Appley.)

SUGGESTED READING

This list and the others that accompany each section are extra sources of information which will assist the reader in gathering together more detailed information on the subjects of each section. The references following each research paper provide more advanced research material for investigation. The suggested readings are primarily in book form and are more general in scope than the individual research references.

Ardrey, R., *African Genesis*. London: Collins, 1961.

The author develops in great detail the thesis that man was preceded by predatory, club-wielding man-apes. According to Ardrey, modern man harbors an innate propensity for fighting his own kind.

Ardrey, R., *The Territorial Imperative*. New York: Atheneum, 1966.

Written as a sequel to *African Genesis*, Ardrey examines recent research in ethology and discusses territorial behavior in animals as it relates to the behavior of men and nation-states.

Carthy, J. D., and Ebling, F. J. (eds.), *The Natural History of Aggression*. New York: Academic, 1964.

This volume summarizes the proceedings of a recent symposium held in London. The papers cover ethological concepts and findings but contributions by anthropologists, sociologists, and psychologists are also included. Many interesting discussions follow specific presentations and some of the leading authorities in the field of animal behavior are represented.

Fletcher, R., *Instinct in Man*. New York: International Universities, 1957.

This book is a very detailed account of instinct (nature) theory as it relates to the behavior of man. It includes an account of early views on the subject as well as an attempt to synthesize contemporary findings into a unified theory of instinct.

Freud, S., "Thoughts for the Times on War and Death." *The Complete Psychological Works of Sigmund Freud*, Vol. XIV, edited by James Strachey. London: Hogarth Press, 1957.

This paper is one of the earliest discussions by Freud on the problem of aggression and war. Many of these ideas are later expanded in his famous letter to Einstein (see Reading 1).

Lorenz, K., *On Aggression*. New York: Harcourt, Brace and World, 1966.

Lorenz has written one of the most influential books on this subject. He develops, among others, the idea that man has no built-in inhibitors for his innate aggressiveness; hence, the difficulty in controlling himself. Considerable scientific evidence from ethological studies is applied to the behavior of man.

Storr, A., *Human Aggression*. New York: Atheneum, 1968.

This book attempts to relate the general findings of instinct (nature) theory to the behavior of man. Considerable psychiatric evidence is explored and interesting chapters on the relationship of aggression to behavior disorders are included. Storr also comments at length on suggestions for the reduction of hostility (see Reading 19).

1   Why War?

—SIGMUND FREUD

Vienna, September, 1932.

Dear Professor Einstein,

When I heard that you intended to invite me to an exchange of views on some subject that interested you and that seemed to deserve the interest of others besides yourself, I readily agreed. I expected you to choose a problem on the frontiers of what is knowable to-day, a problem to which each of us, a physicist and a psychologist, might have our own particular angle of approach and where we might come together from different directions upon the same ground. You have taken me by surprise, however, by posing the question of what can be done to protect mankind from the curse of war.[1] I was scared at first by the thought of my—I had almost written "our"—incapacity for dealing with what seemed to be a practical problem, a concern for statesmen. But then I realized that you had raised the question not as a natural scientist and physicist but as a philanthropist: you were following the promptings of the League of Nations just as Fridtjof Nansen, the polar explorer, took on the work of bringing help to the starving and homeless victims of the World War. I reflected, moreover, that I was not being asked to make practical proposals but only to set out the problem of avoiding war as it appears to a psychological observer. Here again you yourself have said almost all there is to say on the subject. But though you have taken the wind out of my sails I shall be glad to follow in your wake and content myself with confirming all you have said by amplifying it to the best of my knowledge—or conjecture.

You begin with the relation between Right and Might.[2] There can be no doubt that that is the correct starting-point for our investigation. But may I replace the word "might" by the balder and harsher

From Chapter XXV of Volume 5 of *The Collected Papers of Sigmund Freud,* edited by Ernest Jones, M.D., Basic Books, Inc., Publishers, New York, 1951. Used with permission of Sigmund Freud Copyrights, Ltd., The Institute of Psycho-Analysis, London, and The Hogarth Press, Ltd., London.

word "violence"? To-day right and violence appear to us as antitheses. It can easily be shown, however, that the one has developed out of the other; and, if we go back to the earliest beginnings and see how that first came about, the problem is easily solved. You must forgive me if in what follows I go over familiar and commonly accepted ground as though it were new, but the thread of my argument requires it.

It is a general principle, then, that conflicts of interest between men are settled by the use of violence. This is true of the whole animal kingdom, from which men have no business to exclude themselves. In the case of men, no doubt, conflicts of *opinion* occur as well which may reach the highest pitch of abstraction and which seem to demand some other technique for their settlement. That, however, is a later complication. To begin with, in a small human horde, it was superior muscular strength which decided who owned things or whose will should prevail. Muscular strength was soon supplemented and replaced by the use of tools: the winner was the one who had the better weapons or who used them the more skilfully. From the moment at which weapons were introduced, intellectual superiority already began to replace brute muscular strength; but the final purpose of the fight remained the same—one side or the other was to be compelled to abandon his claim or his objection by the damage inflicted on him and by the crippling of his strength. That purpose was most completely achieved if the victor's violence eliminated his opponent permanently—that is to say, killed him. This had two advantages: he could not renew his opposition and his fate deterred others from following his example. In addition to this, killing an enemy satisfied an instinctual inclination which I shall have to mention later. The intention to kill might be countered by a reflection that the enemy could be employed in performing useful services if he were left alive in an intimidated condition. In that case the victor's violence was content with subjugating him instead of killing him. This was a first beginning of the idea of sparing an enemy's life. But thereafter the victor had to reckon with his defeated opponent's lurking thirst for revenge and sacrificed some of his own security.

Such, then, was the original state of things: domination by whoever had the greater might—domination by brute violence or by violence supported by intellect. As we know, this régime was altered in the course of evolution. There was a path that led from violence to right or law. What was that path? It is my belief that there was only one: the path which led by way of the fact that the superior strength of a single individual could be rivalled by the union of several weak ones. *"L'union fait la force."* Violence could be broken by union, and the power of those who were united now represented law in contrast to the violence of the single individual. Thus we see that right is the might of a community. It is still violence, ready to be directed against any individual who resists it; it works by the same methods and follows the same purposes. The only real difference lies in the fact that what prevails is no longer the violence of an individual but that of a community. But in order that the transition from violence to this new right or justice may be effected, one psychological condition must be fulfilled. The union of the majority must be a stable and lasting one. If it were only brought about for the purpose of combating a single dominant individual and were dissolved after his defeat, nothing would have been accomplished. The next person who thought himself superior in strength would once more seek to set up a dominion by violence and the game would be repeated *ad infinitum*. The community must be maintained permanently, must be organized, must draw up regulations to anticipate the risk of rebellion and must institute authorities to see that those regulations—the laws—are respected and to superintend the execution of legal acts of violence. The recognition of a community of interests such as these leads to the growth of emotional ties between the members of a united group of people—communal feelings which are the true source of its strength.

Here, I believe, we already have all the essentials: violence overcome by the transference of power to a larger unity, which is held together by emotional ties between its members. What remains to be said is no more than an expansion and a repetition of this.

The situation is simple so long as the community consists of a number of equally strong individuals. The laws of such an association will determine the extent to which, if the security of communal life is to be guaranteed, each individual must surrender his personal liberty to turn his strength to violent uses. But a state of rest of that kind is only theoretically conceivable. In actuality the position is complicated by the fact that from its very beginning the community comprises elements of unequal strength—men and women, parents and children—and soon, as a result of war and conquest, it also comes to include victors and vanquished, who turn into masters and slaves. The justice of the community then becomes an expression of the unequal degrees of power obtaining within it; the laws are made by and for the ruling members and find little room for the rights of those in subjection. From that time forward there are two factors at work in the community which are sources of unrest over matters of law but tend at the same time to a further growth of law. First, attempts are made by certain of the rulers to set themselves above the prohibitions which apply to everyone—they seek, that is, to go back from a dominion of law to a dominion of violence. Secondly, the oppressed members of the group make constant efforts to obtain more power and to have any changes that are brought about in that direction recognized in the laws—they press forward, that is, from unequal justice to equal justice for all. This second tendency becomes especially important if a real shift of power occurs within a community, as may happen as a result of a number of historical factors. In that case right may gradually adapt itself to the new distribution of power; or, as is more frequent, the ruling class is unwilling to recognize the change, and rebellion and civil war follow, with a temporary suspension of law and new attempts at a solution by violence, ending in the establishment of a fresh rule of law. There is yet another source from which modifications of law may arise, and one of which the expression is invariably peaceful: it lies in the cultural transformation of the members of the community. This, however, belongs properly in another connection and must be considered later.

Thus we see that the violent solution of conflicts of interest is not avoided even inside a community. But the everyday necessities and common concerns that are inevitable where people live together in one place tend to bring such struggles to a swift conclusion and under such conditions there is an increasing probability that a peaceful solution will be found. Yet a glance at the history of the human race reveals an endless series of conflicts between one community and another or several others, between larger and smaller units—between cities, provinces, races, nations, empires—which have almost always been settled by force of arms. Wars of this kind end either in the spoliation or in the complete overthrow and conquest of one of the parties. It is impossible to make any sweeping judgement upon wars of conquest. Some, such as those waged by the Mongols and Turks, have brought nothing but evil. Others, on the contrary, have contributed to the transformation of violence into law by establishing larger units within which the use of violence was made impossible and in which a fresh system of law led to the solution of conflicts. In this way the conquests of the Romans gave the countries round the Mediterranean the priceless *Pax Romana*, and the greed of the French kings to extend their dominions created a peacefully united and flourishing France. Paradoxical as it may sound, it must be admitted that war might be a far from inappropriate means of establishing the eagerly desired reign of "everlasting" peace, since it is in a position to create the large units within which a powerful central government makes further wars impossible. Nevertheless it fails in this purpose, for the results of conquest are as a rule short-lived: the newly created units fall apart once again, usually owing to a lack of cohesion between the portions that have been united by violence. Hitherto, moreover, the unifications created by conquest, though of considerable extent, have only been *partial*, and the conflicts between these have called out more than ever for violent solution. Thus the result of all these warlike efforts has only been that the human race has exchanged numerous, and indeed unending, minor wars for wars on a grand scale that are rare but all the more destructive.

If we turn to our own times, we arrive at the same conclusion which you have reached by a shorter path. Wars will only be prevented with certainty if mankind unites in setting up a central authority to which the right of giving judgement upon all conflicts of interest shall be handed over. There are clearly two separate requirements involved in this: the creation of a supreme agency and its endowment with the necessary power. One without the other would be useless. The League of Nations is designed as an agency of this kind, but the second condition has not been fulfilled: the League of Nations has no power of its own and can only acquire it if the members of the new union, the separate States, are ready to resign it. And at the moment there seems very little prospect of this. The institution of the League of Nations would, however, be wholly unintelligible if one ignored the fact that here was a bold attempt such as has seldom (perhaps, indeed, never on such a scale) been made before. It is an attempt to base upon an appeal to certain idealistic attitudes of mind the authority (that is, the coercive influence) which otherwise rests on the possession of power. We have seen that a community is held together by two things: the compelling force of violence and the emotional ties (identifications is the technical name) between its members. If one of the factors is absent, the community may possibly be held together by the other. The ideas that are appealed to can, of course, only have any significance if they give expression to important affinities between the members, and the question arises of how much strength such ideas can exert. History teaches us that they have been to some extent effective. For instance, the

Panhellenic idea, the sense of being superior to the surrounding barbarians—an idea which was so powerfully expressed in the Amphictyonic Council, the Oracles and the Games—was sufficiently strong to mitigate the customs of war among Greeks, though evidently not sufficiently strong to prevent warlike disputes between the different sections of the Greek nation or even to restrain a city or confederation of cities from allying itself with the Persian foe in order to gain an advantage over a rival. The community of feeling among Christians, powerful though it was, was equally unable at the time of the Renaissance to deter Christian States, whether large or small, from seeking the Sultan's aid in their wars with one another. Nor does any idea exist to-day which could be expected to exert a unifying authority of the sort. Indeed it is all too clear that the national ideals by which nations are at present swayed operate in a contrary direction. Some people are inclined to prophesy that it will not be possible to make an end of war until Communist ways of thinking have found universal acceptance. But that aim is in any case a very remote one to-day, and perhaps it could only be reached after the most fearful civil wars. Thus the attempt to replace actual force by the force of ideas seems at present to be doomed to failure. We shall be making a false calculation if we disregard the fact that law was originally brute violence and that even to-day it cannot do without the support of violence.

I can now proceed to add a gloss to another of your remarks. You express astonishment at the fact that it is so easy to make men enthusiastic about a war and add your suspicions that there is something at work in them—an instinct for hatred and destruction—which goes halfway to meet the efforts of the warmongers. Once again, I can only express my entire agreement. We believe in the existence of an instinct of that kind and have in fact been occupied during the last few years in studying its manifestations. Will you allow me to take this opportunity of putting before you a portion of the theory of the instincts which, after much tentative groping and many fluctuations of opinion, has been reached by workers in the field of psychoanalysis?

According to our hypothesis human instincts are of only two kinds: those which seek to preserve and unite—which we call "erotic", exactly in the sense in which Plato uses the word "Eros" in his *Symposium* or "sexual", with a deliberate extension of the popular conception of "sexuality"—and those which seek to destroy and kill and which we group together as the aggressive or destructive instinct. As you see, this is in fact no more than a theoretical clarification of the universally familiar opposition between Love and Hate which may perhaps have some fundamental relation to the polarity of attraction and repulsion that plays a part in your own field of knowledge. But we must not be too hasty in introducing ethical judgements of good and evil. Neither of these instincts is any less essential than the other; the phenomena of life arise from the concurrent or mutually opposing action of both. Now it seems as though an instinct of the one sort can scarcely ever operate in isolation; it is always accompanied—or, as we say, alloyed—with a certain quota from the other side, which modifies its aim or is, in some cases, what enables it to achieve that aim. Thus, for instance, the instinct of self-preservation is certainly of an erotic kind, but it must nevertheless have aggressiveness at its disposal if it is to fulfil its purpose. So, too, the insinct of love, when it is directed towards an object, stands in need of some contribution from the instinct for mastery if it is in any way to obtain possession of that object. The difficulty of isolating the two classes of instinct in their actual manifestations is indeed what has so long prevented us from recognizing them.

If you will follow me a little further, you will see that human actions are subject to another complication of a different kind. It is very rarely that an action is the work of a *single* instinctual impulse (which must in itself be compounded of Eros and destructiveness). In order to make an action possible there must be as a rule a combination of such compounded motives. This was perceived long ago by a specialist in your own subject, a Professor G. C. Lichtenberg who taught physics at Gottingen during our classical age—though perhaps he was even more remarkable as a psychologist than as a physicist. He invented a Compass of Motives, for he wrote: "The motives that lead us to do anything might be arranged like the thirty-two winds and might be given names in a similar way: for instance, 'bread-bread-fame' or 'fame-fame-bread'." So that when human beings are incited to war they may have a whole number of motives for assenting—some noble and some base, some which are openly declared and others which are never mentioned. There is no need to enumerate them all. A lust for aggression and destruction is certainly among them: the countless cruelties in history and in our everyday lives vouch for its existence and its strength. The satisfaction of these destructive impulses is of course facilitated by their admixture with others of an erotic and idealistic kind. When we read of the atrocities of the past, it sometimes seems as though the idealistic motives served only as an excuse for the destructive appetites; and sometimes—in the case, for instance, of the cruelties of the Inquisition—it seems as though the idealistic motives had pushed themselves forward in consciousness, while the destructive ones lent them an unconscious reinforcement. Both may be true.

I fear I may be abusing your interest, which is after all concerned with the prevention of war and not with our theories. Nevertheless I should like to linger for a moment over our destructive instinct, whose popularity is by no means equal to its importance. As a result of a little speculation, we have come to suppose that this instinct is at work in every living creature and is striving to bring it to ruin and to reduce life to its original condition of inanimate matter. Thus it quite seriously deserves to be called a death instinct, while the erotic instincts represent the effort to live. The death instinct turns into the destructive instinct when, with the help of special organs, it is directed outwards, on to objects. The organism preserves its own life, so to say, by destroying an extraneous one. Some portion of the death instinct, however, remains operative *within* the organism, and we have sought to trace quite a number of normal and pathological phenomena to this internalization of the destructive instinct. We have even been guilty of the heresy of attributing the origin of conscience to this diversion inwards of aggressiveness. You will notice that it is by no means a trivial matter if this process is carried too far: it is positively unhealthy. On the other hand if these forces are turned to destruction in the external world, the organism will be relieved and the effect must be beneficial. This would serve as a biological justification for all the ugly and dangerous impulses against which we are struggling. It must be admitted that they stand nearer to Nature than does our resistance to them for which an explanation also needs to be found. It may perhaps seem to you as though our theories are a kind of mythology and, in the present case, not even an aggreeable one. But does not every science come in the end to a kind of mythology like this? Cannot the same be said to-day of your own Physics?

For our immediate purpose then, this much follows from what has been said: there is no use trying to get rid of men's aggressive inclinations. We are told that in certain happy regions of the earth, where nature provides in abundance everything that man requires, there are races whose life is passed in tranquillity and who know neither coercion nor aggression. I can scarcely believe it and I should be glad to hear more of these fortunate beings. The Russian Communists, too, hope to be able to cause human aggressiveness to disappear by guaranteeing the satisfaction of all material needs and by establishing equality in other respects among all the members of the community. That, in my opinion, is an illusion. They themselves are armed to-day with the most scrupulous care and not the least important of the methods by which they keep their supporters together is hatred of everyone beyond their frontiers. In any case, as you yourself have remarked, there is no question of getting rid entirely of human aggressive impulses; it is enough to try to divert them to such an extent that they need not find expression in war.

Our mythological theory of instincts makes it easy for us to find a formula for *indirect* methods of combating war. If willingness to engage in war is an effect of the destructive instinct, the most obvious plan will be to bring Eros, its antagonist, into play against it. Anything that encourages the growth of emotional ties between men must operate against war. These ties may be of two kinds. In the first place they may be relations resembling those towards a loved object, though without having a sexual aim. There is no need for psycho-analysis to be ashamed to speak of love in this connection, for religion itself uses the same words: "Thou shalt love thy neighbour as thyself." This, however, is more easily said than done. The second kind of emotional tie is by means of identification. What ever leads men to share important interests produces this community of feeling, these identifications. And the structure of human society is to a large extent based on them.

A complaint which you make about the abuse of authority brings me to another suggestion for the indirect combating of the propensity to war. One instance of the innate and ineradicable inequality of men is their tendency to fall into the two classes of leaders and followers. The latter constitute the vast majority; they stand in need of an authority which will make decisions for them and to which they for the most part offer an unqualified submission. This suggests that more care should be taken than hitherto to educate an upper stratum of men with independent minds, not open to intimidation and eager in the pursuit of truth, whose business it would be to give direction to the dependent masses. It goes without saying that the encroachments made by the executive power of the State and prohibition laid by the Church upon freedom of thought are far from propitious for the production of a class of this kind. The ideal condition of things would of course be a community of men who had subordinated their instinctual life to the dictatorship of reason. Nothing else could unite men so completely and so tenaciously, even if there were no emotional ties between them. But in all probability that is a Utopian expectation. No doubt the other indirect methods of preventing war are more practicable, though they promise no rapid success. An unpleasant picture comes to one's mind of mills that grind so slowly that people starve before they get their flour.

The result, as you see, is not very fruitful when an unworldly theoretician is called in to advise on an urgent practical problem. It is a better plan to devote oneself in every particular case to meeting the danger with whatever means lie to hand. I should like, however, to discuss one more question, which you do not mention in your letter but which specially interests me. Why do you and I and so many other people rebel so violently against war? Why do we not accept it as another of the many painful calamities of life? After all, it seems to be quite a natural thing, to have a good biological basis and in practice to be scarcely avoidable. There is no need to be shocked at my raising this question. For the purpose of an investigation such as this, one may perhaps be allowed to wear a mask of assumed detachment. The answer to my question will be that we react to war in this way because everyone has a right to his own life, because war puts an end to human lives that are full of hope, because it brings individual men into humiliating situations, because it compels them against their will to murder other men, and because it destroys precious material objects which have been produced by the labours of humanity. Other reasons besides might be given, such as that in its present-day form war is no longer an opportunity for achieving the old ideals of heroism and that owing to the perfection of instruments of destruction a future war might involve the extermination of one or perhaps both of the antagonists. All this is true, and so incontestably true that one can only feel astonished that the waging of war has not yet been unanimously repudiated. No doubt debate is possible upon one or two of these points. It may be questioned whether a community ought not to have a right to dispose of individual lives; every war is not open to condemnation to an equal degree; so long as there exist countries and nations that are prepared for the ruthless destruction of others, those others must be armed for war. But I will not linger over any of these issues; they are not what you want to discuss with me, and I have something different in mind. It is my opinion that the main reason why we rebel against war is that we cannot help doing so. We are pacifists because we are obliged to be for organic reasons. And we then find no difficulty in producing arguments to justify our attitude.

No doubt this requires some explanation. My belief is this. For incalculable ages mankind has been passing through a process of evolution of culture. (Some people, I know, prefer to use the term "civilization".) We owe to that process the best of what we have become, as well as a good part of what we suffer from. Though its causes and beginnings are obscure and its outcome uncertain, some of its characteristics are easy to perceive. It may perhaps be leading to the extinction of the human race, for in more than one way it impairs the sexual function; uncultivated races and backward strata of the population are already multiplying more rapidly than highly cultivated ones. The process is perhaps comparable to the domestication of certain species of animals and it is undoubtedly accompanied by physical alterations; but we are still unfamiliar with the notion that the evolution of civilization is an organic process of this kind. The *psychical* modifications that go along with the process of civilization are striking and unambiguous. They consist in a progressive displacement of instinctual aims and a restriction of instinctual impulses. Sensations which were pleasurable to our ancestors have become indifferent or even intolerable to ourselves; there are organic grounds for the changes in our ethical and aesthetic ideals. Of the psychological characteristics of civilization two appear to be the most important: a strengthening of the intellect, which is beginning to govern instinctual life, and an internalization of the aggressive impulses, with all its consequent advantages and perils. Now war is in the crassest opposition to the psychical attitude imposed on us by the process of civilization, and for that reason we are bound to rebel against it; we simply cannot any longer put up with it. This is not merely an intellectual and emotional repudiation; we pacifists have a *constitutional* intolerance of war, an idiosyncrasy magnified, as it were, to the highest degree. It seems, indeed, as though the lowering of aesthetic standards in war plays a scarcely smaller part in our rebellion than do its cruelties.

And how long shall we have to wait before the rest of mankind become pacifists too? There is no telling. But it may not be Utopian to hope that these two factors, the cultural attitude and the justified dread of the consequences of a future war, may result within a measurable time in putting an end to the waging of war. By what paths or by what side-tracks this will come about we cannot guess. But one thing we *can* say: whatever fosters the growth of civilization works at the same time against war.

I trust you will forgive me if what I have said has disappointed you, and I remain, with kindest regards,

Sincerely yours,
Sigm. Freud

FOOTNOTES

[1] Throughout this paper, Freud's use of "the war" refers to World War I.

[2] [In the original the words *"Recht"* and *"Macht"* are used throughout Freud's letter. It has unfortunately been necessary to sacrifice this stylistic unity in the translation. *"Recht"* has been rendered indifferently by "right", "law" and "justice"; and *"Macht"* by "might", "force" and "power".]

2   The Predatory Transition from Ape to Man

—RAYMOND A. DART

Of all beasts the man-beast is the worst
To others and himself the cruellest foe.
Richard Baxter (1615-1691): *Christian Ethics*

INTRODUCTION

"It is the last stage in the evolution of Man", said Elliot Smith (1926), "that has always excited chief interest and has been the subject of much speculation.

"These discussions usually resolve themselves into the consideration of such questions as whether it was the growth of the brain, the acquisition of the power of speech, or the assumption of the erect attitude that came first and transformed an ape into a human being. The case for the erect attitude was ably put by Dr. Munroe in 1893.[1] He argued that the liberation of the hands and the cultivation of their skill lay at the root of Man's mental supremacy."

"If the erect attitude was to explain all, why did not the gibbon

From the *International Anthropological and Linguistic Review*, 1 (1953), 201-219. Reprinted with the permission of the author.

become man in Miocene times or earlier?" enquired Elliot Smith. He believed that the steady growth and specialization of the brain had been the fundamental factor in leading Man's ancestor step by step upward; the erect attitude had been brought about because the brain by its previous expansion had made a new posture and skilled movements possible and of definite use in the struggle for existence.

Elliot Smith recognised as self-evident the fact that the hands, once liberated, would lead to further brain development in the process of becoming skilled. Indeed, as he pointed out, a primate as primitive as Tarsius can assume the erect attitude and use its hands for prehension rather than progression. In Oligocene times a Catarrhine Monkey, in becoming an Anthropoid Ape, had lost its tail, developed its brain and become more skilled, and this increased skill had manifested itself by a greater facility in walking in the upright attitude. This type of anthropoid has persisted with relatively slight modifications in the present-day Gibbon!

"But if the earliest Gibbons were already able to walk upright, how is it" Elliot Smith asked "that they did not begin to use their hands thus freed from the work of progression on the earth, for skilled work, apart from tree climbing, of biological usefulness for these competent hands to do?"

It was not until *Australopithecus africanus* (Dart 1925a) was found that any secure answer could be returned to this rhetorical question. Up till then no evidence was available concerning confirmed bodily habits amongst anthropoids, other than tree-climbing and occasional bipedal movements, which they could adopt. But here was represented an ultra-simian group of creatures living in the deepest south of Africa on the unfriendly eastern fringe of the Kalahari desert in caverns on the cliffside of a plateau overlooking the widest possible expanse of open plain devoid of forests, fruits and nuts but abounding in a greater degree with dangerous beasts than any other country in the world.

In the breccia accompanying *Australopithecus* eggshells, crab-shells and turtle shells, and bones of birds, small insectivores, bats, rodents, baboons and buck revealed evidence of the carnivorous diet of these creatures which appeared to be little if at all advanced beyond that of the baboons, which live today in the same locality. So I claimed from the outset and in 1929 that they were flesh-eating, shell-cracking and bone-breaking, cave-dwelling apes; and that in them the carnivorous habits of South African baboons (which, under stress of drought, as in that year, attack flocks of sheep and goats and carry off not only the young but the adults for food and in coastal areas go down to the sea and eat shellfish) had already become stereotyped.

Elliot Smith (op. cit. 1926) had immediately realized after the Taungs discovery that the determining factors in the separation in Miocene times, between Man's ancestors on the one hand and those of the Gorilla and Chimpanzee on the other, were the differential environmental and dietary conditions to which they became exposed after parting company. He said that in the group leading to man the brain growth had reached a stage where the more venturesome members—stimulated by some local failure of their arboreal food supply or by sheer curiosity—sought new sources of food on hill and plain wherever they could obtain the sustenance they needed; and that *Australopithecus africanus*, being found on the Kalahari Desert fringe at Taungs, provided objective evidence of fossil apes with habits in sharp contrast to those living arborealized anthropoids, who are tied down to a predominantly vegetarian diet admidst tropical forest conditions.

I had called the Taungs cave lair a kitchen-midden (1926) and Broom (1946) was inclined to agree with this opinion, because all the baboon skulls he had found had depressed fractures on the top of the head, such as I had demonstrated (1934). He suggested that the Australopithecinae could only have captured antelopes by hunting in packs, surrounding them at waterholes and killing them with sticks or stones. But neither Broom nor Schepers (1946) paid particular attention to the means whereby the man-apes had subsisted in South Africa; nor could the question be said to have arisen in an acute form before it had been discovered (1947) that the Limeworks site in Makapansgat Valley was an australopithecine and not (as I had presumed from the large size of the bones in 1925b) a human deposit. It then became clear (Dart 1948 and 1949), that the animals slain by the Australopithecinae are neither small nor slow: they are huge and active, such as the grotesque and extinct tree-bear (or Chalicothere), the extinct horse (Hipparion), the extinct giraffe (Griquatherium), the elephant, the rhinoceros, hippopotamus, pigs and fourteen or more species of antelopes (eight of which appear to be extinct) from the largest like the kudu to the smallest like the duiker and gazelle, and even carnivores like the lion, hyaenas (two species), hunting dog and jackal.

Other principal features in which the Makapansgat deposit resembles those left by mankind are that these large bones are not gnawed but split and crushed to extract the marrow; and the double-ridged extremities of the arm bones (humeri) of antelopes are cracked and fractured through having been used as bludgeons. The broken ridges of these humeri correspond with double-furrowed fractures found in baboon (and even in australopithecine) skulls found at all three man-ape sites: Taungs, Sterkfontein and Makapansgat.

Thus the chief cultural tools of the Australopithecinae were clubs formed by the long limb bones of antelopes; but it is clear that small punctured, round and triangular holes in skulls have been formed by thrusting home the sharp ends of broken long bones, or the horns of antelopes, employed as daggers, or by smashing blows with the upper canine teeth of carnivores when using their upper jaws as picks. The lower canine teeth of carnivores and pigs have usually been wrenched out of the mandible as though they had been used as natural curve kukri blades. The angles of ungulate mandibles are generally broken off and their rows of teeth chipped as though they had been damaged when employed as choppers and saws respectively. The upper jaws of ungulates are usually broken off so that they fit nicely in one's hand while the chipped character of the teeth edges shows that they have previously served the role of scrapers.

The australopithecine deposits of Taungs, Sterkfontein and Makapansgat tell us in this way a consistent, coherent story not of fruit-eating, forest-loving apes, but of the sanguinary pursuits and carnivorous habits of proto-men. They were human not merely in having the facial form and dental apparatus of humanity; they were also human in their cave life, in their love of flesh, in hunting wild game to secure meat and in employing implements, whether wielded and propelled to kill during hunting or systematically applied to the cracking of bones and the scraping of meat from them for food. Either these Procrustean proto-human folks tore the battered bodies of their quarries apart limb from limb and slaked their thirst with blood, consuming the flesh raw like every other carnivorous beast; or, like early man, some of them understood the advantages of fire as well as the use of missiles and clubs. At any rate there are no known features on the cultural side other than the deliberate manufacture of tools and the systematic employment of fire, which separate these proto-men at the present time from early man.

The carnivorous dietary habits of the South African man-apes have shown the skilled work of biological usefulness, which their competent hands were called upon to do; and have demonstrated the direct relationship which exists between the carnivorous habits and acquiring the upright posture. The cerebral powers of these creatures were necessarily enhanced to carry out these striking, hurling and thrusting manual feats that are utterly beyond the skill of living apes; but it is absurd to imagine now that a large brain is essential to perform such berserk deeds.

Nobody knows what constitutes a human brain from the point of view of absolute size and weight. Individuals of apparently normal intelligence today need not have brains weighing more than 788 grams (i.e. the equivalent of a cranial volume of 830 cc. or only 175 cc. more than the greatest cranial capacity reported in a living ape). But human beings of subnormal intelligence can live long and fairly serviceable lives with a brain of anthropoidal dimensions. Keith (1928) reminded us of "Joe", the Lancashire shepherd, whose brain investigated by Cunningham weighed 500 grams (i.e. less than the Taungs infant) and had a fissural pattern simpler than a chimpanzee's. Yet "Joe" was, and functioned as, a human being! He was five foot nine inches high and died at the age of sixty; he could speak, having a considerable number of words and framing sentences; he knew a sixpenny from a fourpenny bit and could count his fingers but did not apprehend divisions of time. The latter part of his life was spent in the county asylum where he tended the sheep, keeping them within prescribed limits, with great vigilance for days together.

We do not transform such human brains as Joe's into simian brains by callling them anthropoidal, or microcephalic, or abnormal; nor do men and women possessing small brains necessarily have less capacity than larger-brained human beings for standing, walking, running, leaping, striking and hurling and carrying out the other physical and physiological activities, that distinguish human beings from apes. Such people may not be as clever or creative as other men but they are men and not apes. Recently in the Northern Transvaal three such small-brained (circa 500 cc.) adults were found in a single Bantu family, discharging simple yet tolerably useful and definitely human roles in their local environment (Grieves and Hughes 1953). As Keith (op. cit.) said, such brains "represent in a disturbed and somewhat distorted manner, an actual stage in the evolution of the human brain"! The last quarter of a century has shown that that stage in the evolution of the human brain was the australopithecine stage. A microcephalic mental equipment was demonstrably more than adequate for the crude, carnivorous, cannibalistic, bone-club wielding, jawbone-cleaving Samsonian phase of human emergence.

THE CARNIVOROUS HABIT IN MANKIND

Wherever found all prehistoric and the most primitive living human types are hunters, i.e. flesh-eaters. Human carnivorous habits have an omnivorous range extending from grubs and insects on the one side to the most formidable of big game on the other. Secondly, man's taste for flesh is so great that human beings, whether in prehistoric *(Pithecanthropus-Sinanthropus)* or recent times and whether driven by need or not, have practised either real or ritualistic cannibalism. "One of the strongest reasons for considering *anthropophagy* as having widely prevailed in pre-historic ages is the fact of its being deeply ingrained in savage and barbaric religions" (*Encyc. Brit.*, 14 ed., art. *Cannibalism*).

Carlyle frankly recognised this as a fundamental human feature when he said (*Sartor Resartus*, Bk. I. p. 29) "Reader, the heaven inspired melodius Singer . . . has descended, like thyself, from that same hair-mantled, flint-hurling Aboriginal *Anthropophagus*." Through Herodotus, Strabo and other writers, the Greeks and Romans were familiar with contemporary peoples who, like the Scythian Massagetae northeast of the Caspian Sea, regularly killed old people and ate them. Marco Polo and other travellers informed Europeans of cannibalistic practices amongst the wild tribes of China, Tibet, and elsewhere. Human flesh was habitually exposed for sale in the market place in West Africa; some tribes sold the corpses of dead relatives for consumption as food. Cannibalism prevailed until recently over a great part of West and Central Africa, New Guinea, Melanesia (especially Fiji), Australia, New Zealand, and Polynesian Islands, Sumatra and other East Indian Islands, in South America and formerly in North America. Cannibalism from necessity is found not only among Fuegians or Red Indian tribes, but also among civilized races, as the records of seiges and shipwrecks show (vide *Encyc. Brit.*, 14 ed., art. *Cannibalism*).

The loathsome cruelty of mankind to man forms one of his inescapable, characteristic and differentiative features; and it is explicable only in terms of his carivorous, and cannibalistic origin. As Robert Hartman (1885) said, "It is well-known that both rude and civilized peoples are capable of showing unspeakable, and as it is erroneously called, inhuman cruelty towards each other. These acts of cruelty, murder and rapine are often the result of the inexorable logic of national characteristics, and, unhappily, are truly human, since nothing like them can be traced in the animal world. It would, for instance, be a grave mistake to compare a tiger with a blood-thirsty executioner of the Reign of Terror, since the former only satisfies his natural appetite in preying upon other animals. The atrocities of the trials for witchcraft, the indiscriminate slaughter committed by the Khonds, the dismemberment of living men by the Battas, find no parallel in the habits of animals in savage state. And such a comparison is, above all, impossible in the case of the anthropoids, which display no hostility towards men or other animals unless they are first attacked. In this respect the anthropoid ape stands upon a higher plane than many men."

The blood-bespattered, slaughter-gutted archives of human history from the earliest Egyptian and Sumerian records to the most recent atrocities of the Second World War accord with early universal cannibalism, with animal and human sacrificial practices or their substitutes in formalized religions and with the world-wide scalping, head-hunting, body-mutilating and necrophilic practices of mankind in proclaiming this common bloodlust differentiator, this predaceous habit, this mark of Cain that separates man dietetically from his anthropoidal relatives and allies him rather with the deadliest of Carnivora.

Darwin (1871, p. 146) recognised this sinister aspect of human evolution to a degree when he said "The same high mental faculties which first led man to believe in unseen spiritual agencies, then in fetishism, polytheism and ultimately in monotheism, would infallibly lead himself, as long as his reasoning powers remained poorly developed, to various strange superstitions and customs. Many of these are terrible to think of—such as the sacrifice of human beings to a blood-loving god; the trial of innocent persons by the ordeal of poison or fire; witchcraft, etc.,—yet it is well occasionally to reflect on these superstitions, for they show us what an infinite debt of gratitude we owe to the improvement of our reason, to science and to our accumulated knowledge." Yet, although he cited Roman gladiatorial shows, scalping, head-hunting, infanticide, slavery, love of inflicting torture and indifference to suffering, as indications of a low-state of moral sense amongst civilized and primitive peoples, Darwin did not deduce from these observations, that man had arisen from a predaceous anthropoid stock. Still whether cognizant of the wider implications of his comments or not, he made this statement (op. cit., p. 78): "If it be an advantage to man to stand firmly on his feet and to have his head and arms free, of which, from his pre-eminent success in the battle of life, there can be no doubt, then I can see no reason why it should not have been advantageous to the progenitors of man to have become more and more erect or bipedal. They would thus have been better able to defend themselves with stones and clubs, to attack their prey, or otherwise obtain food."

Thus Darwin dared to picture not merely early men but also their progenitors as hunters! What is "prey"? According to the *Concise Oxford Dictionary* (3rd ed., 1934) it is "animal hunted or killed by carnivorous animal for food". The predaceous habit is therefore "living by preying", i.e. hunting down and killing animals for food. On this thesis man's predecessors differed from living apes in being confirmed killers: carnivorous creatures, that seized living quarries by violence, battered them to death, tore apart their broken bodies, dismembered them limb from limb, slaking their ravenous thirst with the hot blood of victims and greedily devouring livid writhing flesh. Further, if Darwin's reasoning was correct, man's erect posture is the concrete expression of signal success in this type of life. It emerged through and was consolidated by the defensive and offensive stone-throwing and club-swinging technique necessitated by attacking and killing prey from the standing position.

This purposive industrial specialization of the hands in accurate hitting and throwing, as I pointed out (1949b), was the only persistent stimulus capable of transferring the body weight from the clambering knuckles and the sitting ischial tuberosities of apes to the only suitable base for such torsional work as man's body performs, namely, the bi-columnar-hexapodal mechanism of the completely extensible lower limbs, each linked to its tripodal foot by its powerful and exceedingly mobile ankle joint. Anatomically and physiologically this implies not so much the development, as Reynolds (1931) thought, of the capacity to leap sideways, but, as Darwin prophetically indicated, of the capacity to stand still and to transform the immobilized feet into a rock-like base from which the whirling body can operate as a whole.

Living anthropoids, lacking bipedal fixity, having developed so little accuracy in wielding or hurling objects to strike other creatures either in offense or defense, that they all rely on their teeth or nails (to which human beings also revert in extremity) rather than on throttling fingers or the impacts of nocuous objects held in their hands, when struggling at close quarters. Man, on the contrary, makes such persistent use of his hands and his whole torsional body strength in the erect posture that he can use his fists deftly and accurately as weapons, either open as in slapping and cuffing, or closed as in boxing and pounding. He is the only fisted creature on earth. Nowhere, also, is man so destitute of intelligence as to be ignorant either of the added effectiveness, both in offense and defense, of objects such as sticks and stones held in his hands, or of the increased advantages, in terms of force and accuracy, of standing firm and still and of using the added height of his erect posture and the force of gravity in applying the blows of his bare fists or weapon-carrying hands, and in hurling projectiles.

This accuracy in hitting and hurling, which apes lack but which men universally possess, is an inherited instinct. It demands no greater intelligence than human microcephalic idiots, with less than australopithecine endocranial capacities, can command (see Keith, op. cit.). What it does require is a short and enlarged pelvis, such as the Australopithecinae possessed, capable of rotating on two columnar limbs about powerful ankles, above feet that have planted heels and big toes capable of adhesion to the ground. With this type of understructure, the elongated human flanks can bend laterally or rotate upon the pelvis as well as flex and extend, while the arms rotate; and the head upon its lengthened neck can move freely and co-ordinatedly as a whole in any direction on the planted torso. All of these co-ordinated movements of the slender upright body, head, neck and arm are required for the performance of accurate hitting and hurling. The divergence in skill between apes and men in hitting and hurling depends upon the acquisition by human beings of a brain capable of co-ordinating with hand and eye movements a series of postural body reflexes demanding a lithe body with tremendous torsional capacity about hips and upon mobile ankles but acquired instinctively by all normal human beings during the first year of infancy.

Such a brain was perhaps already developed and capable of providing skilled work of biological usefulness other than tree-climbing, hitting and hurling, for competent hands pivoted upon a torsional body to do, even in Miocene times or even earlier when the entire anthropoid stock was experienced to a greater or lesser degree in bipedal behaviour. Indeed, it has been suggested (Wells 1931, p. 25) that "the existing great apes have become secondarily quadrupedal in the course of their evolution."

Before the ancestral bipedal brachiator there lie three courses in respect of evolving hands. If he stays in the trees like the gibbon and orang the upper limbs must become the chief body supporters and propulsants; so the arms become long and develop hook-like hands specialized for swinging from the branches. The elongated arms of the brachiating gibbon and orang are also shared by the partially-terrestrialized chimpanzee and gorilla. These hands can be adapted to ground support in the brachiating anthropoid only by the biological absurdity of using the dorsal instead of the palmar aspects of their hands in progression. Consequently during their temporary desertion of the trees the chimpanzee and gorilla are enabled to aid foot-walking by employing the knuckles of their hands as props. The third alternative before the brachiator in departing from the forest is to carry one or more pieces of the deserted tree with him, not to hang therefrom but to use them as props or crutches if necessary and to swing about himself in the club or bludgeon fashion adopted by our presumptive bipedal progenitors. At one extreme the gibbon has specialized in whirling himself about the branches of trees; at the other extreme man has specialized in whirling the branches of the trees about himself.

These bludgeon-whirling activities became significant when there was skilled work apart from tree-climbing of biological usefulness such as hunting for these competent hands to do; the seizing of victims, prey or quarry; the thirst for blood and the hankering after flesh for food; the carnivorous diet.

THE CARNIVOROUS PROPENSITY—A PRIMITIVE PRIMATE CHARACTERISTIC

The recognition of the carnivorous habit as a distinctive australopithecine trait has developed so gradually that it is fruitful to analyse the probable reasons why this basic fact has been obscured and the widespread reluctance even today to accept its implications for the understanding of human nature.

In the first place the apparent preference of primates generally for nuts, fruits and vegetable substances as food and their arboreal habitat has naturally led to the assumption that anthropoids were originally frugivorous. The primate stock was not originally frugivorous; as they arose from Cretaceous insectivorous ancestors, the need for animal proteins in primate diet has an extremely ancient history. The partiality of many primates for fruits has obscured the extent to which most of them are still omnivorous and thus to an appreciable extent carnivorous.

The Tarsier is insectivorous; the Aye-aye family of Madagascan lemurs *(Daubentonia* or *Cheiromys)* live on grubs, for which meat diet they are so specialized that their very large ears can hear them boring under the bark of trees and their teeth are modified rodent-like to gnaw through the bark to catch their succulent prey. The Asiatic lemur, the Slender Loris of India and Malaya, clings to branches and feeds on leaves and fruit, insects, small birds and mice. The squirrel monkeys *(Saimira* or *Chrysothrix)* ranging from the Amazon to Costa Rica are also chiefly insectivorous. The Old World monkeys are generally omnivorous; but some, like the macaques of Malaya, are so attracted by crustacean invertebrate food that they subsist on crabs; while the monkeys of Sierra Leone relish oysters. The most primitive living anthropoid, the Gibbon, lives on fruit, leaves, grasses, shoots and roots; but it also feeds on insects, birds' eggs, birds and other small vertebrates.

South Africa is not richly furnished with forests bearing nuts and fruits. Consequently the terrestrialized Chacma baboons, as Sclater (1900) pointed out, are omnivorous. They live on prickly-pear fruit and leaves, wild fruits, berries, bulbs and aloe-stalk pith. But they avidly devour insects, scorpions, centipedes and lizards, rob the nests of wild bees; and cause great annoyance to farmers not only by devastating orchards and fruit gardens, but also by sucking and devouring ostrich eggs. He further pointed out how they had taken to killing and disembowelling lambs and kids for the sake of the curdled milk in their stomachs.

Almost a century earlier Wood had referred in his *Zoography* (v. I, 1807) to the orchard and vineyard raids of baboons, to their robbing the nests of birds and smashing their eggs if unpalatable. A. C. White *(The Call of the Bushveld)* has mentioned their consumption of scorpions, tarantulas and other creeping creatures found under the flat stones they overturn, the tremendous amount of damage they do to farmers' crops, their bird-nesting and their killing the young of many animals.

Shortridge (1934) refers to baboons turning over stones for lizards, scorpions, beetles etc., he compares their diet to some extent with that of a Bushman and states "I once found a half-grown leopard that had been killed by baboons. The back of the skull had been bitten away, and the brains sucked out or otherwise extracted—but there were few additional signs of injury."

Stevenson-Hamilton (1912) corroborates the baboon diet of scorpions, centipedes, beetles, bees' nests, and birds' or other eggs and adds thereto ants and locusts. He knew that baboons in captivity occasionally accept a small piece of raw meat and he had been told of a troop killing a half-grown klipspringer. He was also familiar with their "attacking parturient and newly-delivered ewes and young lambs of domestic sheep, in order to obtain milk from the udders and stomachs respectively"; but he does not regard them as flesh-eaters.

On the other hand, Mr. H. B. Potter, the internationally known Game Conservator of Zululand, whose headquarters were at the Hluhluwe Game Reserve for most of his professional life, believes that baboons destroy more bird and animal life than all the other vermin put together. Again, whereas the attacking of ewes and young lambs is regarded by many observers as a recent development due to drought or other drastic interferences with their normal food supply, Captain Potter's repeated observations of organized hunting by baboons within the Reserve, indicate that it is not a chancy but a seasonal occurence. The following is his written communication to me upon the matter which I quote in full because of its general importance and the authoritative character of its source.

ZULULAND GAME RESERVE AND PARKS BOARD

Hluhluwe Game Reserve
P. O. MTUBATUBA
24th June, 1949.

Professor Raymond A. Dart,
University of the Witwaterstrand,
Hospital Street,
Johannesburg.

Dear Professor Dart,

I am sorry I have not replied to your letter before. I have been away a lot lately and these are our busiest months.

I have no prejudice against Baboons but one must go on one's personal observations. In this Game Reserve there are some 4 or 5 hundred Chacma Baboons and very few days pass without encountering some of them.

I have personally witnessed on 6 or 7 different occasions spread over the last 20 years baboons actually surrounding a half grown antelope, mostly reedbuck and mountain reedbuck. On what would appear to be a given signal the baboons—sometimes from 15 to 20 in number—have closed in on their victim and as a matter of interest I have mostly refrained from shooting so that the results could be correctly recorded. In almost every case the buck has been caught and literally rent from limb to limb, and all that remains when the affair is all over are the skull and leg bones and pieces of skin. Everything else has been devoured. Old and young baboons in the troop all seem to have taken part in these orgies.

I have noticed that these "hunts" usually take place during the winter months, possibly due to the fact that other food is then scarce. I have actually seen old baboons enter the fowl runs near my home. They have caught and eaten full-grown fowls—pure bred stock at that. On one occasion we saw a Leghorn cock with half of its back eaten away by an old dog baboon which had killed 4 other fowls before we heard the screaming. A postmortem revealed the bones and meat of fowls in the baboon's stomach. Fowl eggs had to be carefully guarded and I have actually seen an old female baboon take eggs from under a sitting hen; 11 out of 15 eggs—one a day—were taken in this manner until the thief was shot.

It is my firm belief that baboons are responsible for the loss of a large percentage of birds' eggs, especially game birds' eggs. I do not consider that more than 5 percent of the eggs laid produce young birds before they reach maturity. Barren braces or very small coveys of francolin, and in some cases guinea fowl, all support my above contention.

I have not witnessed anything in the nature of ambushing game at waterholes on the part of baboons but I have seen lizards, rodents, scorpions and even one legewan (iguana) killed and eaten by them.

In conclusion, I consider that baboons do more damage to bird life (eggs and young birds) and also to young buck of the smaller species than all other vermin put together.

I would very much like to meet you and have a chat about baboons in particular and game in general.

With kind regards,
Yours sincerely,
(Sgd). H. B. Potter
Game Conservator, Zululand

Eugene Marais (*My Friends the Baboons*, 1939) reported the slaughter of lambs and also of half grown sheep in the Heidelberg district of the Transvaal and said that there too "not only sheep but pigs, ducks, fowls, guinea fowls and turkeys are on the list of their booty in the Suikerbosrand." Marais, like Sclater, at the beginning of the twentieth century attributed these carnivorous propensities of baboons to a habit which had spread northwards from the Karroo within the last eighty years; but a feature which has been recorded by so many individual observers and has manifested itself seasonally amongst baboons equivalently in the Cape, Transvaal and Natal over more than half a century is entitled to be regarded as a consistent dietary phenomenon. Baboons anywhere become hunting mammals under the pressure of seasonal or environmental need.

The emergence of carnivorous food habits is therefore not singular but appropriate to a capable and discriminating omnivorous anthropoid ape. The habits of baboons show that the predatory tendency becomes seasonal and even systematic in a much less capable primate than *Australopithecus* in the exacting terrestrial environment of Southern Africa.

We have seen that, despite the insectivorous origin of primates such as South African baboons and of arborealized anthropoids such as the gibbon, there has been a general reluctance to accept the repeatedly demonstrated facts that baboons spontaneously hunt wild game and to recognize the implications for human evolution of primitive man's multi-millennial dependence upon animal slaughter for his food. So compulsive a characteristic could not possibly have emerged in mankind without its having characterized his predecessors; but man has been forced by the growth of population to discover in fish and in cereals alternative chief sources of sustenance and concurrently there emerged strict taboos and legislation against animal foods in particular. From the Biblical story of the Garden of Eden to the cult of vegetarianism (a term applied *circa* 1847 to the present practice of excluding fish, flesh and fowl from human diet, and in some sects, milk, eggs and cheese, or even all cooked foods) civilized man has been steadily and increasingly separated from his carnivorous past.

It is paradoxical to reflect that intensification of vegetarian indoctrination has occurred spontaneously during the past century which has revealed man's ancestral subsistence upon flesh. Vegetarian societies (the oldest of which in England was in Manchester) have been established all over the world but chiefly in the United Kingdom, America, Germany, France, Austria, Holland, Czechoslovakia, Scandinavia and Australia. Food taboos are, however, as ancient as human society! In the western religious world Seventh-Day Adventists and some Bible Christians, in the eastern division the worshippers of Vishnu and the Swami Narang and Vishnu sects, among others, preach abstinence from flesh food. The Salvation Army and the Doukhobors encourage it; and a number of orders in the Roman Catholic Church (e.g. Trappists) and in the Hindu faith (e.g. Dadupanthi Sadus) are pledged abstainers (vide *Encyc. Brit.*, 14 ed., art. *Vegetarianism*).

The real objection to the free use of meat, however, for civilized human food arises from the fact that animals compete with man for utilizable foods. The cow, pig and fowl consume 12-14 lb. dry fodder for each 1 lb. of dry human food produced (as meat, eggs, etc.), the sheep 24 lb. and the ox 64 lb. (vide *Encyc. Brit.* 14th ed., art. *Diet and Dietetics*). "Animals are no longer necessary for transport and an acre of cultivable ground will produce from two to twenty times as much human food in vegetables and fruit as it will in terms of meat. Meat forms already less than 4 per cent of the estimated 2,000 trillion calorie consumption of mankind annually. Where populations are most concentrated in India, China and Japan meat has virtually vanished from human diet and in the long run vegetarians alone will inherit the earth." (Dart 1951).

Whether the factors relentlessly imposing a meatless existence on mankind today be ritualistic or economic, they are contrary to his original nature; they assist however in explaining the incredulity many groups of human beings share about the carnivorous phase in the transition from ape to man and the past reluctance to envisage the implications of the multimillennial proto-human and human predaceous habit for the understanding of man's essential nature and many of his taboos, customs and beliefs.

REFERENCES

Broom, R. (1946) *see* Broom & Schepers.

Broom, R. & Schepers, G. W. H. (1946), The South African fossil apemen, the Australopithecinae, *Mem. Transv. Mus.* no. 2, 144 p. illus.

Carlyle, T. (1833-34), *Sartor resartus*, Bk. I, chap. I.

Dart, R. A. (1925a), Australopithecus africanus: the man-ape of South Africa, *Nature*, 115, 195-199, illus.

——(1925b), A note on Makapansgat: a site of early human occupation, *S. Afr. J. Sci. 22*, 454.

——(1926),   Taungs and its significance, *Nat. Hist. N. Y. 26*, 315-327, illus.

————(1929), A Note on the Taungs skull,   S. Afr. J. Sci. 26, 648-58, illus.

————(1934), The dentition of Australopithecus africanus, Folia anat. Japon. 12, 207-221, illus.

————(1948),   An Australopithecus from the Central Transvaal, S. Afr. Sci. I, 200-201, illus.

————(1949a), The predatory implemental technique of Australopithecus, Amer. J. Phys. Antnrop. n.s. 7, 1-38.

————(1949b), Innominate fragments of Australopithecus prometheus,   Amer. J. Phys. Anthrop. n.s. 7, 301-334, illus.

————(1951), In foreword to   Botany for medical students by Edward R. Roux, Johannesburg, Juta, 201 p., illus.

Darwin, C. (1871), The descent of man, London, J. Murray.

Elliot Smith, G. (1926), Evolution of man: essays, London: Oxford Univ. Press, 159 p., illus.

Encyclopaedia Britannica, 14th ed., c. 1929. v. 4 Cannibalism; v. 7 Diet and Dietetics; v. 23 Vegetarianism.

Grieves, J. K. & Hughes, A. R. (1953), Medium grade microencephaly in a Bantu family, S. Afr. J. Med. Sci. 18, 19-30.

Hartman, R. (1885), Anthropoid apes, Internat. sci. series; v. 53, London: Kegan Paul, 326 p.

Keith, A. (1928), The human body, Home Univ. Lib., 6th imp., Butterworth, 256 p.

Marais, E. N. (1939), My friends the baboons, London, Methuen, 124 p.

Munroe, R. (1893), Presidential address, Section H. Anthropology, Rep. Brit. Ass. for the Adv. of Sci. 63, 885-895, v. 63.

Reynolds, E. (1931), The evolution of the human pelvis in relation to the mechanics of the erect posture, Pap. Peabody Mus. 11, no. 5.

Sclater, W. L. (1900), The Mammals of South Africa, London, Porter 1900-1901, 2 v., illus.

Shortridge, G. C. (1934), The Mammals of South West Africa, London, Heinemann, 2 v., illus.

Stevenson-Hamilton, J. (1912), Animal life in Africa, London, Heinemann, 539 p., illus.

Wells, L. H. (1931), The foot of the South African native, Amer. J. Phys. Anthrop. 15, 185-289.

White, A. C. (1948), The call of the Bushveld, Bloemfontein, White, 261 p., illus.

Wood, W. (1807), *Zoography; or, The beauties of nature displayed,* London, Cadell & Davies, 3 v.

EDITOR'S NOTE TO THE ABOVE[2]

We agree in the main with Dr. Dart's paper. But when he says that man's "accuracy in hitting and hurling, . . . is an inherited instinct," we feel doubt, because even five year-old children are often far from accurate in hitting and hurling, and only *some* adults acquire that accuracy by practice. How can it then be an inherited instinct?

Professor Dart's thesis that the South-African apemen, at the stage they were found, were omnivorous, must be considered as proven.

Of course, they were only the ancestors of the modern Bushmen and Negroes, and of *nobody else*.

That there was an omnivorous (seemingly quite short) stage between the frugivorous ape stage and the human one, must also be considered as a fact, not only for the African apemen, but also at least for some of the primitive men in Asia and Europe.

However, this omnivorous, and often carnivorous stage, may have been very short for some of the European races (who go back to different ape-ancestors), which is proven by the intolerance to meat eating (and much more so to exclusive meat diets), observed in modern man, who suffers from many diseases which immediately disappear under a vegetarian diet.

That apes, such as the gorilla, do not thrive on a mixed diet, we demonstrated in Vol. I, No. 2-3 (Alan H. Kelso de Montigny, "Strange doings at a New York Zoo", pp. 156-157).

The terrible cases of arthritis and grave abscesses found among the fossil remains of the Neandertaloid-Armenoid hybrids of Krapina, who ate meat and were cannibals, prove again that they had not been eating meat for a very long time, because their diseases prove their inadaptedness. If they had been eating it for 100,000 years, they would have been adapted to it, like the pig, which is a true omnivore, while we are not.

We also will realize that the meat eating baboons are a far cry from the apes (gorilla, chimpanzee, gibbon, orangutan) who are the only true vegetarians.

The Editors

FOOTNOTES

[1] Presidential Address to Section H. *Report of the British Association.*

[2] These comments were made by the editor of the journal which originally published Dart's paper. We suspect that it will be helpful in stimulating further discussion of Dart's research, and in contributing more information relevant to the central theme of this book.

3   Ritualized Fighting

——KONRAD LORENZ

If we put together, into the same container, two sticklebacks, lizards, robins, rats, monkeys or boys, who have not had any previous experience of each other, they will fight. If we do the same with two animals of different species, there will be peace—unless, of course, there is a prey-predator relationship between them. Intraspecific aggression, or aggression for short, is found in the vast majority of vertebrates and in many invertebrates. There cannot be any doubt about the important functions it achieves in the interest of the survival of the species.

The first, and probably the most important, of these functions is the spacing out of individuals of one species over the available habitat; in other words, the distribution of "territories." Another is the selection of the "better man" by rival fighting, relevant in conjunction with defence of the family or the society by the male;* a third is the establishing of a social rank order which is of particular importance in social animals in which learning is highly developed, so that the individual experience of the aged leader is of great advantage to the community. There are, of course, some other, less important functions of aggression in its simple unmixed form, but they need not concern us here. There are, however, some highly important functions which aggression performs in the system of mutual interactions between independently variable motivations governing animal and human behaviour. Of these we shall talk later.

From J. D. Carthy and F. J. Ebling (Eds.), *The Natural History of Aggression.* London: Academic, 1964. Reprinted with the permission of the Institute of Biology, London, England.

The selection pressure exerted by all these functions has caused aggression to evolve independently in a great number of different animals, the same performance being achieved by very different means. Very often weapons which originally served functions other than those of intraspecific aggression were pressed into its service. All these weapons were made to kill, or at least to inflict as serious an injury as possible, as they were primarily adapted to overcome a prey or to defend the organism against a predator. In both cases, the damage done to a non-conspecific animal is entirely to the advantage of the species, while the injuries suffered by the combatants in intraspecific fighting clearly are not. Moreover, they are not really necessary for the achievement of any of the functions already mentioned; these are served just as well by a sound thrashing as by a killing. There seem to be very few subhuman species of animals in which intraspecific fights lead regularly to serious wounds and death, which, in these cases, must be regarded as a sacrifice made by the species in order to attain the advantages of aggressive behaviour.

However, very few animals go to these extremes. The only case to my certain knowledge in which the attack of a rival often results in the immediate death of the fellow-member of the species, concerns a lizard, *Lacerta mellisselensis*, in which G. Kramer[1] repeatedly saw one male break another's back by a single bite and twist of the head. In Indian elephants, according to J. H. William's reliable reports,[2] it is also a frequent occurrence that a tusker is mortally wounded in natural rival fighting. Circumstantial evidence makes it probable that mass fighting in rats and other rodents may also lead to the death of individuals under natural conditions. For the rest, I believe that our experience with captive animals and, of course, with our own species, misleads us into overestimating the loss of individuals incurred by intraspecific fighting in most animals.

In the great majority of species measures have been taken to render aggression less dangerous. They may consist of defence movements so well adapted to the species-specific form of attack that the latter is almost invariably parried. (Whoever has witnessed a serious clash between two lions will have been surprised by the total absence of gaping wounds after much roaring, flashing of fangs and pounding of armed paws.) They may also consist of merely passive armour, the fat-padding and the upright mane on the neck of a Przewalsky stallion being good examples. However, in very many cases the necessity of making aggression less dangerous has led to changes in the behaviour patterns of fighting itself, in other words to "ritualization". I cannot go into detailed discussion of the rather complex concept which we associate with this term. Its definition is necessarily an injunctive one, and it is a matter of taste on which of its part-constitutive characters one chooses to put most stress. So I shall not disagree with anyone who contends that the evolutionary changes in fighting, of which I am going to speak now, are not very typical examples of what we call ritualization. However, the ritualized redirected activities of which I am going to talk later on, definitely are.

One line of differentiation which tends to alleviate the damage done to individuals without decreasing the survival value of aggression, is the development of so-called threat behaviour. It invariably arises out of a conflict between the motivations of attack and escape, and in its most primitive forms it may consist of the simple superposition of motor patterns simultaneously activated by both. In many spiny-rayed fish, the aggressive urge to swim at the adversary and the escape drive trying to effect the opposite often results in both combatants turning broadside on at close quarters with heads averted. At the same time aggression induces them to show their most beautiful display colours, and fear causes them to erect their unpaired fins to the utmost. In this broadside-on posture, there are primarily no appreciable elements of ritualization. But as the attitude automatically displays the full size of the individual to the eyes of the rival, there is obviously a strong selection pressure at work to make the fish appear as large as possible. Unpaired fins have become enlarged, their borderlines enhanced by conspicuous colourations and their erection exaggerated to the point at which the danger of tearing the fin membranes arises. A new motor pattern was "invented," namely the lowering of the radii branchiostegi in a vertical plane so that the fish's visible contours were enlarged by the spread of the gill membrane; the latter was adorned by new, striking colour patterns. With these additions, all pointing unequivocally to its communicative function, the broadside-on display definitely takes on the character of a ritualized activity.

Even more widely distributed among teleosteans than the broadside-on display is the tail-beat, very probably a derivation of the former. Standing parallel to the rival with head averted, the fish delivers, in the direction of the adversary, a stiff beat with the maximally spread caudal fin. In some species, in which both fish stand very closely together in display, the rival may be actually swept away by the stream of water thus produced. Even with the mildest form of tail-beat, the lateral line organs of the reactor must receive a pressure wave indicative of the actor's strength.

Another motor pattern of threat, not quite so widely spread among bony fishes as the two aforementioned, also arose from the conflict between aggression and fear, and also, very probably, evolved independently in several families of fish. It consists in a halfhearted, tentative bite at the other fish's head, very different indeed from the wild ramming thrust launched in uninhibited aggression. The ritualization of this movement is contained more in its releasing mechanism than in its motor co-ordination: it can only be performed if the reactor retaliates in kind. In other words, any intentional movement preparing for this kind of bite is instantly suppressed if the other fish continues to stand in the attitude of broadside-on display. Never, never does a fish bite into the opponent's unprotected flank. As a result, the movement under discussion is fully performed only when both fish do so simultaneously, so that they bite at each other's mouth, hence the descriptive term "mouthfighting". From this movement, two lines of differentiation lead up to more ritualized forms. In the genus *Tilapia* both fighters open their mouths as far as possible and push against each other; in *T. mossambiqua* and other species of the genus, the inside of the mouth, particularly the toothed areas, are brightly coloured and displayed a few seconds before the fishes start to push. In other Cichlids, each fish grabs at the other's jaw, takes a firm hold and then pulls for all he is worth. It must be borne in mind that such grasping with the jaws, and particularly pulling at a strongly resisting object, does not occur in any other situation.

All three behaviour patterns just described precede actual fighting. All of them very obviously serve to "size up" the opponent, to measure the fighting potential of one rival against that of the other before damage is inflicted. A small fish may swim up to a bigger one and display broadside-on, but will collapse and flee the moment the other unfolds his unpaired fins and shows his size and colours. If the difference in size and strength between the rivals is slight, matters may proceed to tail-beating and, if still slighter, to mouthfighting. The combatants must indeed be very equally matched if the observer is to see an actual, damaging fight, which only takes place when the introductory motor patterns have failed to lead to a decision. The damaging fight, in most bony fishes, consists in delivering ramming thrusts at the adversary's flank. As both fishes try to do so simultaneously, this results in their rapid circling round each other, technically termed the "merry-go-round".

In fish, as indeed in most animals, more highly ritualized patterns of fighting have evolved along analogous lines. Almost invariably the threatening behaviour, particularly that part of it, in which physical strength of the fighters is measured *and, at the same time, exhausted*, has developed more and more, *postponing* the breaking out of damaging motor patterns further and further until, in some species, the latter have become vestigial or have disappeared altogether. In this evolutionary process, changes in the thresholds of the different motor patterns play an important role. We know from the comparative study of closely allied species that quantitative changes in the frequency with which certain homologous motor patterns appear, are the most recent steps in the evolution of behaviour. We know that rareness of occurrence generally means that the motor pattern in question has a higher threshold than other "cheaper" and more frequently shown patterns activated by the same kind of excitation. It is a justified asssumption that the differences in the thresholds of the single motor patterns offered a good point of attack to the selection pressure tending to eliminate damage to the individual while preserving the survival functions of fighting.

A beautiful serial stepladder of differentiation can be demonstrated in the fighting behaviour of the many species of Cichlids known to us. It shows very plainly that, in the least differentiated forms, the thresholds of the ramming thrust and the merry-go-round are only very slightly higher than those of broadside-on display and tail-beating. As fighting excitation does not rise smoothly but rather in a saw-toothed curve, the sequence of thresholds leading from display to damaging fight is not directly observable and, in some particularly excitable species, hostilities may even be opened by a ramming thrust that makes some scales fly. In species with a more highly ritualized form of fighting there is a more regular succession of motor patterns showing very clearly their correlation with the several consecutive stages of mounting excitation. With higher ritualization, these stages become progressively more distinct from each other. This has to do with a phenomenon called *typical intensity*, to which Desmond Morris[3] was the first to draw our attention. In most innate motor patterns there is an infinite gradation between slight initial hints of the movement and its full performance. As Erich von Holst[4] has shown, this scale of intensities is very probably caused by slight differences in the thresholds of single neural motor elements: with rising excitation more and more of them are recruited into activity. In some ritualized movements functioning as releasers, the selection pressure exerted by the demand for unambiguity of the signal has tended to diminish, if not to abolish, the *variability of form caused by differences of intensity*. Very probably it has done so by pushing up the lowest thresholds and lowering the highest ones, thus narrowing down the intensely-correlated variability of the motor patterns. This, indeed, is one of the many amazing analogies existing between phyletic and cultural ritualization. Practically all behaviour rendered "formal" by the latter is characterized by being what we call "measured". The measure thereby implied is that of typical intensity. [Some of the films which I am going to show will, I hope, demonstrate this point convincingly.] [5]

We are better informed about the evolution of ritualized fighting in Cichlids than in any other group of animals I know of. But it is to be inferred that forms of intraspecific aggression which achieve their survival functions without doing damage to individuals, have evolved independently, but along analogous lines, in very many other vertebrates. Among lizards of the genus *Lacerta* there are, as already mentioned, species with the most deadly damaging fighting known among subhuman vertebrates, but there are other species of the same genus in which the whole combat consists of a strictly fair wrestling match. In *Lacerta agilis* the males first stand parallel to each other facing in opposite directions, head to tail, each animal flattening its body by active movements of its ribs in such a manner, that it becomes high and narrow, thus increasing its contours as seen from the rival's viewpoint. The green flank as well as a lateral part of the yellow belly are flattened into a vertical plane, offering a striking colour display. After a short time of this highly ritualized threat, both males tilt their heads slightly towards each other, actually offering the heavily armoured occiput to be bitten. Usually it is the weaker animal that takes hold first and tries to shake the opponent. If the latter is much heavier, it often gives up and goes into a submissive attitude even before getting bitten itself. Otherwise there is a regular alternation of bites, every lizard desisting after a ritually determined time and offering, in turn, its occiput to be grabbed by the opponent. Neither G. Kitzler,[6] who very thoroughly investigated the lizards here mentioned, nor myself ever saw a *Lacerta agilis* bite another in any place other than the occiput, nor did we ever observe any other form of damaging fight. The same is true in *Lacerta strigata major*, in which, most surprisingly, the males grab each other at the proffered knee and, thus linked into a circular arrangement, dance round each other in a wrestling match strongly reminiscent of a certain Swiss national sport, the "Hosenwrangeln".

Some reptiles have developed forms of ritualized fighting in which the mouth and teeth are not used at all. The males of the Marine Iguana, *Amblyrhynchus cristatus*, push against each other with their heads which are heavily armoured with large protuberant scales; those of many kinds of poisonous snakes of the order Viperidae perform a complicated wrestling match. The selection pressure causing the complete disappearance of biting seems to have been a different one in both of these cases.

I. Eibl-Eibesfeldt[7] is very probably right in supposing that the extremely sharp teeth of *Amblyrhynchus*, adapted to scrape hard growths from submerged rocks, would do too much damage if used in the frequent territorial fighting of males. This is borne out by the fact that the females of the species, indeed more deadly than the males, hurt each other grievously when, after a short introductory bout of head shoving, they proceed to a damaging fight using their teeth. This, however, they do only once a year when contending for nesting sites, and that only in the one subspecies *A.c. venustissima* which lives on an island where egg-laying sites are scarce. Very probably in vipers the story is a different one. Though immune to their own species' poison, these creatures cannot afford to risk their highly vulnerable hunting weapons in rival combat.

All ritualized fighting which has completely done away with damage inflicted on the combatants but which achieves a decision by the mere exhaustion of one of them, can fulfil its essential survival function only on the condition that the vanquished individual is as effectively and as permanently subdued as if it has suffered serious wounds. Although this is a perfectly logical postulate, one is again and again surprised to observe how completely the loser of a ritualized fight is intimidated and how long he retains the memory of the victor's superiority. It is to be supposed that a very special mechanism must be necessary to make the experience of a lost battle so impressive, in spite of the lack of any bodily damage.

I now come to an entirely different type of ritualized aggressive behaviour which, though most specialised and found in much fewer organisms, seems to me to be of greater importance to our understanding of social behaviour in the higher vertebrates and in Man. It is a fact worthy of deep meditation that, for all we know, the bond of personal friendship was evolved by the necessity arising for certain individuals to cease from fighting each other in order to combat more effectively other fellow-members of the species. It is easy to visualize behaviour mechanisms accomplishing a concerted unanimous attack of many individuals on a non-conspecific enemy. Very many animals are known in which a specific "mobbing" signal elicits an intense and universal attack on a potential predator. Such a response can be blindly mechanical and does not necessitate individual recognition between the creatures participating in it. The problem to be solved by evolution becomes much more difficult when the "enemy" to be attacked is himself a fellow-member of the species. The mates of a pair of Cichlids, defending their territory and their brood against hostile neighbours, stand in dire need of all the aggression they can muster, but they must not fight each other, in spite of the fact that each of them, in its striking aggressive colouration and intense threatening behaviour, offers nearly as good stimulation to attack behaviour as does the neighbour that has to be repelled.

Evolution has found a really brilliant solution for this difficult problem. We know from many other observations that aggression, though evoked by one object, can be easily directed towards another, if inhibitory factors prevent its discharge in the direction of the primarily eliciting stimulation. N. Tinbergen has called this process redirection.[8] In the case of our Chichlids, there are two inhibitory factors tending to deflect attack from the mate and redirect it at the territory neighbour. One is sexual motivation counteracting aggression, and the other, much more interesting one, is habituation, one mate getting habituated to the other as an individual. This, of course, presupposes a faculty for personal recognition. I here defy the philosopher to tell me that a fish does not have a "persona": if it has enough personality to be recognized by another fish as an individual, I regard this as sufficient justification for speaking of personal recognition, especially as it is in a very definite "role"—Latin *persona*—in which the partner is recognized.

In most Cichlid fishes, the obviously immense survival value of attacking the territory neighbour in preference to the mate has caused the redirection mechanism accomplishing this to be ritualized into a reliably rigid ceremony. The motor elements involved retain the form of the patterns of threat behaviour already described, except that they are welded into a fixed sequence and invariably directed, as a final goal, against an object other than the mate. One partner approaches the other in the attitude of broadside-on display and may even deliver a tail-beat or two directed at its mate. But while doing so it does not come to a full stop, as it would if threatening an enemy; on the contrary, it lays a visible and very expressive emphasis on the fact that it is moving and keeps moving towards another goal. Immediately after this ceremony, the fish rushes off in the direction of the territory border, quite literally looking for trouble. The reliability of the mechanism deflecting aggression from the mate still hinges, in most of the Cichlid species, on the possibility of discharging an attack against a hostile neighbour.

B. Oehlert, my daughter-in-law, found that she could keep Orange Cichlids, *Etroplus maculatus*, in permanent marital peace only if she kept two pairs, separated by a clear glass pane, in one tank. It sounds like a joke that her attention was regularly drawn to algal growth rendering the pane opaque by the observation of male *Etroplus* beginning to treat their females in an unkind manner. Cleaning of the glass at once re-directed aggressive behaviour towards the neighbour and restored peace between mates. The beautiful energetic economy of the behaviour mechanism under discussion lies in circumventing the necessity to suppress aggression: far from being suppressed, the aggressive drive aroused by the presence of the mate is actually exploited to perform the all-important function of territory defence.

There are, however, more far-reaching consequences to the ritualization of re-directed attack. Whenever ritualization has the effect of welding together, into one obligatory sequence of movements, a number of hitherto independent elements of behaviour, the whole sequence assumes the character of an independent fixed motor pattern with all its physiological properties. It is linked to its own releasing mechanism and, in case the stimulus situation activating it is withheld, it activates its own appetitive behaviour directed at its consummation. Now the releasing mechanisms of all ceremonies derived from redirected aggression contain, as their most important element, an individually acquired familiarity with the area to be defended; so do the mechanisms releasing ritualized redirection of attack depend on the acquired familiarity and acquaintance with an individual fellow-member of the species. In other words, any organism possessing a ceremony derived from re-directed aggression is bound to the individual object of this activity with very similar and just as strong bonds as a highly territorial animal is bound to its home. Monika Meyr-Holzapfel,[9] in trying to find an objective expression to describe, in animal sociology, what we would simply call a friend in common parlance, has coined the term "the animal with the home valence". I don't think I could find a more honorific title for my wife.

Indubitably, ritualized aggressive behaviour is at least one root of bond behaviour. The latter can be defined as the keeping together in space of two or more individuals by a set of responses which each of them selectively elicits in the other. We know neither whether all bond behaviour has arisen out of aggression, nor whether ritualized redirection of aggression is its only origin. Both are certainly true for ducks and geese, which have been extensively and intensively studied by ethologists in the last decades. The bond of lifelong individual friendship keeping together wild geese and determining, by its immense strength, the whole structure of their society, is demonstrably based on the so-called triumph ceremony which, also demonstrably, originated in a way strictly analogous to that of the ritualized redirected attack in Cichlids I have tried to describe.

There may be other independent ways in which bond behaviour has evolved, but wherever it does, it seems to have done so as a means of controlling aggression, that is to say on the basis of aggressive behaviour pre-existing. In the Canidae for instance, in the dog-like carnivores, all gestures and ceremonies of greeting, love and friendship are obviously derived from the expression movements denoting infantile submission. It is quite conceivable that appeasement ceremonies, with high ritualization, have become independently autochthonous motor patterns whose performance constitutes as great a need for the organism as does that of ritualized redirected aggression in the case of the geese's triumph ceremony.

How much of the primarily motivating aggression may still be contained in ritualized redirected attack, or, for that matter, in any behaviour patterns affecting bond behaviour, cannot be deduced from their similarity to or dissimilarity from threat and fighting, but must be investigated separately in every single case. In the triumph ceremony of geese there is certainly quite a lot of autochthonous aggression, as can be demonstrated in the quasi-pathological case of homosexual gander pairs. In these, bond behaviour is much more intense than it ever is in normal heterosexual pairs, occasionally reaching a true ecstatic climax. As in other known cases, abnormally high intensity of ritualized activity causes true aggression in Freud's sense, that is to say to a recrudescence of the phylogenetically older, unritualized behaviour patterns. In other words, ritualized redirection suddenly breaks down and the partners proceed to fight with a fury never otherwise observed in goose combat. In Cichlids, even under normal circumstances, the danger of redirection failing and attack being launched at the mate, is forever present. In homicide as every policeman knows, the loving spouse is the most likely suspect, the word "loving" emphatically not being used ironically.

The strongest reason, however, which makes me believe that all bond behaviour has evolved, by way of ritualization, on the basis of intraspecific aggression, lies in an unsuspected correlation between both. We do not know, as yet, of a single organism showing bond behaviour while being devoid of aggression; in a way, this is surprising, as, at a superficial appraisal, one would expect bond behaviour to evolve rather in those highly gregarious creatures which, like many fish and birds, live peacefully in large schools or flocks, but this obviously never happens. The great assemblies of these animals are always strictly "anonymous", even in the birds of such high organization as starlings, as G. Kramer[10] has conclusively shown. The dependence of bond behaviour in intraspecific aggression is most strikingly demonstrated in those species in which a regular seasonal change takes place between aggressiveness and schooling or flocking. In these cases, whether they concern fish or birds, all individual ties are dissolved immediately when the organism changes from its aggressive to its non-aggressive phase. Also, there seems to be a strong positive correlation between the strength of intraspecific aggression and that of bond behaviour. Among birds, the most aggressive representatives of any group are also the staunchest friends, and the same applies to mammals. No more faithful friendship is known in this class than that which S. Washburn and I. de Vore[11] have shown to exist among wild baboons, while the symbol of all aggression, the wolf, whom Dante calls the "bestia senza pace", has become "man's best friend", and that not on the grounds of properties developed in the course of domestication.

Of course, the relationship between bond and aggression is entirely one-sided. We have reason to believe that intraspecific fighting evolved millions of years earlier than bond behaviour, as indeed all present-day reptiles show the first, while being entirely devoid of the second. But, to the best of our knowledge, bond behaviour does not exist except in aggressive organisms. This certainly will *not* be news to the student of human nature, to the psychiatrist and the psychoanalyst. The wisdom of the old proverbs as well as that of Sigmund Freud has known for a very long time indeed how closely human aggressiveness and human love are bound together.

In conclusion of this highly condensed and correspondingly in-complete presentation of ritualized intraspecific aggression and of its most important consequences, I want to add a few words which, I hope, may stimulate discussion. When Julian Huxley, nearly thirty years ago, coined the word ritualization, he used it, without any quota-tion marks, for a phylogenetic process as well as for a cultural one; in other words, the conception he associated with the term was purely functional. The functional analogies of phylogenetic and cultural ritual-izations are indeed so profound and reach into such amazing details that a concept embracing both is fully justified. Indubitably, one of the most important functions phyletic ritualization has to perform in the interest of a species' survival is the one discussed in this paper, the controlling of intraspecific aggression. It is to be hoped that cultural ritualization will prove able to do the same with that kind of intra-specific aggression in Man which threatens him with extinction.

There cannot be any doubt, in the opinion of any biologically-minded scientist, that intraspecific aggression is, in Man, just as much a spontaneous instinctive drive as in most other higher vertebrates. The beginning synthesis between the findings of ethology and psycho-analysis does not leave any doubt either, that what Sigmund Freud has called the "death drive", is nothing else but the miscarrying of this instinct which, in itself, is as indispensable for survival as any other. In this symposium there has been a most satisfying agreement, on this point, between psychiatrists, psychoanalysts and ethologists.

However, it comes very hard to people not versed in biological thought to concede that Man, with a capital M, still does possess in-stincts in common with animals. That particular kind of pride which proverbially comes before a fall prevents men from understanding the workings of their own instincts, including that of aggression. As it is causal insight alone which can ever give us the power to influence chains of events and to direct them to our own ends, it is highly dangerous to assume the ostrich attitude in respect to the nature of human instincts. Science is often accused of endangering humanity by endowing it with excessive power over nature. This reproach would be justified only if scientists were guilty of not having included Man among the objects of their research. They have indeed done so, but they have earned no thanks in return. Men like to think of themselves as something outside and above nature. They dislike hearing what a small part of nature they really are and they hate the thought of being subject to its universal laws. They burned Giordano Bruno when he told them that their planet was only one particle of dust in one small dust cloud among innumerable other, bigger ones. When Charles Darwin discovered that they are descended from animals, they would fain have burned him, too; they did their best to silence him in other ways. When Sigmund Freud undertook to investigate the deeper springs motivating human social behaviour, by methods which, though implying the study of subjective phenomena, still were those of inductive natural science, he was accused of lack of reverence, of materialistic blindness to all values and even of pornographic tendencies. Humanity defends its self-conceit with all means, fair and foul, and it seems sadly necessary to preach that kind of humility which is the prerequisite for recognizing the natural laws which govern the social behaviour of men.

To put it very crudely: If we know enough about the functions of our intestinal tract to enable medical men to cure many of its disorders, we owe this ability, amongst other things, to the fact that men were never prevented by excessive respect from investigating the physiological causality prevailing in the workings of their bowels. If, on the other hand, humanity is so obviously powerless to stem the pathological disintegration of its social structure and if it behaves, as a whole, in no way more intelligently than any species of animals would under the same circumstances, this alarming state of affairs is largely due to that spiritual pride which prevents men from regarding themselves and their behaviour as parts of nature and as subject to its universal laws.

FOOTNOTES

* Without such a correlated other function, rival fighting can lead to intraspecific selection which very often effects evolutionary changes not in the interest of the survival of the species.

[1] Beobachtungen uber Paarungsbiologie und soziales Verhalten von Mauereidechsen (Observations on the sexual behavior and social restraints of lizards). Z. Morphol. Oekol. Tiere, 1937, Vol. 32, no. 4.

[2] Elephant Bill. London, 1950.

[3] British zoologist and author of The Naked Ape (London: Jonathan Cape, 1967). See "Typical intensity" and its relation to the problem of ritualization, Behaviour, 1957, 11: 1-12.

[4] See, for example, Holst, E. v., and Saint-Paul, U. v., On the functional organisation of drives. Animal Behavior, 1963, 11, 1-20, translated from Naturwissenschaften, 18, 409-422.

[5] This point was emphasized at the time of the symposium by the use of some filmed presentations prepared by the author.

[6] Die Paarungsbiologie einiger Eidechsen (The sexual behavior of lizards). *Z. Tierpsychol*, 1942, 4: 353-402.

[7] Der Kommentkampf der Meerechse (Amblyrhynchus cristatus Bell) nebst einigen Notizen zur Biologie dieser Art (Ritualized fights of iguanas and some notes on the biology of the species). *Z. Tierpsychol.*, 1955, 12: 49-62. See also Eibl-Eibesfeldt, *Ethology: The Biology of Behavior*. New York: Holt, Rinehart, & Winston, 1970, pp. 317-318.

[8] *Redirection* as used here has the same meaning as the psychoanalytic term *displacement*. For further information on the term see Tinbergen, N., Comparative studies of the behaviour of gulls (*Laridae*): A progress report. *Behaviour*, 1959, 15: 1-70.

[9] Soziale Beziehungen bei Saugetieren (Social relations of mammals). In F. E. Lehmann (Ed.), *Gestaltungen sozialen Lebens bei Tier und Mensch*. Bern: Francke, 1958, pp. 86-109.

[10] Uber individuell und anonym gebundene Gemeinschaften der Tiere und Menschen (On individual and anonymous bonds in communities of animals and men). *Studium Gen.*, 1950, 3: 564-572.

[11] The social life of baboons. In C. H. Southwick (Ed.), *Primate Social Behaviour*. Princeton, N. J.: Van Nostrand, 1962.

PART 2   Nurture Theory

The evidence concerning the biosocial
nature of man, as we know it today, does
not support the notion of an aggressive,
death, or destructive instinct in man ...
So far as the development, by evolutionary
means, of aggressive tendencies in man is
concerned, the idea can be thoroughly
dismissed.
    Ashley Montagu

Before beginning to read this section on nurture theory, it might be useful for the reader to close the book for a moment and consider the variety of stimuli that are impinging on his senses. For example, there may be a radio playing in the background, sunlight shining through the window, trees swaying in a breeze, or there may be the smell of tar as men repair the streets. All these stimuli have some effect on behavior. Collectively these environmental factors may be called "nurture variables." The word "nurture" refers to those external events that influence behavior. These may be such things as falling out of bed, love and consideration provided by parents, or stimulation supplied by social agencies such as schools, the church, and so forth. It would be impossible in this short space to discuss all of the variables that influence our behavior. More relevant to our discussion are variables leading to or anticipating behavior that may be labeled "aggressive."

    Most of us at one time or another have heard of people who were "born bad." In the history of psychology there have been many reports of families or individuals within families who, one way or another, have behaved strangely, presumably as a result of "bad blood." Such cases were discussed as early as 1875 by Richard Dugdale, who reported on a family known as the Jukes. Dugdale covered seven generations of the Jukes including 540 blood relatives and 169 of those related by marriage. Dugdale's report was somewhat sketchy in places and when he failed to find support for his contention of the genetic cause of crime, he stretched the truth to some extent to fill in the gaps. But the unusually high percentage of familial participation in criminal acts seemed curiously related to some inherited characteristics. Another famous family, the Kallikaks, were studied by Henry Goddard in a report published in 1912. Goddard's contention was that the Kallikaks' regrettable history resulted from the union of Martin Kallikak, Sr., with a feeble-minded girl. The offspring of this union gained a reputation for criminal behavior. More recently, Richard Speck, the convicted murderer of eight student nurses in Chicago, was found to have a strange combination of chromosomes, known as the XYY syndrome, which has been interpreted as evidence that chromosomal makeup may influence criminal behavior. This interpretation remains a possibility although recently studies have reported conflicting results. In all these cases, however, environmental causes cannot be dismissed. The behavior of the Jukes and the Kallikaks may have been *learned* rather than *inherited*, and it is likely that the circumstances surrounding their birth and upbringing contributed a great deal to their later behavior.

What, then, are some of the variables affecting our lives that are not of a genetic nature? William Mason, in a series of studies with monkeys, has found that monkeys reared with mother and peer contact behave quite differently from those reared without these contacts. The former usually mature into sexually potent and physically normal adults. A monkey raised in a laboratory with little or no maternal or peer contact does not develop into a healthy adult and a number of bizarre behavioral anomalies are associated with the effects of maternal and peer deprivation. For example, an adult isolate cannot "read" the nonverbal signals of his conspecific peers and does not know when or how to be submissive. Similarly, isolates do not know how to react to the sexual signals of females of their own kind. Self-biting is a frequent occurrence with these animals and they resort to other strange behaviors when excited. These animals are highly aggressive and their fights are longer and more severe than in normal animals. We can say, therefore, that an important environmental variable is missing in the case of the isolate-reared monkey. There are many instances in which human babies are raised under similar conditions, and it is not difficult to understand how serious a problem this can be for the individual and society.

It is difficult to ascertain which variables are most critical in the normal development of a human being and thus which factors may be responsible for the appearance of aggressive or hostile behavior. Some of the more interesting ideas of how aggressive behavior originates are discussed in the articles that follow. One of these is known as the "frustration-aggression hypothesis." Essentially this hypothesis states that frustration generally leads to aggression, sometimes of a concealed nature. Conversely, aggression is also an indicator for a frustration of some kind. First proposed by Neal Miller and his associates in 1939, the hypothesis has undergone considerable modification and it has proved to be most useful in generating further research. Leonard Berkowitz, for example, has compiled large quantities of data from his research which, in part, are concerned with selected aspects of this hypothesis (Reading 9).

It is relatively easy to identify environmental or nurture variables *per se*, but it is somewhat difficult to delimit them; that is, to establish where nature variables end and nurture variables begin. At what point does biology turn into sociology? Some studies have suggested that behavior in the absence of an observable stimulus, the so-called "vacuum response," is evidence that the buildup of action-specific energy has caused the response. But how do we know that some external stimulus, unknown to us, has not triggered the response? We are partly at the mercy of our limits of observation. Thus the dichotomy between nature and nurture variables becomes, at best, somewhat arbitrary. However, predominantly social or cultural influences can be identified, at least if we choose to subject a phenomenon, such as aggression, to that level of analysis.

It is important to realize that the arguments presented in this section do not exclude the possible effects of a person's genetic history. It is quite conceivable that a person could inherit, for example, a quick temper and therefore be subject to more frequent acts of rage. But the predominant emphasis in this section is on the social origins of aggressive behavior. Included in this category is the way that aggression is maintained and transmitted and the events that influence its mode of expression. A quick temper must have an environment for its expression, and the cultural milieu in which it develops may or may not allow it to reach its full potential.

SOCIAL LEARNING THEORY AND AGGRESSION

A theory of importance to our topic—social learning theory—has been developed by Bandura and Walters (1959, 1963). As we have not included it in this section it is appropriate to discuss it briefly.

Social learning theory encompasses more than the topic of

aggression. However, the two experimenters have spent much of their time during the past few years in investigating aggression in children. In essence social-learning theory (as it relates to aggression) states that an individual may learn aggression by *observation* and *imitation* of an aggressive model. Thus the rewarding consequences of the behavior (aggression) are transmitted to the observer *vicariously*. As the experimenters point out, the consequences of the observed behavior are crucial for the observer. Children who observe an aggressive model rewarded by the response will display more aggression than those who observe an unrewarded or punished model. John Wayne may be cited as an example of a successful model for aggression. The "bad guys" who are continually defeated by him are, one might say, examples of unrewarded or punished models.

In addition, it should be mentioned that "real-life" experiences play an important part in shaping aggression. If for example, the child observes successful aggression on television, but fails to gain rewards when he himself uses aggression, he is unlikely to develop aggressive behavior to the same degree as the child who is successful at practicing it. Vicarious rewards can suggest the pattern of behavior, but the observer must learn for himself how well the response(s) will work for him. Thus *reinforcement* plays a leading role in the development of aggression as portrayed in the social-learning theory. Naturally, this theory is relevant to our subject and should be taken into consideration when the papers that follow are examined. The role of imitation and reinforcement on the development of aggressive responses by animals, for example, is frequently overlooked in animal behavior experiments.

One final issue which is germane to the present discussion is the controversy over violence in television programs. At the root of this is an old argument that is related to our basic issue of nature vs. nurture. We may phrase the question as follows: Does the observation of violent behavior *instigate* violent behavior on the part of the observer, or does it *reduce* the probability of aggression by means of vicarious experience? If one believes in the notion of *catharsis*, then one expects that aggressive energy would be dissipated by observation of aggression. However, we have seen that according to the data of social-learning experiments, aggression may be imitated after observation and increased if that imitation is reinforced. Which point of view is correct? This example illustrates how theoretical issues can have important consequences for the lay public. An interesting contribution to this research problem is found in Reading 8.

REFERENCES

Bandura, A., and Walters, R., *Adolescent Aggression*. New York: Ronald, 1959.

———, *Social Learning and Personality Development*. New York: Holt, Rinehart and Winston, 1963.

Dollard, J., Doob, L. W., Miller, N. E., Mowrer, O. H., and Sears, R. R., *Frustration and Aggression*. New Haven: Yale University Press, 1939.

Mason, W. A., The effects of environmental restriction on the social development of rhesus monkeys. In: Southwick, C. H. *Primate Social Behavior*. New York: Van Nostrand Rheinhold, 1963, 161-173.

Montagu, M. F. A., Chromosomes and crime. *Psychology Today,* October 1968, 2:42-48.

SUGGESTED READINGS

Bandura, A., and Walters, R., *Adolescent Aggression*. New York: Ronald, 1959

Antisocial aggression in adolescents is discussed within the context of the authors' extensive research on modeling and imitation by children. Research data are included and many practical problems and treatment procedures are discussed.

Berkowitz, L., *Aggression: A Social Psychological Analysis*. New York: McGraw-Hill, 1962.

A theoretical framework is developed from research in social psychology. Human studies are emphasized, although findings with animal subjects are included when relevant. A very comprehensive treatment of various nature theories introduces the discussion.

Chagnon, N. A., *Yanomamö: The Fierce People*. New York: Holt, Rinehart and Winston, 1968.

An anthropological field study. Chagnon details the way of life of a fierce tribe in South America. The author provides much useful information regarding the function and transmission of a violent way of life.

Dollard, J., Doob, L. W., Miller, N. E., Mowrer, O. H., and Sears, R. R., *Frustration and Aggression*. New Haven: Yale University Press, 1939.

This book has probably generated more research than any other on the subject of human aggression. The research is dated, but the times in which it was written are illuminated by many applications of the hypothesis to then current events.

Fried, M., Harris, M., and Murphy, R. (eds.), *War: The Anthropology of Armed Conflict and Aggression*. New York: Natural History Press, 1968.

This book contains many interesting contributions and is interdisciplinary in approach. The papers and discussions were originally part of a symposium held by the American Anthropological Association in 1967.

Montagu, M. F. A. (ed.), *Man and Aggression*. New York: Oxford University Press, 1970.

Devoted to discrediting the theories of Robert Ardrey and Konrad Lorenz, this collection of articles from many well-known contributors criticizes nature theory as it is represented in the work of Lorenz and Ardrey. Alternative points of view are discussed.

4   Why Do Animals Fight?

The fact that all animals fight has attained immense importance in our day, because it is used as an argument in favor of the doctrine that men also should fight one another, that warfare ought not to be abolished. This doctrine I shall speak of for convenience as militarism; which is far preferable to calling it Nietzscheism or Trietschkeism or Prussianism, for all such names are invidious and more or less unjust. Militarists are at work in every nation, and in every nation they emphasize what they call the "biological" argument for war. They paint lurid pictures of "nature red in tooth and claw," dwelling on the many instances of rapacity, cruelty and destruction which undoubtedly occur in nature. They claim that theirs is a true picture of the life of animals, and also of the natural life of man. Their argument looks plausible, and it furnishes entertaining reading for the populace. But I believe it to be fallacious, partly because it exaggerates the cruel facts in nature, but far more because it misinterprets their meaning.

The attempts to refute the biological argument for war, so far as I have seen them, have been inadequate, some of them even absurd. Some pacifists have claimed that "No animal fights its own kind." Now, if the word pacifist means a person who longs to see war abolished, and who is willing to labor to the very best of his ability toward that end, I am myself a pacifist. But I believe that the cause of truth is more fundamental than the cause of any one man's theory as to how to make peace. Let us tell the truth, regardless of consequences. And the essential truth in this matter is that every animal fights its own kind. If we wish to discover any biological support for a policy of pacification, we must not seek to do it by asking "Do animals fight?" That question is

From the *International Journal of Ethics*, 31, 1921, 264-278. Reprinted with permission of the University of Chicago Press.
Along with Darwin, Whitman, and Heinroth, Wallace Craig is generally credited as being a pioneer of the science of ethology. It is most interesting that he should depart from the widely held notion of a biological instinct for fighting. This early and important paper is frequently overlooked in discussions on animal fighting.

not worth investigation, because the answer is already known. We must change the question and ask "Why do animals fight?"

The militarists are ready with an answer. They claim that the means by which a race progresses, or even maintains its present standard, is the killing of the less fit in the struggle for existence. And if the struggle to the death should cease, the race would degenerate. This presupposes the truth of the theory of the all-sufficiency of natural selection. Now, I shall not allow myself to be drawn into the debate between the rival theorists in genetics. I favor no one proposed solution of the problem of heredity and the method of evolution. On the contrary, I emphasize that this problem is not solved and that, therefore, the militarist has no right to base an argument for war upon one particular theory as to the origin of species. Our knowledge of genetics is not sufficient, and we doubt whether it ever will be sufficient, taken by itself, to settle the debate either for or against the biological argument for war. The pacifist is right in saying that the argument for war is "not proven," but he cannot say that it is positively disproven, by what we know as to the method of evolution. To find the evidence on which to base a positive verdict, we must turn away from the theories of evolution—which are all highly speculative—and examine at first hand the facts as to fighting among animals.

To understand why an animal fights, we must watch its fighting behavior, and also study the relation of its fighting to its other behavior, to its life history—in short, to its whole economy. This cannot be done by the old-fashioned method of surveying the entire animal kingdom, collecting therefrom a mixture of fragmentary and miscellaneous information. On the contrary, our problem can best be attacked by an intensive study of one species, or one group of related species. And if it can be shown that even one flourishing group of animals has evolved into its present prosperous state without its members engaging in internecine strife, that is enough to prove that warfare is not necessary for evolution.

My own specialty, as a student of behavior, is the Columbidae, and I shall use pigeons as my chief examples. But I have studied other birds and mammals sufficiently to be sure that the statements made in this paper have a wide and general application among them. The birds and mammals are the most important animals for our problem. And the pigeons are a properly representative group: because, first, their behavior is typical; they quarrel and fight just about as much, or as little, as do the majority of birds. A healthy pigeon never allows another to trespass on his territory, or in any way interfere with his interests, with impunity. And, secondly, the pigeons are a "dominant" group; that is to say, the pigeon family is found all over the world, it has evolved into a large number of species, and the number of its individual members is enormous. All signs indicate that the pigeon family is in the most flourishing condition and in a state of rapid, progressive evolution. If the members of such a group live and act in a manner contradictory to the militarist theory, this is sufficient to prove that the militarist policy is not necessary for the welfare or the evolution of a race.

A friendly critic has asked me whether the generalizations presented in this article are true of carnivores such as the lion. In reply I would say that I believe they are. The published accounts of leonine behavior, so far as I have read them, indicate that our knowledge of the lion is fragmentary and incomplete. But the lion is only a large cat. Feline behavior is best known in the common cat. All the generalizations set forth in the present article are true of cats. Some of them are less true of cats than of pigeons, but, on the other hand, some of them are more true of cats than of pigeons. For example, the felines "space out" much farther than pigeons do, and consequently come less into conflict with each other. The terrible powers of the carnivores are exercised chiefly upon their prey, not upon their own kind. I never knew of a cat being killed by another cat in a fight. And our problem concerns only fighting between animals of the same species.

With reference especially to the higher vertebrates, we shall maintain and defend the two following theses: I. Fundamentally, among animals, fighting is not sought nor valued for its own sake; it is resorted to rather as an unwelcome necessity, a means of defending the agent's [1] interests. II. Even when an animal does fight, he aims, not to destroy the enemy,[1] but only to get rid of his presence and his interference.

I. The animal fights in order to gain or to retain possession of that which is of value to him, such as food, mate or nest. With animals, as with men, the cause of a quarrel is very commonly a coveted territory. My dog drives away other dogs from my house and yard. In general, each agent drives away other animals from his own nesting place, his chosen place for sleeping at night, his place for basking in the sun, or other territory which he can appropriate and use. Two animals fight only when their interests conflict. This is the fundamental fact in regard to infra-human fighting.

Animals do not enjoy fighting for its own sake. Unless his anger is aroused, the agent's behavior indicates that he has no appetence[2] for the fighting situation; he does not seek it; when in it he does not endeavor to prolong it; and he reveals by his expressions that he does not enjoy it. On the contrary, fighting belongs under the class of negative reactions or aversions;[2] it is a means of getting rid of an annoying stimulus. As McDougall says, the stimulus of the instinct of pugnacity is the thwarting of some other instinct. If the animal's instincts are not thwarted, if annoying stimuli do not thrust themselves upon him, he will never fight. He fights only when he is attacked, or threatened, or his interests are interfered with. He does not go in active search of a fight, except in play, in which no injury is done to either side. Of course it is true that when necessity compels him to fight, he shows eagerness to attack, and joy when he achieves the victory; these are necessary in order that he may be a good defender of his rights. But when we try the experiment of keeping a bird in such a peaceful environment that he never tastes the jobs of battle (except in harmless play), he shows no sense of loss, but is manifestly happier than the birds that fight. The pigeon, unless his temper is aroused, has no appetence for a battle. He has appetence for a great many other objects; as, water, food, mate, nest: if kept without such appeted objects he shows distress, tries to get out of his cage, and in every way makes clear to us that he is seeking the appeted object. But when he is kept without enemies, he never manifests the least appetence for them. This is true of all birds and mammals, so far as I know them. I am sure that it is true of the common fowl, although the cock is one of the most celebrated of fighters. It is a popular error to suppose that a bird such as the game cock, which shows unyielding bravery when in a fight, must necessarily be an aggressive bird, seeking the job of battle for its own sake. The fact is that in the poultry yard the game cock is not a quarrelsome bird; if neighboring cocks do not unwarrantably thrust their presence upon him he lets them alone and attends to his own affairs.

At this point the question naturally arises, Why do animals fight so much as they do? For it is undeniable that under certain conditions there is a very great amount of fighting among them. In seeking an answer to this question it is important to notice the circumstances under which the excessive amount of fighting takes place. One of these circumstances is that of caged animals which are crowded so closely together that they constantly fall afoul of one another. Pigeons, if thus crowded in quarters that are too small for them, fight to a degree that is cruel and distressing. Since each pair of birds insistently drive trespassers away from their nest and from a certain territory around their nest, if the pigeon-keeper crowds the nest-boxes too closely together constant fighting must inevitably result. Such a state of affairs is not natural. It does not exist in any species in a state of nature. And it has given to some theorists an exaggerated notion as to the frequency of fighting among animals. In a great many wild species the amount of fighting is, I am sure, extremely slight.

Indeed, anyone who has watched wild animals with a philosophical eye must have been struck by the beauty and delicacy of the adjustments through which they avoid collisions with their fellows. Thus, in a flock of flying geese each goose maintains with astonishing precision the standard distance between himself and the goose ahead of him. When a great number of robins are searching for worms on a large lawn, each keeps with a pretty faithful constancy a distance of two or more yards between himself and any other robin. This tendency to "keep one's distance" is so widespread among animals and so various in its manifestations that it constitutes a study in itself. It is known to naturalists as "spacing out." Birds "space out" their nests with similar regularity, the distance varying with the species. The bank swallow burrows within a foot or two from the burrow of his neighbor; the kingfisher prefers a distance which I should estimate at about a half-mile. The interesting fact about these myriad cases of spacing out is that nearly all of them (of course, not quite all) are adjusted without fighting. Evidence indicates that in a vast number of cases the animal seeking a nesting-place, and finding one that is already occupied, peacefully passes on and looks elsewhere. These quiet cases of adjustment are likely to be overlooked by the observer; the cases of fighting are conspicuous and thus seem to be a greater proportion than they really are.

Notwithstanding all these facts, there remain cases in which two animals, even among those that are in a free and natural state, may be observed to quarrel and fight throughout a whole day, and to renew the quarrel when they wake the next morning. The reasons for such excessive fighting may be stated briefly as follows:

The adjustment of conflicting interests requires intelligent co-operation and some degree of social organization. Most animals, because of the low state of their intelligence and of their social organization, have a narrowly limited power of co-operating. That is why their differences must be settled so often by fighting. The following illustration, though simple, shows the point clearly. If we set up a new pigeon cote containing several compartments, each with its own door, and allow the pigeons to choose compartments for themselves, it may happen that two males will choose different doors from the very first, in which case they may live side by side in peace. But it may happen that on trying the new dove cote both males become enamored of the same door and each tries to enter it and make it his own. If these birds were endowed with reason, one of them would address the other in this wise: "Friend, in this dove cote there is plenty of room for you and me and for our families. Let us agree that you shall use the right door and I the left." But since pigeons are not endowed with reason they cannot make such a conceptual agreement; if both birds have chosen the same door they can adjust the difficulty only by fighting for it. In short, the reason why animals fight is that they are too stupid to make peace.

That this is the true explanation is indicated by the fact that if we lend the birds our reasoning power, if we act as arbitrator and settle their disputes for them, they gladly accept our adjustments and live in peace. Thus, in the case of the two pigeons fighting for possession of the same compartment in the dove cote, if I take one of the birds and keep him in a different compartment until it becomes "home" to him, then I can let him out, he will return to his new home and leave his neighbor at peace in the one which he had first chosen. This is an illustration of what we said before (p. 68), that if a man who keeps a flock of birds acts as a "benevolent despot" among them, administering justice and successfully resolving all conflicts of interest, the birds under his rule show no desire to fight, and are happier than those that do fight.

The amount of quarreling among animals varies (other conditions being equal) with the degree of their stupidity. Some individual pigeons are much more quarrelsome than others. The truculence of some individuals is due to the fact that they were reared under unnatural conditions which kept them in constant brawls. If as adults they are allowed to live a free, normal life, they outgrow their excessive quarrelsomeness. As their experience increases they tend more and more to adjust their differences with their neighbors without fighting. This fact is of great importance for our problem; because it proves that as viewed by the birds themselves fighting is not a good, it is a necessary evil.

The law that aggressiveness does not pay is conspicuously true among animals. The contentious pigeon, as every fancier can testify, brings disaster not only to his neighbors but also to his own cherished home and family. Hence it is not surprising that in the course of evolution the birds have become as gentle and peaceful as we know them to be, and that even the individual bird strives, with his limited intelligence, to adjust the difficulties that arise between himself and his neighbors and to avoid actual fighting.

II. These last observations naturally lead up to our second thesis, which is, that even when an animal does fight he aims, not to destroy the enemy, but only to get rid of his presence and his interference. This point is important, for if the militarist theory were correct, that the function of fighting in the economy of nature is the elimination of the "weaker" individuals or groups, then we ought to find that the behavior of the fighting animal is directed toward the extinction of the enemy's line of descent, as, by the destruction of the reagent himself and also his eggs and young. But the behavior of animals does not conform to this theory. Especially among the higher vertebrates, observation shows that they do not follow the militarist policy, and without it they live and thrive and progress.

The only animals which in any degree follow the militarist policy are, I believe, those of parasitic habit. Some of the ants, bees and wasps have become "robbers," systematically destroying their congeners, even those of their own species, in order to appropriate their stores of larval food. It is a sad and discouraging fact that wherever in the animal kingdom some members have developed sociality and co-operation and thrift, their very prosperity has furnished opportunity for the breeding of a race of despicable parasites. But there is a grain of consolation in the fact which has long been known and has recently been emphasized by Professor Wheeler,[3] that the parasitic habit leads, in the course of time, to the destruction of the parasite itself. A specialist on the hymenoptera could probably write a positive refutation of the militarist theory, by showing that those animals which do follow a militarist policy thereby lead to their own extinction.

But I am not a specialist on insects. Among the higher vertebrates, parasitic behavior is found only in rare cases, as in that of the European Cuckoo. We are told that the young cuckoo works systematically to push its nest-mates out of the nest and thus destroy them. Such a systematic attempt to destroy its rivals is, I am sure, not to be found in any non-parasitic species. The young cuckoo exhibits the behavior that would be found in other animals if they followed a militarist policy, and makes clear to us by contrast that they do not follow a militarist policy. The nonparasitic bird or mammal aims, as we said before, not to destroy his rival, but only to free himself from his presence and his interference.

To prove this thesis we shall present an analysis of the animal's fighting behavior. We shall inquire especially as to the behavior of that individual animal who is the better fighter of the two in the contest, the one destined to be the winner if the battle be fought out. With such an individual as agent, the reagent may do one of three things: namely, (a) flee; (b) submit; (c) persist in fighting. We shall describe the behavior of the agent in these three cases.

(a) If the reagent flees, the agent does not pursue him indefinitely and seek to destroy him. On the contrary, he pursues him only far enough to eliminate him from the field of interest about which the battle is being fought. Thus, when a dog has driven an intruding dog out of the territory over which he claims sovereignty, he does not pursue him farther (unless there be some other cause for the pursuit), but by barking and other expressions of triumph and satisfaction he shows that his end has been fully accomplished. So it is with all animals that I have ever observed driving others away from their territory or their mates or their food. That is all that the agent cares to accomplish. He shows no tendency to pursue the enemy to the death.

That this generalization is true and fundamentally important is evidenced by the fact that in many groups of animals, indeed in probably all animals in some degree, there has been an evolutionary change from destructive forms of fighting to forms of fighting which are merely expressive or ceremonial, which drive away the reagent by threatening or warning him without doing him any injury. This is a part of what Hocking[4] has named "the dialectic of pugnacity." The pigeon warns his enemy by a display which is highly ceremonious, consisting of elaborate cooing and gesturing. The celebrated mock-battles of the capercaillie and of a great many other birds furnish other instances of the same trend in evolution. All birds and all mammals are endowed with instincts to threaten the enemy, to make feints, to hiss or growl or roar, or to vent their anger in other expressions which serve to warn the reagent and often cause him to flee without a blow being struck. It is important to notice that these attempts to drive away the enemy are used first, before the physical combat. If the aim were to destroy the enemy he would be attacked silently and by stealth: but in most species he is not so attacked; he is first warned, and given every opportunity to withdraw from the field. In a great majority of the conflicts among animals, the ceremonial combat is all that is needed and all that is used: the reagent may withdraw as soon as he is threatened; or he may at first make a counterdisplay, but withdraw on discovering that the agent is more determined than he. The physical combat is resorted to only after the ceremonial has been tried and has failed to settle the dispute.

(b) If the enemy submits, the agent ceases fighting. In pigeons this is witnessed again and again. In the heat of battle the agent may rush upon his enemy, jump on his back, peck him with all his might, and pull out his feathers. But if the reagent lies down unresisting, the agent's blows quickly diminish into gentle taps, he jumps off his prostrate foe, walks away, and does not again attack the enemy so long as he is quiet. This behavior is typical, and it proves that the pigeon is devoid of any tendency to destroy his rival.

Further study of this behavior indicates that it is not merely negative, not merely the absence of an impulse to destroy. The bird has a positive impulse to quit fighting a non-resisting bird of his kind. One explanation of this impulse is to be found in the mode of instinctive sex recognition. When a male meets a stranger belonging to his own species, provided this male has not learned by experience to discriminate the sexes, the only discrimination he shows is this: if the stranger fights, the agent treats it as a male; if the stranger refuses to fight, the agent treats it as a female; if the stranger first fights, then submits, the agent treats it first as a male, then as a female. This mode of sex recognition is so widespread in the animal kingdom that it seems to be fundamentally ingrained in the nature of the male. Audubon tells us that when he watched a battle between two wild turkeys, when one of them had been defeated, he was surprised to see that the victor, instead of injuring him, showed toward him the amorous behavior which is generally accorded to a female. Audubon need not have been surprised. Behavior of this sort is now known to be characteristic of a great many animals ranging all the way from the lower invertebrates to the Primates. Thus we see that in the male animal there is a fundamental trait which tends to prevent him, and in most cases does prevent him, from doing any injury to a non-resisting member of his species.

We showed under (a) that the enemy who flees is not pursued to the death. We have now shown under (b) that the enemy who submits commonly finds his life spared. The non-destruction of the enemy is more certain under (a) than under (b). An enemy who has fled from the field of operations is perfectly safe. It is but natural that an enemy who has submitted is not perfectly safe; the pugnacity may smolder, and may break out again into violent fighting. The facts are too complex to be treated in this short article. Suffice it to say that in those rare cases in which an animal, or a group of animals, kills a non-resisting member of the same species, such killing is in various ways exceptional or accidental;[5] it is not a policy, not a common and regular form of behavior, and very far from being a systematic pursuit.

(c) When the reagent refuses to flee and refuses to submit, the agent is obliged to resort to physical force. It is extremely interesting to notice that even in this case the physical force used is often of a form which serves merely to rid the agent of his enemy without doing him any hurt. Thus, when the common pigeon quarrels with his neighbors on one of the high ledges on which they like to perch, his principal method of dealing with his opponent is to seize him by the nape of the neck, drag him to the edge of the ledge, and hurl him off into space. The bird that is thus hurled off spreads his wings and flies without injury.

From what has been said thus far, it is clear that when a pigeon deals with a rival pigeon, his behavior is directed first toward inducing the reagent to flee voluntarily, then toward forcing him off the field. Only when these means have failed, when the reagent refuses to flee, refuses to submit, and is too powerful to be hurled off the ledge, only then does the agent endeavor to the utmost to injure his opponent. Then the two pigeons meet in a grim, silent, unrelenting, physical struggle. This brutally physical struggle appears in extreme contrast to the more common pigeon fights, which are highly ceremonious. Yet even these fiercest struggles, unless they are protracted for a very long time, do not result in the death of either combatant. And at any time when either combatant feels that he has had "enough," he needs only to leave the field in possession of the victor; he thereby saves himself from further injury. For even after the most prolonged and painful battle it still remains true that if the enemy flees or submits the agent ceases fighting.

In conclusion, we can give only a few brief statements as to the bearing of the facts of animal behavior upon human problems and upon philosophy.

1. No bird or mammal follows a policy of non-resistance. And we find no trend of evolution toward a policy of absolute non-resistance. Even in the peaceful settlement of his disputes each animal asserts his rights by expressing his determination, and often by making feints or threats. Further, each individual and each social group is prepared to resort to force, and to exert force to the utmost, if necessary, in order to defend its interests. Defensive fighting pays.

2. On the other hand, aggressive fighting does not pay. Among animals, as among men, fighting is a wasteful and harmful means toward the attainment of the ends sought by the contestants. In adaptation to this fact, we find that both in the history of the race and in that

of the individual there is a trend away from destructive forms of fighting, toward the adjustment of disputes by harmless means. Progress in this direction involves the development of intelligence, of self-control, and of a technique for the adjustment of difficulties. Such progress has been made by all the higher animals, in varying degree. Birds and mammals strive to control their angry impulses. They cope with the problem of pacification, for their interests depend upon it. We human beings, when we strive toward world peace, are only travelling farther in the same line of progress in which our infra-human ancestors took the first steps. Whether we shall ever achieve world peace, I do not predict; because I do not know whether we can ever develop our understanding, our self-mastery and our political technique to a degree of perfection sufficient to cope with the immense difficulties of world organization.

3. The third conclusion, which we shall present here chiefly as a criticism of militarism, is, in its broader aspects, a criticism of the whole "biological"[6] philosophy of certain schools of current thought. This conclusion is that a distinctively "biological" need for fighting does not exist. The reasons why animals fight are substantially identical with the reasons why men fight. These reasons all pertain to the subject matter of our first two conclusions, and they may be summed up in one sentence. Animals and men fight because they must conserve their interests, and their technique for the adjustment of conflicting interests is too imperfect to adjust all cases of conflict. This, we believe, is the only true argument for war. The militarist denies that this is the whole argument. He says that if we ever should achieve a world organization which would adjust all conflicts peaceably and abolish war, the race would degenerate, because war is a "biological" necessity. His "biological" argument takes two forms: viz., (a) the definite; (b) the indefinite.

(a) The definite "biological" argument is based upon the theory of the all-sufficiency of natural selection. In reply to this argument, I do not need to offer any opinion as to the method of evolution. All I need to say is that, whatever the method of evolution may be, it cannot be in contradiction with the facts of animal behavior. And the facts of animal behavior prove that fighting to the death is not necessary for the welfare or for the evolution of a race.

(b) The indefinite form of the argument pictures a mysterious, irresistible "biological" necessity or "destiny," which threateningly overhangs our human efforts and will cause the extinction, or at least the degeneration, of any people that lives up to its humanitarian ideals. In answer to this, I have given evidence to prove that the animals also are humanitarians. The higher animals strive to avoid destructive fighting, and some large and important groups of species have so far achieved this result that they have reduced fatalities to a negligible quantity. These great groups of animals have been evolving their pacific regime, and thriving under it, for millions of years, and are today in a state of progressive evolution, the very opposite of degeneration. Therefore, we may brand as false and contrary to the evidence of facts, the militarist statement that degeneration threatens to overtake us if we should put into practise our humanitarian ideals.

FOOTNOTES

[1] We shall use the term "agent" for the animal whose behavior we are studying. His opponent we shall speak of as the "reagent" or the "enemy." The word enemy is thus used, as it is used by writers on the science of war, in a non-invidious sense.

[2] Craig. Appetites and Aversions as Constituents of Instincts, *Biol. Bull.*, 1918, Volume 34.

[3] Wheeler, W. M. The Parasitic Aculeata, a study in evolution, *Proc. Amer. Philos. Soc.*, 1919, Volume 58.

[4] Hocking, W. E. *Human Nature and Its Remaking*. New Haven, 1918. Chapter XXIV.

[5] See, e.g., W. H. Hudson. *The Naturalist in La Plata*. London, 1892. Chapter XXII.

[6] This word is put in quotation marks to make clear that our criticism is aimed, not at biology proper, but at the misuse of biological theory by certain philosophers and writers on human affairs.

5   An Anthropological Analysis of War

—BRONISLAW MALINOWSKI

I. WAR THROUGH THE AGES

In any symposium of social sciences on war a place might be rightly claimed for anthropology, the study of mankind at large. Obviously the anthropologist must not appear merely as an usher, heralding the ad-

From the *American Journal of Sociology*, 46, 1941, 521-550. Reprinted with permission of the University of Chicago Press.

vent of war in the perspective of human evolution; still less as the clown of social science, amusing the symposium with ancedotes on cannibalism or head-hunting, on preposterous magical rites or quaint war dances.

Anthropology has done more harm than good in confusing the issue by optimistic messages from the primeval past, depicting human ancestry as living in the golden age of perpetual peace. Even more confusing is the teaching of those who maintain or imply that war is an essential heritage of man, a psychological or biological destiny from which man never will be able to free himself.[1]

There is, however, a legitimate role for the anthropologist. Studying human societies on the widest basis in time perspective and spatial distribution, he should be able to tell us what war really is. Whether war is a cultural phenomenon to be found at the beginnings of evolution; what are its determining causes and its effects; what does it create and what does it destroy—these are questions which belong to the science of man. The forms, the factors, and the forces which define and determine human warfare should, therefore, be analyzed in a correct anthropological theory of war.

All these problems have their practical as well as theoretical bearing. As a member of a symposium on war, inspired by pragmatic as well as philosophical interests, the anthropologist himself must be fully acquainted with the present circumstances of warfare and the practical problems which arise out of our contemporary crisis. There is no time to be wasted on fiddling while Rome burns—or, more correctly, while Rome assists Berlin in burning the world.

Dictated by common sense, indispensable to sound statesmanship, running through abstract and philosophic reflection, persistent in and above the battle cries of intrenched armies and scheming diplomacies, the main problem of today is simple and vital: shall we abolish war or must we submit to it by choice or necessity? Is it desirable to have permanent peace and is this peace possible? If it is possible, how can we implement it successfully? There is obviously a price and a great price to be paid for any fundamental change in the constitution of mankind. Here, clearly, the price to be paid is the surrender of state sovereignty and the subordination of all political units to world-wide control. Whether this is a smaller or greater sacrifice in terms of progress, culture, and personality than the disasters created by war is another problem, the solution of which may be foreshadowed in anthropological arguments.

I think that the task of evaluating war in terms of cultural analysis is today the main duty of the theory of civilization. In democratic countries public opinion must be freed from prejudice and enlightened as regards sound knowledge. The totalitarian states are spending as much energy, foresight, and constructive engineering on the task of indoctrinating the minds of their subjects as in the task of building armaments. Unless we scientifically and ethically rally to the counterpart task, we shall not be able to oppose them. At the same time the full cultural understanding of war in its relation to nationality and state, in its drives and effects, in the price paid and advantages gained, is necessary also for the problem of implementing any fundamental change.

The problem of what war is as a cultural phenomenon naturally falls into the constituent issues of the biological determinants of war, its political effects, and its cultural constructiveness. In the following discussion of pugnacity and aggression we shall see that even pre-organized fighting is not a simple reaction of violence determined by the impulse of anger. The first distinction to emerge from this analysis will be between organized and collective fighting as against individual, sporadic, and spontaneous acts of violence—which are the antecedents of homicide, murder, and civic disorder, but not of war. We shall then show that organized fighting has to be fully discussed with reference to its political background. Fights within a community fulfil an entirely different function from intertribal feuds or battles. Even in these latter, however, we will have to distinguish between culturally effective warfare and military operations which do not leave any permanent mark either in terms of diffusion, of evolution, or of any lasting historical aftereffect. From all this will emerge the concept of "war as an armed contest between two independent political units, by means of organized military force, in the pursuit of a tribal or national policy."[2] With this as a minimum definition of war, we shall be able to see how futile and confusing it is to regard primitive brawls, scrimmages, and feuds as genuine antecedents of our present world-catastrophe.

II. WAR AND HUMAN NATURE

We have, then, first to face the issue of "aggressiveness as instinctual behavior"; in other words, of the determination of war by intrinsically biological motives. Such expressions as "war is older than man," "war is inherent in human nature," "war is biologically determined" have either no meaning or they signify that humanity has to conduct wars, even as all men have to breathe, sleep, breed, eat, walk, and evacuate, wherever they live and whatever their civilization. Every schoolboy knows this and most anthropologists have ignored the facts just mentioned. The study of man has certainly evaded the issue concerning the relation between culture and the biological foundations of human nature.[3]

Put plainly and simply, biological determinism means that in no civilization can the individual organism survive and the community continue without the integral incorporation into culture of such bodily functions as breathing, sleep, rest, excretion, and reproduction. This seems so obvious that it has been constantly overlooked or avowedly omitted from the cultural analyses of human behavior. Since, however, the biological activities are in one way determinants of culture, and since, in turn, every culture redefines, overdetermines and transmits many of these biological activities, the actual interrelation and interdependence cannot be left outside anthropological theory. We shall have briefly to define in what sense certain phases of human behavior are biological invariants and then apply our analysis to aggression and pugnacity.

Every human organism experiences at intervals the impulse of hunger. This leads to search for food, then to the intake, that is, the act of eating, which, in its wake, produces satiety. Fatigue demands rest; accumulated fatigue, sleep; both followed by a new state of the organism which the physiologist can define in terms of the conditions of the tissues. The sex impulse, more sporadic in its incidence and surrounded by more elaborate and circumstantial cultural determinants of courtship, sex taboos, and legal rulings, nevertheless leads to a definite joint performance—that of conjugation, which again is followed by a state of temporary quiescence as regards this impulse. Conjugation may start a new biological sequence of events: conception, pregnancy, and childbirth, which must occur regularly within any community if it is to survive and its culture to continue.

In all these simple and "obvious" facts there are a few theoretical principles of great importance. Culture in all its innumerable varieties redefines the circumstances under which an impulse may occur, and it may in some cases remold the impulse and transform it into a social value. Abstinences and long-drawn fasts may slightly modify the workings of the organism as regards sex and hunger. Vigils and prolonged periods of intensive activity make rest and sleep determined not merely by organic but also by cultural rulings. Even the most regular and apparently purely physiological activity of breathing is linked up with cultural determinants—partly in that housing and sleeping arrangements somewhat condition the amount of oxygen available and the rate of breathing and partly in that the act of breathing, identified with life itself, has been the prototype of a whole set of practices and beliefs connected with animism. What, however, can never be done in any culture is the full elimination of any of these vital sequences, imposed on each culture by human nature. We can condense our argument into the form of the simple diagram:

impulse→bodily reaction→satisfaction

We can say that the least variable as regards any cultural influences of the three phases is the central one. The actual intake of air or food, the act of conjugation, and the process of sleep are phenomena which have to be described in terms of anatomy, physiology, biochemistry, and physics. The second important point is that both links—between impulse and bodily reaction and between that and the satisfaction—are as clear-cut physiological and psychological realities as is the bodily reaction itself. In other words, each culture has integrally to incorporate the full vital sequence of the three phases. For each of those tripartite vital sequences is indispensable to the survival of the organism, or, in the case of sexual conjugation and pregnancy, to the survival of the community. However complicated and substantial might be the cultural responses to the basic needs of man—responses such as courtship, marriage, and family in relation to sex; economic arrangements within the household, food producing activities and the tribal or national commissariat, in response to hunger—they are in one way biologically determined, in that they have to incorporate each integral vital sequence with all its three phases and links between them, intact and complete.

Can we regard pugnacity and aggressiveness and all the other reactions of hostility, hate, and violence as comparable to any vital sequence so far discussed? The answer must be an emphatic negative. Not that the impulses of aggression, violence, or destruction be ever absent from any human group or from the life of any human being. If the activity of breathing be interrupted by accident or a deliberate act of another individual, the immediate reaction to it is a violent struggle to remove the obstacle or to overcome the human act of aggression. Kicking, biting, pushing, immediately start; a fight ensues, which has to end with the destruction of the suffocated organism or the removal of the obstacle. Take away the food from the hungry child or dog or monkey and you will provoke immediately strong hostile reactions. Any interference with the progressive course of sexual preliminaries—still more, any interruption of the physiological act—leads in man and animal to a violent fit of anger.

This last point, however, brings us directly to the recognition that the impulse of anger, the hostilities of jealousy, the violence of wounded honor and sexual and emotional possessiveness are as productive of hostility and of fighting, direct or relayed, as is the thwarting in the immediate satisfaction of a biological impulse.

We could sum up these results by saying that the impulse which controls aggression is not primary but derived. It is contingent upon circumstances in which a primary biologically defined impulse is being thwarted. It is also produced in a great variety of nonorganic ways, determined by such purely cultural factors as economic ownership, ambition, religious values, privileges of rank, and personal sentiments of attachment, dependence, and authority. Thus, to speak even of the *impulse* of pugnacity as biologically determined is incorrect. This becomes even clearer when we recognize, by looking at the above diagram, that the essence of an impulse is to produce a clear and definite bodily reaction, which again produces the satisfaction of the impulse. In human societies, on the contrary, we find that the impulse of anger is in almost every case transformed into chronic states of the human mind or organism—into hate, vindictiveness, permanent attitudes of hostility. That such culturally defined sentiments can lead, and do lead, to acts of violence, simply means that acts of violence are culturally, not biologically, determined. Indeed, when we look at the actual cases of violent action, individual, or collective and organized, we find that most of them are the result of purely conventional, traditional, and ideological imperatives, which have nothing whatsoever to do with any organically determined state of mind.

It is interesting to find that when the argument for a biological or psychological determinism of aggressiveness as something inherent in man's animal nature is put forward, examples from prehuman behavior are easy to find. It is easy to show that dogs, apes, baboons, and even birds fight over females, food, spatial or territorial rights. The study of immature children in primitive tribes, or in our own nurseries, discloses that the argument by violence is very often used and has to be constantly watched over and regulated by adults.[4] This, indeed, might have suggested to any competent observer that the elimination of violence and of aggression, and not its fostering, is the essence of any educational process.

When we are faced with the question where, how, and under what circumstances, acts of purely physiological aggression occur among human adults, we come again to an interesting result. Cases of sound, normal people attacking, hurting, or killing one another under the stress of genuine anger do occur, but they are extremely, indeed, negligibly, rare. Think of our own society. You can adduce an indefinite number of cases from a mental hospital. You can also show that within very specialized situations, such as in prisons or concentration camps, in groups cooped up by shipwreck or some other accident, aggression is fairly frequent. Such a catastrophe as a theater on fire or a sinking boat has sometimes, but not always, the effect of producing a fight for life, in which people are trampled to death and bones broken through acts of violence, determined by panic and fear. There are also cases in every criminal record, primitive or civilized, of homicidal injuries or bruises which occur under outbursts of anger and hatred, or a fit of jealousy. We see that "aggressiveness" within the framework of an adult cultural group is found under the headings of "panic," "insanity," "artificial propinquity," or else that it becomes the type of antisocial and anticultural behavior called "crime." It is always part and product of a breakdown of personality or of culture. It is not a case of a vital sequence which has to be incorporated in every culture. Even more, since it is a type of impulsive sequence which constantly threatens the normal course of cultural behavior, it has to be and is eliminated.

III.   THE HARNESSING OF AGGRESSION BY CULTURE

Another interesting point in the study of aggression is that, like charity, it begins at home. Think of the examples given above. They all imply direct contact and then the flaring-up of anger over immediate issues, where divergent interests occur, or, among the insane, are imagined to occur. Indeed, the smaller the group engaged in co-operation, united by some common interests, and living day by day with one another, the easier it is for them to be mutually irritated and to flare up in anger. Freud and his followers have demonstrated beyond doubt and cavil that within the smallest group of human co-operation, the family, there frequently arise anger, hatred, and destructive, murderous impulses. Sexual jealousies within the home, grievances over food, service, or other economic interests occur in every primitive or civilized household. I have seen myself Australian aborigines, Papuans, Melanesians, African Bantus, and Mexican Indians turning angry or even flaring into a passion on ocassions when they were working together, or celebrating feasts, or discussing some plans or some issues of their daily life. The actual occurrence, however, of bodily violence is so rare that it becomes statistically negligible. We shall see shortly why this is so.

Those who maintain that "natural aggressiveness" is a permanent cause of warfare would have to prove that this aggressiveness operates more as between strangers than between members of the same group. The facts taken from ethnographic evidence give an entirely different answer. Tribal strangers are above all eliminated from any contact with one another. Thus, the Veddas of Ceylon have arrangements by which they can transact exchange of goods and give symbolic messages to their neighbors—the Tamils and Singhalese—without ever coming face to face with them. The Australian Aborigines have an elaborate system of intertribal avoidances. The same applies to such primitive groups as the Punans of Borneo, the Firelanders, and the Pygmies of Africa and Malaysia.[5]

Besides the avoidances there are also to be found clear and legalized forms of contact between tribes. In Australia and in New Guinea, all over the Pacific, and in Africa we could find systems of intertribal law, which allow one group to visit another, to trade with them, or to collaborate in an enterprise. In some regions an intrusion on the part of a stranger, against the rules of intertribal law, and breaking through the normal dividing line, was dangerous to the intruder. He was liable to be killed or enslaved; at times he served as the *pièce de resistance* in a cannibal repast. In other words, the execution of such a trespasser was determined by tribal law, by the value of his corpse for the tribal kitchen, or of his head to the collection of a head-hunting specialist. The behavior of the murderers and of the murdered has, in such cases, obviously nothing to do with the psychology of anger, pugnacity, or physiological aggressiveness. We have to conclude that, contrary to the prevailing theoretical bias, aggression as the raw material of behavior occurs not in the contact between tribal strangers but within the tribe and within its component co-operative groups.

We have seen already that aggression is a by-product of co-operation. This latter organizes human beings into systems of concerted activities. Such a system, or institution, as we can call it, is the family. A small group of people are under the contract of marriage. They are concerned with the production, education, and socialization of children. They obey a system of customary law, and they operate conjointly a household—i.e., a portion of environment with an apparatus of implements and consumers' goods. The clan and the local group, the food-producing team and the industrial workshop, the age grade and the secret society are one and all systems of concerted activities, each organized into an institution.[6]

Let us try to understand the place of aggressiveness within an institution. There is no doubt at all that, within these short-range cooperative and spatially condensed forms of human organization, genuine aggressiveness will occur more readily and universally than anywhere else. Impulses to beat a wife or husband, or to thrash children, are personally known to everybody and ethnographically universal. Nor are partners in work or in business ever free of the temptation to take each other by the throat, whether primitive or civilized. The very essence of an institution, however, is that it is built upon the charter of fundamental rules which, on the one hand, clearly defines the rights, prerogatives, and duties of all the partners. A whole set of minor and more detailed norms of custom, technique, ethics, and law also clearly and minutely lay down the respective functions as regards type, quantity, and performance in each differential activity. This does not mean that people do not quarrel, argue, or dispute as to whether the performance or prerogatives have not been infringed. It means, first and foremost, that all such disputes are within the universe of legal or quasilegal discourse. It also means that the dispute can always be referred, not to the arbitrament of force, but to the decision of authority.

And here we come upon the fact that the charter—the fundamental customary law—always defines the division of authority in each institution. It also defines the use of force and violence, the regulation of which, is, indeed, the very essence of what we call the social organization of an institutionalized group. The patriarchal family supplies the father with the right to rule and even with the implements of violence. Under mother-right the father has to submit, to a much larger extent, to the decisions and influences of his wife's family, notably of her brother. Within the institution of the clan, quarrels and dissensions are very stringently proscribed, for the clan in many cultures acts as the unit of legal solidarity. The myth of the perfect harmony of all clansmen, however, had to be exploded.[7] Nevertheless, quarrels within the clan are rapidly and effectively eliminated by the definite, centralized, and organized authority vested in the leaders and the elders. The local group not only has the right to co-ordinate the activities and the interests of its component households and clans; it also has the means of enforcing its decisions if violence has to be used or prevented. The tribe, as the widest co-ordinating group, has also its legal charter, and it has often also some executive means for the enforcement of decisions bearing upon quarrels, disputes, and feuds within the group.

It is characteristic once more that most fighting on the primitive level occurs between smaller units of the same cultural group. The members of two families or two clans or two local groups may come to blows. We have instances of such fighting among the Veddas, the Australian Aborigines, and other lowest primitives.[8] Such intratribal fighting is always the result of the infraction of tribal law. A member of a clan or a family is killed. A woman is abducted or an act of adultery committed. Only in the rarest of cases, a spontaneous brawl or fight ensues immediately. For there exist rules of tribal law which define the way in which the dispute has to be fought out. The whole type of fighting between families, clans, or local groups is conventional, determined in every detail by beliefs and elements of material culture, or by values and agreements. The collective behavior in such fighting, which is characteristic of the primitive level of lowest savages, is guided at every step and is controlled by factors which can be only studied with reference to the social organization, to customary law, to mythological ideas, as well as to the material apparatus of a primitive culture.[9]

When there is a strong rivalry between two groups, and when this leads to a general state of mind—generating frequent outbursts of anger and sentiments of hatred over real divergences of interest—we find an arrangement in which occasional fights are not only allowed, but specially organized, so as to give vent to hostile feelings and reestablish order after the feelings have been overtly expressed. Such occasional tournament fights take sometimes pronouncedly peaceful form. The public songs of insult, by which the Eskimo even up their differences and express hatred, grievances, or hostility, are a well-known example of this. In Central Europe the institution of Sunday afternoon drinking and fighting fulfils the function of an organized and regulated exchange of insults, blows, at times injuries and casualties, in which accumulated resentments of the week are evened up. We have a good description of such regulated fights within the group among the Kiwai Papuans, among the Polynesians, and among the South American Indians.

Anthropological evidence, correctly interpreted, shows, therefore, that there is a complete disjunction between the psychological fact of pugnacity and the cultural determination of feuds and fights. Pugnacity can be transformed through such cultural factors as propaganda, scare-mongering, and indoctrination into any possible or even improbable channels. We have seen the change in France: the pugnacity of yesterday has overnight become a lukewarm alliance, and the friendship of the most recent past may, at any moment, flare up into the pugnacity of tomorrow. The raw material of pugnacity does admittedly exist. It is not in any way the biological core of any type of organized violence, in the sense in which we found that sex is the core of organized family life, hunger of commissariat, evacuation of sanitary arrangements, or the maintenance of bodily temperature a biological factor around which center cultural adjustments of clothing and housing. Anger and aggressiveness may flare up almost at any moment in the course of organized co-operation. Their incidence decreases with the size of the group. As an impulse, pugnacity is indefinitely plastic. As a type of behavior, fighting can be linked with an indefinitely wide range of cultural motives.

Everywhere, at all levels of development, and in all types of culture, we find that the direct effects of aggressiveness are eliminated by the transformation of pugnacity into collective hatreds, tribal or national policies, which lead to organized, ordered fighting, but prevent any physiological reactions of anger. Human beings never fight on an extensive scale under the direct influence of an aggressive impulse. They fight and organize for fighting because, through tribal tradition, through teachings of a religious system, or of an aggressive patriotism, they have been indoctrinated with certain cultural values which they are prepared to defend, and with certain collective hatreds on which they are ready to assault and kill. Since pugnacity is so widespread, yet indefinitely plastic, the real problem is not whether we can completely eliminate it from human nature, but how we can canalize it so as to make it constructive.

IV.   TRIBE-NATION AND TRIBE-STATE

In our study of anthropological evidence in so far as it throws light on modern warfare, we are in search of genuine primitive antecedents of fighting, such as occurred in historical times, and of fighting as it has become transformed in the modern world wars. The use of violence, clearly, has to receive a fuller sociological treatment. Nationalism and imperialism, and even totalitarianism—in my opinion, a phenomenon of cultural pathology—must be supplied with their evolutionary background and their ethnographic antecedents.

We have already seen that when two clans or two local groups fight with each other within the framework of the same tribal law, we deal with cases of legal mechanisms, but not with antecedents of war. We have to face now the question of how to define, in terms of social organization and of culture, the groups which can legitimately be regarded as pursuing some prototype of international policy, so that their battles can be considered as genuine precursors of warfare.

The concept of tribe and of tribal unity would naturally occur to every anthropologist or student of social science. An ethnographic map of the world shows, on every continent, well-defined boundaries which separate one tribe from the other. The unity of such a tribe consists *de facto* in the homogeneity, at times identity, of culture. All tribesmen accept the same tradition in mythology, in customary law, in economic values, and in moral principles. They also use similar implements and consume similar goods. They fight and hunt with the same weapons and marry according to the same tribal law and custom. Between the members of such a tribe communication is possible because they have similar artifacts, skills, and elements of knowledge. They also speak the same language—at times divided by some dialectical varieties—but generally allowing free communication. As a rule, the tribe is endogamous, that is, marriage is permitted within its limits but not outside. Consequently, the kinship system usually welds the whole tribe into a group of related and mutually co-operative, or potentially antagonistic, clans. The tribe in this sense, therefore, is a group of people who conjointly exercise a type of culture. They also transmit this culture in the same language, according to similar educational principles, and thus they are the unit through which the culture lives and with which a culture dies.

In the terminology here adopted, we can say that the tribe as a cultural entity can be defined as a federation of partly independent and also co-ordinated component institutions. One tribe, therefore, differs from the other in the organization of the family, the local groups, the clan, as well as economic, magical, and religious teams. The identity of institutions, their potential co-operation due to community of language, tradition, and law, the interchange of services, and the possibility of joint enterprise on a large scale—these are the factors which make for the unity of a primitive, culturally homogeneous group. This, I submit, is the prototype of what we define today as nationality: a large group, unified by language, tradition, and culture. To the division as we find it between primitive culturally differentiated tribes there correspond today such divisions as between Germans and Poles, Swedes and Norwegians, Italians and French. In our modern world these divisions do not always coincide with the boundaries of the state. Hence, all the contemporary political problems of nationalism, imperialism, the status of minorities, and of irredentist groups are covered by the principle of national self-determination. All such problems hinge obviously on the relation between nation and state.

The principle of political unity or statehood can also be found— on a primitive level—in creating divisions as well as in establishing unity. We know already that authority, as the power to use physical force in the sanctioning of law, exists even at the lowest level of development. We have seen that it is the very essence of the constitution of organized systems of activities, that is, institutions. We have seen that it also functions as the basis for a wider territorial control of the relations between institutions. At the lowest level we found the local group as the widest co-ordinating unit with political prerogatives. If we were to survey the political conditions at a somewhat higher level of development, we would find in most parts of the world, in Melanesia and Polynesia, in Africa and parts of America, that political power is wielded by much larger regional groups, united on the principle of authority and, as a rule, equipped with military organization, the duty of which is partly internal policing, partly external defense or aggression. Much of my own field work has been done in the Trobriand Islands, where such politically organized regions were to be found and where a clear prototype of a politically organized state could be seen at work.

We have thus introduced another concept for which the word "tribe" is also used in anthropology. I submit that the distinction between political and cultural units is necessary. To implement it terminologically, I suggest that we coin the two expressions "tribe-nation" and "tribe-state." The tribe-nation is the unit of cultural co-operation. The tribe-state has to be defined in terms of political unity, that is, of centralized authoritative power and the corresponding organization of armed force. It is clear from all that has been said that the tribe-nation is an earlier and more fundamental type of cultural differentiation than the tribe-state. The two do not coincide, for we have many instances of the tribe-state as a subdivision of the tribe-nation. The Maori of New Zealand, the Trobriand Islanders, the Zulu before European advent, as well as many North American tribes, could be quoted as examples of this. Among them the tribe-nation embraces many tribe-states. On the other hand, we could adduce from East and West Africa examples in which two or more tribe-nations are united within the same tribe-state. I have in mind the kingdoms of Unyoro and Uganda, such political units as the Masai or the Bemba, all of whom have "subject minorities" within their dominion.[10]

The two principles of statehood and nationality must, therefore, be kept apart in theory, even as they are different in cultural reality. Nevertheless, there has always existed a convergence of the two principles and a tendency toward the coalescence of the two groups—the nation and the state. In Europe this tendency, under the name of nationalism, has made its definite appearance in political aspirations and as a cause of wars and rebellions ever since the French Revolution and the Napoleonic Wars. Its main exponents were Germany, Poland, and Italy, where the disjunction of the two principles had been most pronounced. Many historians regard nationalism in this sense as an entirely new phenomenon of recent European history. In reality nationalism is probably as old as an early appearance of political power. On the one hand, a primitive nation, that is, a tribe carrying a homogeneous culture, is best protected against outside disturbances by being organized into a tribe-state. On the other hand, the strongest tribe-state is the one which coincides with the tribe-nation, since political organization is even under primitive conditions most solidly based on the association with the group who are fully co-operative through the possession of one language, one system of customs and laws, one economic machinery, and one type of military equipment.

V. WAR AND PRIMITIVE POLITICS

We can now return to the role played by fighting in the early crystallization of statehood and nationality. As a working hypothesis, we might suggest that once a strong local group developed a military machine, it would use this in the gradual subjugation of its neighbors and extension of its political control. Ethnography supplies us with the evidence that fighting between local groups of the same culture does exist. It also supplies us with a clear picture of conditions in which fairly extensive political units, which form states within a larger nation, are in existence. The study of the *status quo* and of fragments of history among the Maori of New Zealand, among several African tribes, as well as all we know about the pre-Columbian history of Mexico and Peru, points to the fact that once armed military operations start in a region, they tend to the formation of the nation-state. The archeologist and historian concerned with the Mediterranean world might show that analogous developments produced the Roman state, some of the Greek political units, and the empires of Egypt, Babylonia, Assyria, and Persia. Wars of nationalism, therefore, as a means of unifying under the same administrative rule and providing with the same military machine the naturally homogeneous cultural group, that is, the nation, have always been a powerful force in evolution and history.

Warfare of this type is culturally productive in that it creates a new institution, the nation-state. Obviously, since the political unit extends to embrace the cultural one, both assume a different character. The co-ordination of any subdivisions of such a group, whether regional or institutional, become standardized and organized. Moreover, a nation-state usually assumes a much more pronounced control over economics and man-power, over contributions to the tribal exchequer and public services rendered. It can also enforce its decisions, that is, sanction administrative activities and customary law. It is legitimate, therefore, to regard fighting of this type as a genuine antecedent of certain historical wars. For fighting here functions as an instrument of policy between two tribe-states, and it leads to the formation of larger political groups, and finally, of the tribe-nation.

It is necessary to remember that organized fighting at higher stages of savagery or barbarism does not always present this politically significant character. Most of the fighting at this stage belongs to an interesting, highly complicated, and somewhat exotic type: raids for head-hunting, for cannibal feasts, for victims of human sacrifice to tribal gods. Space does not allow me to enter more fully into the analysis of this type of fighting. Suffice it to say that it is not cognate to warfare, for it is devoid of any political relevancy; nor can it be considered as any systematic pursuit of intertribal policy. Human man-hunting in search of anatomic trophies, the various types of armed body-snatching for cannibalism, actual or mystical, as food for men and food for gods, present a phase of human evolution which can be understood in terms of ambition, thirst for glory, and of mystical systems. In a competent analysis of warfare as a factor in human evolution, they must be kept apart from constructive or organized systems of warfare.[11]

So far we have dealt with fighting organized on political principles and performing a political function, and we have dealt briefly with sportive types of human man-hunting. Where does the economic motive enter into our problem? It is conspicuously absent from the earliest types of fighting. Nor are the reasons difficult to find. Under conditions where portable wealth does not exist; where food is too perishable and too clumsy to be accumulated and transported; where slavery is of no value because every individual consumes exactly as much as he produces—force is a useless implement for the transfer of wealth. When material booty, human labor, and condensed wealth—i.e., precious metals or stones—become fully available, predatory raids acquire a meaning and make their appearance. Thus, we have to register a new type of fighting: armed expeditions for loot, slave wars, and large-scale organized robbery. We could quote examples from East and Southeast Africa, where cattle raiding was a lucrative industry associated with war. Among the northwestern tribes of Africa, slavery is found perhaps in its simplest type and furnishes one of the main motives of intertribal feuds. Nomadic tribes who, as organized robber bands, controlled some of the caravan routes in North Africa and in Asia, developed and used their military efficiency for a systematic levy of tribute, and for loot at the expense of their wealthier sedentary, mercantile or agricultural, neighbors.

We have, in the above analysis, made one or two distinctions, perhaps too sharply, but for the purpose of isolating the principles which lead to the appearance of genuine, purposeful warfare. We have spoken of nationalism as an early tendency leading to political wars and the formation of primitive nation-states. We spoke of organized raids carried out under the economic motive. These types of fighting very often coincide. It is even more important to realize that nationalism, as the tendency of extending political control to the full limits of cultural unity, is never a clear-cut phenomenon. Nationalism seldom stops at the legitimate cultural boundaries of the nation. Whether it be a Hitler or a Chaka, a Napoleon or an Aztec conqueror, a Genghis Khan or an Inca ruler, he will readily and naturally overstep the boundaries of his nation. Nationalism readily turns into imperialism, that is, the tendency of incorporating other nations under the political rule of the military conqueror.

Here we arrive at a new phenomenon which has played an important role in the development of mankind. Conquest, the integral occupation of another cultural area by force, combines all the benefits of loot, slavery, and increase in political power. Conquest is a phenomenon which must have played an enormous part in the progress of mankind at the stage where, in a parallel and independent manner, we had the establishment of large agricultural communities and militarily strong nomadic or cattle-raising tribes. From the conditions found in various parts of the ethnographically observable world and from the records of history we can retrace and reconstruct the main characteristics of culturally constructive conquests. The best ethnographic areas for this analysis are to be found among the East African tribes, where we still can study the symbiosis of invading Hamitic or Nilotic cattle breeders and nomads with sedentary agricultural Bantus. Or, we could turn to some parts of West Africa, where we find extensive monarchies, in which the sedentary agricultural West African Negroes live under the rule of their Sudanese conquerors.* From the New World the histories of the Mexican and Peruvian states embody rich material for this study of conquest.

The most important cultural effect of conquest is an all-round enrichment in national life through a natural division of function between conquerors and conquered and through the development and crystallization of many additional institutions. The conquerors provide the political element; those conquered, as a rule, supply economic efficiency. This also means that the conquerors, in exploiting the subject community, organize a tribal exchequer, institute taxes, but also establish security and communication, and thus stimulate industry and commerce. Under the impact of two different cultures, the customary law of each tribe becomes formulated, and often a compound system of codification is drawn up. Religious and scientific ideas are exchanged and cross-fertilize each other.

War as an implement of diffusion and cross-fertilization by conquest assumes, therefore, an important role in evolution and history. Such war, let us not forget, made a very late appearance in human evolution. It could not occur before such high differentiation in types of culture as that of nomadic pastoralism and sedentary agricultural pursuits. No fruits of victory were obtainable in any economic, political or cultural sense before slavery, loot, or tribute could be effected by violence.

VI. THE CONTRIBUTION OF ANTHROPOLOGY TO THE PROBLEM OF WAR

Glancing back over our previous arguments, we can see that we have arrived at certain theoretical conclusions, new to anthropological theory. It will still be necessary to show where our gains in clarity and definition are related to modern problems.

As regards the theoretical gains, we have shown that war cannot be regarded as a fiat of human destiny, in that it could be related to biological needs or immutable psychological drives. All types of fighting are complex cultural responses due not to any direct dictates of an impulse but to collective forms of sentiment and value. As a mechanism of organized force for the pursuit of national policies war is slow in evolving. Its incidence depends on the gradual development of military equipment and organization, of the scope for lucrative exploits, of the formation of independent political units.

Taking into account all such factors, we had to establish, within the genus of aggression and use of violence, the following distinctions: (1) Fighting, private and angry, within a group belongs to the type of breach of custom and law and is the protozype of criminal behavior. It is countered and curbed by the customary law within institutions and between institutions. (2) Fighting, collective and organized, is a juridical mechanism for the adjustment of differences between constituent groups of the same larger cultural unit. Among the lowest savages these two types are the only forms of armed contest to be found. (3) Armed raids, as a type of man-hunting sport, for purposes of head-hunting, cannibalism, human sacrifices, and the collection of other trophies. (4) Warfare as the political expression of early nationalism, that is, the tendency to make the tribe-nation and tribe-state coincide, and thus to form a primitive nation-state. (5) Military expeditions of organized pillage, slave-raiding, and collective robbery. (6) Wars between two culturally differentiated groups as an instrument of national policy. This type of fighting, with which war in the fullest sense of the word began, leads to conquest, and, through this, to the creation of full-fledged military and political states, armed for internal control, for defense and aggression. This type of state presents, as a rule, and for the first time in evolution, clear forms of administrative, political, and legal organization. Conquest is also of first-rate importance in the processes of diffusion and evolution.

The types of armed contest, listed as (4) and (6) and these two only, are, in form, sociological foundations, and in the occurrence of constructive policy are comparable with historically defined wars. Every one of the six types here summed up presents an entirely different cultural phase in the development of organized fighting. The neglect to establish the differentiation here introduced has led to grave errors in the application of anthropological principles to general problems concerning the nature of war. The crude short-circuiting—by which our modern imperialisms, national hatreds, and world-wide lust of power have been connected with aggression and pugnacity—is largely the result of not establishing the above distinctions, of disregarding the cultural function of conflict, and of confusing war, as a highly specialized and mechanized phenomenon, with any form of aggression.

We can determine even more precisely the manner in which anthropological evidence, as the background of correct understanding and informed knowledge, can be made to bear on some of our current problems. In general, of course, it is clear that since our main concern is whether war will destroy our Western civilization or not, the anthropological approach, which insists on considering the cultural context of war, might be helpful.

Especially important in a theoretical discussion of whether war can be controlled and ultimately abolished, is the recognition that war is not biologically founded. The fact that its occurrence cannot be traced to the earliest beginnings of human culture is significant. Obviously, if war were necessary to human evolution; if it were something without which human groups have to decay and by which they advance; then war could not be absent from the earliest stages, in which the actual birth of cultural realities took place under the greatest strains and against the heaviest odds. A really vital ingredient could not, therefore, be lacking in the composition of primitive humanity, struggling to lay down the foundations of further progress.

War, looked at in evolutionary perspective, is always a highly destructive event. Its purpose and *raison d'etre* depend on whether it creates greater values than it destroys. Violence is constructive, or at least profitable, only when it can lead to large-scale transfers of wealth and privilege, of ideological outfit, and of moral experience. Thus, humanity had to accumulate a considerable stock of transferable goods, ideas, and principles before the diffusion of those through conquest, and even more, the pooling and the reorganization of economic, political, and spiritual resources could lead to things greater than those which had been destroyed through the agency of fighting.

Our analysis has shown that the work of cultural exercise is associated with one of the two widest groups, the tribe-nation. The work of destroying and also of reconstructing in matters cultural is associated with the tribe-state. Here, once more, it will be clear to every social student that, in giving this ethnographic background to the concepts of state and nation, of nationalism and imperialism, we may have contributed to the theoretical clarification of the corresponding modern facts.

What matters to us today, as ever, is human culture as a whole, in all its varieties, racial and religious, national or affected by regional differentiation of interests and of values. Nationhood in its manifold manifestations, today as always, is the carrier of each culture. The state should be the guardian and the defender of the nation, not its master, still less its destroyer. The Wilsonian principle of self-determination was scientifically, hence morally, justified. It was justified to the extent only that each culture ought to have full scope for its development—that is, every nation ought to be left in peace and freedom. Self-determination was a mistake, in that it led to the arming of new nations and more nations, while it ought to have meant only the disarming of dangerous, predatory neighbors. Self-determination can be perfectly well brought about by the abolition of all states, rather than by the arming of all nations.

Thus, the general formula which anthropological analysis imposes on sound and enlightened statesmanship is the complete autonomy of each cultural group, and the use of force only as a sanction of law within, and in foreign relations, a policing of the world as a whole.

VII. TOTALITARIANISM AND WORLD WARS I AND II IN THE LIGHT OF ANTHROPOLOGY

Anthropological analysis of modern conditions cannot stop here, however; nor need it remain satisfied with the important but very general statement just formulated. To vindicate its claim of applicability to the savagery of civilization, as well as to the civilizations of savages, it is necessary to go a step or two further and submit the cultural pathology of today, that is, totalitarian systems and World Wars I and II, to a somewhat more detailed and searching analysis.

World war, that is, total war, is, in the light of our anthropological criteria, as distinct from the historical wars up to 1914 as these were different from head-hunting or slave-raiding. The influence of present warfare on culture is so total that it poses the problem whether the integral organization for effective violence—which we call totalitarianism—is compatible with the survival of culture.

Culture, as we know, is exercised in each of its varieties by the co-operative working of partly independent, partly co-ordinated institutions within the group, which we define as the nation. It has been thus exercised and transmitted from the very beginnings of humanity, right through to the beginnings of this century. The foundations of the industrial, liberal, and democratic era which, as I am writing this, still survives in the United States and in a few Latin-American countries, were laid on the very same structure of institutional differentiation and co-operation by the state, which controlled the development of human civilization as a whole. The principle of totalitarianism, black or red, brown or yellow, has introduced the most radical revolution known in the history of mankind. In its cultural significance it is the transformation of nationhood and all its resources into a lethal, "technocratic" instrument of violence. This becomes a means justified by the end. The end is the acquisition of more power for one state, that is, more scope for organizing violence on a larger scale and for further destructive uses. Thus, the end of totalitarianism, in so far as it gradually saps all the resources of culture and destroys its structure, is diametrically opposed and completely incompatible with the constitution of human societies for the normal, peaceful business of producing, maintaining, and transmitting wealth, solidarity, reason, and conscience, all of which are the real indices and values of civilization.

The war of 1914-18 was, I submit, different in all fundamentals from the historical wars of constructive conquest. In its technique, in its influence on national life, and also in its reference to the international situation it became a *total* war. Fighting goes on now not merely on all the frontiers geographically possible; it is waged on land, on sea, and in the air. Modern war makes it impossible to distinguish between the military personnel of an army and the civilians; between military objectives and the cultural portion of national wealth, and the means of production, the monuments, the churches, and the laboratories. Lines of communication, seats of government, centers of industry, and even centers of administrative, legal, and scientific activity are rapidly becoming targets for destruction, as much as garrisons, fortified lines, and airdromes. This development is not only due to the barbarism of a nation or of a dictator. It is inevitable, for it is dictated by the modern technique of violence.

The total character of war, however, goes much further. War has to transform every single cultural activity within a belligerent nation. The family and the school, the factory and the courts of law, are affected so profoundly that their work—the exercise of culture through autonomous self-contained institutions—is temporarily paralyzed or distorted. It is enough to look at the statistics of mobilization in manpower, in activity, and in public opinion to realize that at present it has become possible to transform some hundred million human beings into one enormous war machine. And it is obvious that when two war machines of this size are launched against each other, the one with the less perfect and total mobilization is bound to succumb.[12]

The stupendous, almost miraculous successes of Hitler's Germany have so dazzled the public opinion of neutrals and belligerents alike, that some of the real lessons have not yet been learned. In the mingled reaction of horror and admiration which followed the *Blitzkrieg* against Poland, the "conquest" of Denmark and Norway, with the implicit subjugation of Sweden, the campaign against the Low Countries, and the shattering collapse of France, many of us had to fight hard against the feeling that, after all, totalitarianism is "a better and bigger" regime than "the decaying demo-plutocracies." Sound anthropological understanding of these facts as cultural phenomena teaches something else. An organized gang of criminals will always gain the upper hand in an armed attack on a bank. The only chance the bank has is, not in fighting the gang, but in having a police force to protect it. And the police will be really efficient if it prevents the formation of an armed gang with such instruments of violence at its disposal, as would make defense impossible, or at least costly and destructive. Prepared aggression will always get the better of unprepared defense. Defense must be prepared so as to prevent aggressiveness, rather than fight it.

And here we come to the most important element in the cultural assessment of totalitarianism. Born out of the first World War, it was, in principle, nothing less and nothing more than the application of the political techniques developed between 1914 and 1918 as the type of political, economic, educational, and propagandist regime, suitable to the carrying-out of a major war.

Nazi Germany developed a system of values which could, through the technique of modern propaganda and under the sanction of a perfectly organized police, be made to become the doctrine of the whole nation. The system of values was based on the superiority of one race, of one nation within this race, and of one organized gang within the nation. Such a doctrine, it can be seen easily, is functionally adapted to the creation of highly artificial, but nevertheless effective, sentiments of superiority, aggressiveness, national egoism and morality which fits perfectly well into a universal barrack-room drill. Parallel to indoctrination, there had to go the complete reorganization of social life. The family, the municipality, the schools, the courts, the churches, and all the institutions of intellectual and artistic production were put directly under the forced and armed control of the state. Never before in humanity has the autonomous working of component institutions of intellectual and artistic production been so completely submitted to state control. Never, that is, has the exercise of culture become so completely paralyzed. This means, in terms of individual psychology, that any differential initiative, any formation of dependent critical judgment, any building-up of public opinion through discussion, controversy, and agreement, has been replaced by a passive acceptance of dictated truths. As regards the social structure of the nation, the control from above has had the effect of replacing spontaneous solidarity between husband and wife, between parents and children, among friends or partners, by a mechanically imposed "spirit of unity," to be accepted regardless of any personal impulse, reasonable judgment, or ruling of conscience.

We know well how the results of individual research, the teaching of the various religions, and the creation of artists have been prescribed, limited, and directed. In religion, notably, we can see that Naziism is trying to substitute its own dogmatic system, its ritual, and its ethics for those of Christianity, as well as for the established ethics of Western civilization and the convictions of scientific judgment.

It is not necessary to inveigh against the totalitarian system; certainly this is not a place for moral indignations or partisan views. Scientific ethics, in any case, must be limited to a clear statement of the consequences of a type of action, whether this be a small-scale enterprise or a world-wide system. The science of man, however, has always the right and the duty to point out what the consequences of a cultural revolution will be. This is the foundation of all applied science. Social science must not be afraid of predicting, anticipating, and developing some ethics of reason. This does not mean that we have the duty or the liberty of condemning certain ends on moral grounds. We can, however, point out, if this is the result of our considered opinion and analysis, that totalitarianism must lead to the destruction of the nation with which it is associated and, later on, to destruction on international scale.

Totalitarianism is an extreme expression in the shift of balance between state and nation. It is extreme because modern means of mechanical mobilization of man-power, economic resources, and spiritual values have become so dangerously effective that it is now possible to refashion whole communities—consisting of hundreds of millions—and to change each of them from a nation, exercising, transmitting, and developing culture, into a belligerent machinery supreme in war, but unsuited, perhaps unable, to carry on the national heritage of culture.

The German nation, once leading in science and in art, rich in a highly differentiated regional folklore, peasant life, and economic diversity, has now been changed into a large-scale barracks. It would be an important historical task to show how much of Germany's greatness was due to the racial, regional, and traditional differences of its component parts. The progressive extinction of this diversity is the price which Germany, as a nation, had to pay in order to make Germany, the state, so powerful. Nationalism in this modern totalitarian form is pernicious because it has become the greatest enemy of the nation itself.

And what is the place of totalitarianism in international policies and politics? It is obvious that humanity is now faced with two alternatives—the final victory, in the long run, of totalitarianism or democracy. No state organized on a peace basis, that is, for the fullest and most effective exercise of civilization, can compete with a state organized for efficiency in war. Nazi victory can be final only if Hitler's nation-state, one and alone, assumes full control of the whole world. If this were probable or even possible, we might well argue that once humanity is submitted to one conqueror, the conditions of creative and constructive conquest will set in, with the usual beneficent results, obtained at a great price, but finally acceptable.

The possibility of a complete victory of one state does not exist. If Germany wins, she will have at least three more totalitarian powers to reckon with—Italy, Russia, and Japan. When Italy falls out and becomes a mere appendage of Hitlerism, the United States of America may have to enter the ranks of totalitarian countries. For, on the assumption that Great Britain is beaten and absorbed into the German-led totalitarian block, as France has become, the United States must continue in isolation. This will mean, again, either embracing totalitarianism or withdrawing into a precarious state of semi-independence in matters political, economic, and cultural. Fortunately, Great Britain is still fighting the battle of liberty and civilization and, as its habit has always been, it may remain beaten in all battles except the final one.

Totalitarism, unless it becomes the universal empire of one single power, is not a source of stability but of age-long periodic world wars. Anthropological analysis supports those who believe that war must be abolished. Nationalism, in the sense of demand for cultural autonomy within each group united by language, tradition, and culture, is legitimate and indispensable to the carrying-out of the very business of culture. Such cultural autonomy of the component parts of present-day humanity is, or was, the principle of the national life of Switzerland

and of the old Austro-Hungarian empire; of the relations between the powerful United States and its Latin-American neighbors, who would be entirely unable to defend their cultural autonomy by force, but enjoy it, with all-round benefits, by the policy of the "good neighbor."

We are now living in a world where fashions come and go, and where the soundest ideals and principles are discredited because they are considered to have become worn out or worn too long. This attitude in itself is almost as pernicious as certain germs of totalitarianism. The student of social science ought to fight against it. I would, therefore, reiterate the beliefs which inspired some of the finest thinkers and best fighters of the last war. I believe that war can legitimately be fought only to end war. I believe that the future peace of mankind is possible only on a principle of a commonwealth of nations. I believe that in a humanity still divided by races, cultures, customs, and languages, a full tolerance in racial relations, in the treatment of nationalities, and national minorities, and in the respect for the individual, is the very mainspring of all progress and the foundation of all stability. The great enemy of today is the sovereign state, even as we find it in democratic commonwealths—certainly as it has developed into the malignant growth of totalitarianism. The real failure of the Wilsonian League of Nations was due to the fact that its very builders refused to pay the price which it obviously imposed. They were not prepared to abrogate one ounce of their national sovereignty, forgetting that this was the very material out of which the League had to be constructed.

Unless we courageously, resolutely, and with due humility, take up the principles, the ideals, and the plans which originated at first in America and were also first denounced by this country, we shall not be able to overcome the major disease of our age. This may be called total war, or totalitarianism, or extreme state sovereignty, or injustice in matters racial, religious, and national. It always results in the substitution of force for argument, of oppression for justice, and of crude, dictated mysticism for faith and reason.

FOOTNOTES

[1] The view of the primeval pacifism of man is associated with the names of Grafton Elliot Smith, W. J. Perry; of Fr. W. Schmidt and the other members of the Vienna school. The studies of R. Holsti, van der Bij, and G. C. Wheeler show that the "lowest savages" did not live in a state of "perpetual warfare." This is substantially correct. It does not, however, justify generalizations such as Elliot Smith's: "Natural man . . . is a good-natured fellow, honest and considerate, chaste and peaceful."

The view that war has been, is, and will remain the destiny of mankind has been elaborated by S. R. Steinmetz and supported by such anthropological authorities as Sir Arthur Keith and Professor Ralph Linton. It has been partly accepted, among other leaders in social science, by Dr. J. Shotwell and Professor Quincy Wright. A balanced and clear as well as essentially sound presentation of the beginnings of warfare and its real determinants is to be found in the article on "War" in the *Encyclopaedia of the Social Sciences*, written by Professor Alvin Johnson [Vol. 15. New York: Macmillan, 1934].

[2] Cf. my article, "The Deadly Issue," *Atlantic Monthly*, CLIX (December, 1936), 659-69.

[3] The above phrases within quotations marks have been taken from current scientific literature concerning war. The theoretical problems of basic human needs and their satisfaction in culture have been fully treated in my article, "Culture," in the *Encyclopaedia of the Social Sciences* [Vol. 4. New York: Macmillan, 1931], and in an essay on "The Group and the Individual in Functional Analysis" published in this *Journal*, XLIV (May, 1939), 938-64.

[4] Cf., for instance, the arguments and factual documentation given in the books by E. F. M. Durbin and John Bowlby, *Personal Aggressiveness and War* (London: K. Paul, Trench, Trubner & Co., 1938), and by Edward Glover, *War, Sadism and Pacifism* (London: G. Allen & Unwin, 1933). Both these books can be taken as examples of the incorrect and insufficient analysis of what aggressiveness really is, and of the tendency to confuse the issues by blaming "human nature" for the present catastrophic incidents of collective, mechanized slaughter, which we like to call "World War II." Good examples without faulty interpretation will also be found in *Frustration and Aggression*, by John Dollard and others, published by the Institute of Human Relations, Yale University (1939). To my colleagues at this Institute, to Dr. John Dollard and Dr. Neal A. Miller, I am greatly indebted for the benefit derived in discussions on aggressiveness and instinctive behavior. Part of the present argument was read as a paper before the Monday Evening Group of the Institute, and the suggestions and criticisms of Professors Mark A. May, Clark L. Hull, and Robert M. Yerkes have been incorporated into this article.

[5] Cf. C. G. and B. Z. Seligman, *The Veddas* (Cambridge, 1911), and G. C. W. C. Wheeler, *The Tribe and Intertribal Relations in Australia* [London: Whiting, 1910]. A full ethnographic analysis of factual data cannot be given in this article. The professional anthropologist will be able to assess the documentary evidence of the references. I hope soon to publish a memoir with full ethnographic material in support of the present argument. I am under a great debt of obligation to the Cross-Cultural Survey, organized by Professor G. P. Murdock at the Institute of Human Relations, Yale University. In this survey evidence concerning war and intertribal relations is fully collected and classified under rubrics 43-44. It is accessible to all students of anthropology. Dr. Stephen W. Reed and Dr. Alfred Métraux have assisted me greatly in discussing the anthropological problems and facts bearing on my approach to war.

[6] I have suggested, in the above-mentioned article "Culture," that this concept of institution is, in anthropological analysis, preferable to that of culture complex. This point will be more fully elaborated in a forthcoming article entitled "The

Scientific Approach to the Study of Man," to appear in a volume entitled *Man and Science*, of the "Science and Culture Series," edited by Dr. R. N. Anshen [New York: Harper & Row, 1942, pp. 207-242].

[7] Cf. my *Crime and Custom in Savage Society* (New York, 1926), among other contributions to this problem.

[8] Perhaps the best and most detailed account of a type of fighting in which one clan functions as a social unit against another is to be found in Lloyd Warner's book on the Murngin entitled *A Black Civilization* (New York: Harper & Bros., 1937). His evidence shows that such armed disputes, though at times destructive and lethal, are carried out with strict rules, over definite issues of clan interests; and are concluded in a peace ceremony, which re-establishes the order of tribal law after an infraction by one of the clan members. All the data, well assembled and classified, can be easily studied in the Cross-Cultural Survey at Yale.

[9] If space would allow, we could show that witchcraft, which is also an important tool of expressing anger or hatred, is a characteristic substitute mechanism. The use of direct violence is eliminated by translating the reaction of anger into a sentiment of hatred, and expressing this, not by any fighting or use of force, but by mystical acts of hostility.

[10] Such conditions can clearly be paralleled from the map of historical and even contemporary Europe. Austro-Hungary was a monarchy in which some fourteen or fifteen nationalities were federated. Germany before the Napoleonic Wars was a nation divided into many small states. Italy was also parceled out and partly subject to foreign rule before its unification in 1871. Poland for one hundred and fifty years was a nation partitioned among three large states. Switzerland is a political entity embracing four component nationalities.

[11] "The Deadly Issue," *Atlantic Monthly*, December, 1936.

* The reader will note that this article was written long before the African independence movements of recent years. However, there are still examples of African nations which are dominated by minority rule (i.e. Rhodesia, S. Africa). THE EDITORS

[12] Fuller data illustrating the complete remolding of all national life during war and as preparedness for war, will be found in Professor Willard Waller's symposium on *War in the Twentieth Century* (New York: Dryden Press, 1940). The four essays on economy, the state, propaganda and public opinion, and on social institutions in war time should, in my opinion, be read carefully by all students of the subject. They show that total war completely transforms the substance of modern culture. The reader, pursuing them in the light of our present analysis, may be able to draw even more pointed conclusions, especially in assessing that totalitarianism is nothing but the constitution of the nation on a war-time basis. That this effect of war is not generally understood or appreciated can be seen from the final essay in Professor Waller's volume, in which Dr. Linton, a competent authority upon all matters anthropological, seems to minimize the destructiveness of war and of its profound influence on culture. Speaking of modern war, he affirms that "its uniqueness, especially as regards potentialities for destruction, has been greatly overrated. . . . The principles upon which successful war must be waged have not changed since the dawn of history, while the destructive intent of war has certainly diminished. . . . In Europe . . . still other factors . . . will keep

intentional destruction to a minimum. . . . It seems safe to predict . . . that . . . there will be no swift victories against large or powerful nations" (What about France? *op. cit.*, pp. 535-38). Such opinions, and they are by no means exclusive to the writer quoted, show how easy it is to miss the real issue. Even if we admitted that the toll of some twenty million human lives taken by the last World War was of no great importance; nor yet the fifty million maimed and rendered useless, we would have to assess the disorganization in economic matters, the lack of security as regards wealth and life, and the general debasement of civic and ethical principles. The real issue, however, discussed in the text, is whether the integral influence of preparedness for, and the use of, violence do or do not disorganize the texture of modern civilization.

6   The Frustration-Aggression Hypothesis

——NEAL E. MILLER
(with the collaboration of
ROBERT R. SEARS,
O. H. MOWRER,
LEONARD W. DOOB
and JOHN DOLLARD)

The frustration-aggression hypothesis is an attempt to state a relationship believed to be important in many different fields of research. It is intended to suggest to the student of human nature that when he sees aggression he should turn a suspicious eye on possibilities that the organism or group is confronted with frustration; and that when he views interference with individual or group habits, he should be on the look-out for, among other things, aggression. This hypothesis is induced from commonsense observation, from clinical case histories, from a few experimental investigations, from sociological studies and from the results of anthropological field work. The systematic formulation of this hypothesis enables one to call sharp attention to certain common characteristics in a number of observations from all of these historically distinct fields of knowledge and thus to take one modest first step toward the unification of these fields.

A number of tentative statements about the frustration-aggression

From *Psychological Review,* 48, 1941, 337-342. Reprinted with permission of the American Psychological Association. This article is a revision of a paper read at the Symposium on Effects of Frustration at the meeting of the Eastern Psychological Association at Atlantic City, April 5, 1941.

hypothesis have recently been made by us in a book.[1] Unfortunately one of these statements, which was conspicuous because it appeared on the first page, was unclear and misleading as has been objectively demonstrated by the behavior of reviewers and other readers. In order to avoid any further confusion it seems advisable to rephrase this statement, changing it to one which conveys a truer impression of the authors' ideas. The objectionable phrase is the last half of the proposition: "that the occurrence of aggression always presupposes the existence of frustration, and contrariwise, that the existence of frustration always leads to some form of aggression."

The first half of this statement, the assertion that the occurrence of aggression always presupposes frustration, is in our opinion defensible and useful as a first approximation, or working hypothesis. The second half of the statement, namely, the assertion "that the existence of frustration always leads to some form of aggression" is unfortunate from two points of view. In the first place it suggests, though it by no means logically demands, that frustration has no consequences other than aggression. This suggestion seems to have been strong enough to override statements appearing later in the text which specifically rule out any such implication.[2] A second objection to the assertion in question is that it fails to distinguish between instigation to aggression and the actual occurrence of aggression. Thus it omits the possibility that other responses may be dominant and inhibit the occurrence of acts of aggression. In this respect it is *inconsistent* with later portions of the exposition which make a distinction between the instigation to a response and the actual presence of that response and state that punishment can inhibit the occurrence of acts of aggression.[3]

Both of these unfortunate aspects of the former statement may be avoided by the following rephrasing: Frustration produces instigations to a number of different types of response, one of which is an instigation to some form of aggression.

This rephrasing of the hypothesis states the assumption that was actually used throughout the main body of the text. Instigation to aggression may occupy any one of a number of positions in the hierarchy of instigations aroused by a specific situation which is frustrating. If the instigations to other responses incompatible with aggression are stronger than the instigation to aggression, then these other responses will occur at first and prevent, at least temporarily, the occurrence of acts of aggression. This opens up two further possibilities. If these other responses lead to a reduction in the instigation to the originally frustrated response, then the strength of the instigation to aggression is also reduced so that acts of aggression may not occur at all in the situation in question. If, on the other hand, the first responses do not lead to a reduction in the original instigation, then the instigations to them will tend to become weakened through extinction so that the next most dominant responses, which may or may not be aggression, will tend to occur. From this analysis it follows that the more successive responses of non-aggression are extinguished by continued frustration, the greater is the probability that the instigation to aggression eventually will become dominant so that some response of aggression actually will occur. Whether or not the successive extinction of responses of non-aggression must inevitably lead to the dominance of the instigation to aggression depends, as was clearly stated in later pages of the book, upon quantitative assumptions beyond the scope of our present knowledge.[4,5]

Frustration produces instigation to aggression but this is not the only type of instigation that it may produce. Responses incompatible with aggression may, if sufficiently instigated, prevent the actual occurrence of acts of aggression. In our society punishment of acts of aggression is a frequent source of instigation to acts incompatible with aggression.

When the occurrence of acts of aggression is prevented by more strongly instigated incompatible responses, how is the existence of instigation to aggression to be determined? If only the more direct and overt acts of aggression have been inhibited, as is apt to be the case because such acts are the most likely to be punished, then the instigation to aggression may be detected by observing either direct or less overt acts of aggression. If even such acts of aggression are inhibited, then a different procedure must be employed. Two such procedures are at least theoretically possible. One is to reduce the competing instigations, such as fear of punishment, and observe whether or not acts of aggression then occur. The other is to confront the subject with an additional frustration which previous experiments have demonstrated would by itself be too weak to arouse an instigation strong enough to override the competing responses inhibiting the aggression in question. If the instigation from this additional frustration now results in an act of aggression, then it must have gained its strength to do so by summating with an already present but inhibited instigation to aggresssion. The presence of the originally inhibited instigation to aggression would be demonstrated by the effects of such summation. Thus the fact that an instigation may be inhibited does not eliminate all possibility of experimentally demonstrating its presence.

At this point two important related qualifications of the hypothesis may be repeated for emphasis though they have already been stated in the book. It is not certain how early in the infancy of the individual the frustration-aggression hypothesis is applicable, and no assumptions are made as to whether the frustration-aggression relationship is of innate or of learned origin.

Now that an attempt has been made to clarify and to qualify the hypothesis, four of the chief lines of investigation which it suggests may be briefly considered.[6]

1. An attempt may be made to apply the hypothesis to the integration and elucidation of clinical and social data. Here the fact that certain forms of aggression are spectacularly dangerous to society and to the individual is relevant. This means that acute personality conflicts are apt to arise from the problem of handling aggression and that the problem of aggression is apt to play an important role in shaping certain great social institutions such as the in-group as an organization against the out-group.

2. An attempt may be made to formulate more exactly the laws determining the different ways in which instigation to aggression will be expressed under specified circumstances. Some of the problems in this field are suggested by the phenomena of displacement of the object of aggression, change in the form of aggression, and catharsis of aggression.

3. An attempt may be made to secure more information concerning the other consequences which frustration may produce in addition to the instigation to aggression. Such an attempt would lead into studies of rational thought and problem solution as suggested in the classical work of John Dewey, and into studies of experimental extinction, trial-and-error learning, substitute response and regression.[7] Work along this line of investigation may deal either with the clinical and social significance of these other consequences of frustration or with the discovery of the laws governing them.

4. An attempt may be made to improve or to reformulate the basic frustration-aggression hypothesis itself. The determination of the laws which allow one to predict exactly under which circumstances instigation to aggression may be expected to occupy the dominant, the second, the third, or some other position in the hierarchy of instigations aroused by a frustrating situation is a most important problem of this type. Another problem is the reduction of the frustration-aggression hypothesis to more fundamental principles and the more accurate restatement of the hypothesis in terms of the more basic principles. One of the steps in this direction would be to scrutinize any

exceptions to the hypothesis as now formulated. Another step would involve a careful study of the early stages of the socialization of the individual in an attempt to analyze the interlocking roles of three factors: first, innate physiological reaction patterns; second, learning mechanisms; and third, the structure of the social maze which poses the learning dilemmas and contains the rewards and punishments. An empirical and theoretical analysis along these lines might lead to a fundamental reformulation giving a closer approximation of the socially and scientifically useful truths imperfectly expressed in the present frustration-aggression hypothesis.

NOTES

[1] J. Dollard, L. W. Doob, N. E. Miller, O. H. Mowrer, and R. R. Sears. *Frustration and aggression.* New Haven: Yale University Press, 1939.

[2] *Op. cit.*, pp. 8-9, 19, 58, 101-102.

[3] *Ibid.*, pp. 32-38; also 27, 39-50, 75-87, 111, 166. In this later exposition a distinction is made not only between instigation to aggression and acts of aggression but also between conspicuous acts of overt aggression and inconspicuous acts of non-overt aggression. It is assumed that the former are more apt to be culturally inhibited by strong punishments than the latter.

[4] *Op. cit.*, p. 40.

[5] The notions used here are similar to those employed by Professor Hull in describing trial-and-error learning. See Hull, C. L. Simple trial-and-error learning—an empirical investigation, *J. comp. Psychol.*, 1939, 27, 233-258.

[6] Both of the first two of these chief lines of investigation have been developed at length in *Frustration and Aggression.* No attempt was made there to elaborate upon either the third or the fourth. Thus that first effort does not purport to be a complete systematization of all principles within a single field, but rather, an exploratory attempt to apply a strictly limited number of principles to several different fields. *Op. cit.*, pp. 18, 26.

[7] These problems are discussed in more detail by Dr. Sears in the next paper of this series, "Non-aggressive responses to frustration." [*Psychological Review*, 1941, 48, 343-346.]

PART 3　Antecedent Conditions: Recent Research

I don't subscribe to the notion that
wars begin in the minds of men; they
begin when somebody steps across a
border or fires a gun, and none of
that is in the mind. If you want to
analyze what starts a war, do so, but
don't begin by looking into the minds
of men. That gets you nowhere.
　　　B. F. Skinner

Since men began writing about violent events and warfare, the study of
such matters has largely been led by philosophers, religious thinkers,
and to some extent, students of metaphysics. Poets and other literary
figures have also found reason to comment on these dismal aspects of
human affairs. The points of view of such traditional disciplines as
these, though each has its separate value, are not generally regarded as
scientific. The scientific study of aggression, as we have mentioned
previously, is a comparatively recent development. When we speak of
scientific study, we mean to emphasize the role played by scientific
controls. These control procedures are meant to insure that our experimental results can be directly attributed to our manipulations.

In any experiment there are generally two major variables
operating. One of these is the independent variable. This is under the
direct control of the experimenter; that is, the experimenter decides
what manipulations are to be made. An example of such a variable would
be the intensity of the foot-shock administered to a series of experimental animals. The second kind of variable is the dependent variable.
This variable, as the name implies, owes its expression to the independent variable. We presume, in other words, that our manipulations of

the independent variable will have some effect on the dependent variable. Therefore, this latter entity may be defined as the behavior that is being measured in the experiment. Referring to our experimental animals again, the number of bites administered by our subjects to a suspended target object would be an example of this second kind of variable. In this case we can say that we have a measure of aggression as defined by the number of bites registered in the presence of some specified level of foot shock. Such a procedure has been used by Ulrich (see Reading 10) and others in demonstrating a relationship between pain and aggression. Again, if one manipulates the independent variable he should see some effect on the dependent variable, all other things being equal. The statement "all other things being equal" is very crucial to any experiment since we must assume that no other variables are operating to produce our results. Extraneous variables, those that are not controlled or manipulated, must be minimized and, when possible, completely eliminated. If we were conducting an experiment and did not control the sex of the animals used as subjects, the temperature in the apparatus, or the noise factors, it would be difficult to ascertain whether the results had not been influenced by these things in some way. Thus control procedures are very important.

Research that is not conducted in the laboratory does not generally utilize controls and consequently strong predictions cannot be made. For example, the psychoanalyst Sigmund Freud seldom if ever used control procedures. Given his working situation as a practicing therapist, this kind of procedural refinement would have been practically impossible to implement. Since he relied on his observations and conversations with patients in order to formulate hypotheses about behavior, the theories he constructed from his findings are largely subjective. This, of course, does not mean that he was wrong, only that other explanations for his findings cannot be ruled out. It is thus quite possible that a number of extraneous variables may have been operating. Since Freud could not directly observe his patients' pasts, he had to rely on their reports. This kind of information may be helpful, but is not always reliable.

The experiment, of course, is not the only way of conducting research. Historians must conduct their research in libraries, just as students do when they are confronted with the necessity of writing a "research" paper. In both of these cases the events one investigates have been, at one time, *behavior*. Presumably, someone witnessed an event and recorded it. If we want to know about this event we must consult the record, much as a scientist must consult old journal articles in order to understand the "history" of the given problem which he is studying. Whatever biases or inaccuracies exist in the historical record can only be corrected by offering alternative histories which, in turn, may be equally biased or inaccurate. Unfortunately, we cannot "experiment" with history; we can only interpret it. This is a shortcoming, of course, and makes predictions somewhat difficult if not impossible. This objection has been raised by B. F. Skinner, who has said:

> ... I don't believe in arguing from history. However, I do think it is interesting to watch what is going on even though I don't like to make predictions on historical evidences.[1]

Similarly, archaeological data is largely historical, hence the term "natural history." We cannot directly *observe* these past events and although we may get some idea about how organisms behaved in earlier times, we cannot bear witness to the behavior itself. We are, therefore, left with the alternatives of making "educated" guesses about these phenomena or making no guesses at all. While history must be examined "second hand," prehistory is even more elusive since, by definition, no recorded observations (with the exception of primitive cave drawings) exist. We are still dealing with evidence, of course, and scientists and historians must approach their subject matter with the tools at their disposal, despite their inadequacies. Until a genuine "time machine" is developed, we'll have to be satisfied with what we can learn from the archives and from the fossilized remnants of beasts and brethren. The written and natural history of aggression are important subjects and should not be overlooked by the serious student of this problem.

In the works presented in this section attempts have made to control and exclude extraneous factors. By following the procedures carefully, we can see the ingenious methods that scientists have devised to pose their research questions. We trust that the reader will find these methods as interesting as the findings.

FOOTNOTE

[1] Evans, R. I. *B. F. Skinner: The Man and His Ideas.* New York: Dutton, 1968.

SUGGESTED READINGS

Berkowitz, L., *The Roots of Aggression*. New York: Atherton Press, 1969.

Studies on various aspects of the frustration-aggression hypothesis, including an interesting introductory chapter by the editor, provide the reader with an excellent source book for an examination of this hypothesis.

Birney, R. C., and Teevan, R. C. (Eds.), *Instinct*. New York: Van Nostrand-Reinhold, 1961.

This is a small paperback book with some very good basic research papers on the instinct problem. Both old and new research is included.

International Journal of Group Tensions. Published quarterly by the International Organization for the Study of Group Tensions, 177 East 77th St., New York 10021.

A new journal publishing cross-disciplinary research on racial, ethnic, religious, social, and political conflicts.

Journal of Conflict Resolution. A quarterly for research related to war and peace. Published by the Center for Research on Conflict Resolution, University of Michigan, Ann Arbor, Michigan, 48104.

An excellent journal which prints both theoretical and experimental papers by behavioral scientists from varied disciplines.

Journal of Peace Research. Published by International Peace Research Institute, Oslo, Norway.

This journal publishes articles on a wide range of subjects but emphasizes papers on international relations.

Journal of Social Issues. Published quarterly by the Society for the Psychological Study of Social Issues, a division of the American Psychological Association, P. O. Box 1248, Ann Arbor, Michigan, 48104.

This journal contains theories and hypotheses developed around research into problems of groups, nations, and society in general. An attempt is made to make the articles as nontechnical as possible, and to apply the research to contemporary problems.

Scott, J. P., *Aggression*. Chicago: University of Chicago Press, 1958.

A comprehensive treatment of the biology of aggression with interesting chapters on social and ecological causes of aggression.

Southwick, C. H., *Animal Aggression*. New York: Van Nostrand-Reinhold, 1970.

This book contains a series of articles on aggression in animals and is a useful supplement to more general treatments of animal aggression such as those found in Lorenz *(On Aggression)* and Ardrey *(African Genesis)*.

Yates, A. J., *Frustration and Conflict*. New York: Van Nostrand-Reinhold, 1965.

This book contains technical research concerning various aspects of the frustration-aggression hypothesis and conflict, with the early selections directly concerned with aggression. Later papers are more specialized but quite relevant to aggression research.

EDITOR'S NOTE

Bulletins and reprints, some of which are free, are available from the Lemburg Center for the Study of Violence. Current and past research papers are important sources of information. The institute is associated with Brandeis University, Waltham, Massachusetts, 02154.

7 Some Patterns in the History of Violence

—FRANK H. DENTON
WARREN PHILLIPS

INTRODUCTION

This paper reports on progress made in a research project aimed at describing systematic trends in the violence between political groups. The report is in two parts. Part one describes the formulation of an empirical test for the existence of (1) a short-term (15-30 years) and (2) a long-term (80-120 years) periodic fluctuation in the historical occurrence of war. Although some background is given about why such a test should be made, the research described in part one is based, largely, on empirical rather than theoretical generalizations. That is, not much attention is given to the "why" of such patterns.

The tests tend to confirm the existence of the expected patterns. Obviously, it is desirable to go beyond the simple observation that an empirical regularity exists to some explanation of the forces leading to that regularity. The available data do provide clues as to possible explanations. This second part of the paper speculates about several possible reasons for these patterns. The explanations are consistent with the data, but their testing must await the collection of historical matter broader in scope than that now available.

TRANSFORMATIONS OF THE INTERNATIONAL SYSTEM

Among the effects of the rise to prominence of general systems models in the social sciences, especially international relations, has been the focusing of attention on transformations of the system. The idea of systematic analysis of changes in the system, although new in its particular form, is based on a long history of speculation about evolution-

Reprinted by permission from *The Journal of Conflict Resolution*, 12 (2), 182-195.

Any views expressed in this paper are those of the author. They should not be interpreted as reflecting the views of The RAND Corporation or the official opinion of any of its governmental or private research sponsors. Papers are reproduced by The RAND Corporation as a courtesy to members of its staff.

ary or cyclical transformations in the social experience. Perhaps influenced by the unrest, violence, and (at times) chaos of this half-century, recent writers such as Spengler, Toynbee, and Sorokin have largely discarded evolutionary "development" in favor of large cyclic movements in history.

Data which would permit the empirical testing of these speculative theories about system changes are not always readily obtainable. In order to perform such tests, data are needed which cover extended periods of time and which are collected in a systematic manner. Records of the incidence of wars represent one of the few types of data meeting these requirements. Some previous research has been done in this area. Sorokin, for one, has attempted in a semi-systematic manner to test for cyclic patterns in war. He expresses a very negative view about the occurrence of cyclical patterns as a result of this examination (1957, pp. 534-604). Richard Rosecrance, in an excellent volume, examined patterns of diplomacy for system changes. Although he was apparently not looking for a cyclic pattern, his findings suggest such a pattern (1963).

Other collections of statistics on war have formed the basis for exploratory examinations for possible time-dependent patterns. Lewis Richardson investigated the possibility of a general trend toward a greater number of wars in the international system. He chose a run of 432 years, divided it in half, and demonstrated that no such trend existed in the data examined. His examination of the frequency of outbreak of war on an annual basis indicated a random fluctuation over time which could be well approximated by a Poisson distribution. Groupings of the data into longer periods (longer than one year and less than 230 years) do, however, indicate fluctuations which cannot be explained as random occurrences (1960, pp. 139-40).

J. E. Moyal (1949) reexamined Quincy Wright's list of wars in order to test two hypotheses. His first hypothesis states that the number of wars in a given period may be correlated with the number of wars in a past period, separated by a fixed interval. Moyal finds that, with time lags of five and 15 years, there is a distinctly significant autocorrelation in the *outbreak* of war.

Moyal's second hypothesis assumed a slow fluctuation in the probability of an outbreak of war. He averaged the number of outbreaks of war using a fifty-year running average. On plotting this running average against the central data, a marked periodic variation in the outbreak of wars was discernible, the half-life being about 100 years.

These findings, as well as those in previous research by one of the authors, suggest that periods of relative peacefulness tend to alternate with relatively warlike periods. Such a regular fluctuation in the intensity of war is of interest in itself; however, these earlier findings also suggest that violence in the international system is to a great extent a reflection of the political state of affairs of the system. More specifically, when the system is characterized by instability and intense political disputes it is also characterized by widespread and intense violence. Thus, to some extent, changes in the state of the international political system can be observed by observing changes in the amount and type of violence in the system. Again, more precisely, it is assumed that the data on war go beyond just indicating the amount of violence; these data reflect something about international politics as well. This assumption, critical to the "explanation" of the observed patterns, is discussed in the brief review of previous research in the next section.

EXPLORATORY ANALYSES

Two exploratory analyses of systematic trends in violence generated findings which are felt to be relevant to the work reported upon in this analysis. The first study defined the spatial-temporal domain of interest as the international system from 1820 until 1949 (Denton, 1966). The second study, also by Denton (unpublished), focused on Latin America between 1820 and 1949. The data for both of these studies were taken from Lewis F. Richardson's collection (1960).

In each case one goal of the analysis was to test the data for a periodic upswing in violence every 15 to 30 years. Since the concept of "upswing in violence" could be represented by one or two large wars as well as by many small wars, the Richardson and Moyal measures of new outbreaks of war did not provide a satisfactory index. Rather a composite "amount of violence" index was formulated from the number of wars, the number of casualties, and the number of participants.[1]

Both studies indicated a definite tendency for a periodic increase in the level of violence about every 25 years.

The level of violence in each of the bars in Figure 1 is found by averaging the level of violence over five-year periods separated by intervals of 15, 20, or 25 years. Thus the first bar in Figure 1, for the 25-year cycle case, is the mean level of violence for the five-year periods: 1820-24, 1845-49, 1870-74, 1895-99, 1920-24, and 1945-49. A pattern with all averages near zero indicates that the amount of war was roughly uniform across time. If a recurring tendency exists for more war at a fixed interval, some averages should deviate significantly

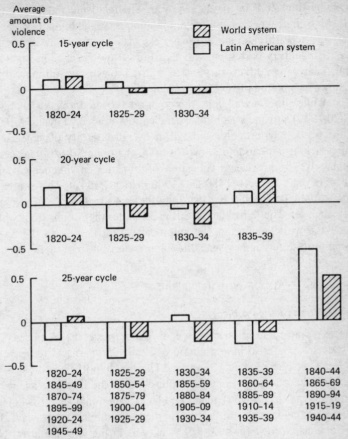

Figure 1. A generation cycle.

from zero. It is readily apparent that these two analyses indicate a tendency for violence to be relatively intense about every 25 years. These findings are surprisingly consistent with Richard Rosecrance's divisions of the temporal dimension of the international political system.[2]

A chi-square test assuming as a null hypothesis an equiprobable distribution of values above and below the means for each of the five intervals indicated a deviation from equiprobability significant at the .15 level for the world system and at the .04 level for the Latin American system.

In addition to the 25-year cycle indicated above, a long-term cyclical effect was suggested. Again, known historical patterns are indicated by the trends in the data on war. The relatively turbulent times following the French revolution show a moderate amount of violence. The stable political situation usually associated with the latter part of the 19th century corresponds to a period of low violence in Figure 2. And finally, the 20th century, often referred to as a revolutionary time, is shown as a period of very intense violence.

While the data for this period covered by Richardson (1820-1949) provide a preliminary indication of a longer cyclic pattern, a test for a long-term rise and fall in overall intensity of violence requires a longer period to allow for more than one 80-120 year cycle. Quincy Wright provides a collection of statistics on violence covering the 420 years between 1480 and 1900 (Wright, 1942). This source of data is used for testing the two hypotheses suggested by the exploratory analyses. The remainder of this paper states the propositions which guide the work; delineates the empirical world; lays down the operational procedures; and presents the findings derived from these tasks.

ANALYSIS OF WRIGHT'S DATA

Formulation of the Problem

The exploratory research reported on above suggests two cycles. The first cycle appears to have a period of about 25 years and to be superimposed on one of longer duration. The two hypotheses to be tested are: (1) There is a tendency for an upswing in the level of violence at about 25-year intervals. It might also be expected that the upswing will possess a gradual rise and decline, perhaps resembling a sine wave. (2) There is a tendency for a 80-120 year cycle in the violence level. This cycle may have a sine-wave pattern. However, a sharp reaction against extreme instability may result in a "saw tooth" effect. That is, the pattern may be a gradual rise followed by a sharp decline.

The data used for testing these hypotheses are taken from Quincy Wright's compilation covering the period 1480-1900. Wright's data include " . . . all hostilities involving members of the family of nations . . . which were recognized as states of war in the legal sense or which involved over 50,000 troops" (1942, p. 636). Thus, relatively small incidents of violence not involving a declaration of war are systematically excluded from the tables, contrary to the Richardson data used in the exploratory analysis. For each of his wars, Wright lists the start and end dates, the number of battles, and the primary partici-

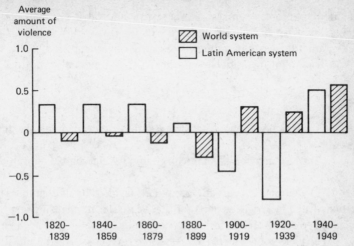

Figure 2. A suggestion of a longer cycle.

pants.[3] Wright also divides wars into four *types:* (1) defense of civilization; (2) civil; (3) balance-of-power, and (4) imperial (1942, p. 641).

Criticisms have been justly leveled at both Richardson's and Wright's work for the lack of concreteness in the explanation of data-collecting procedures.[4] Work is now underway in an attempt to replicate and extend these data in at least two institutions.[5] This study is conceived of in an iterative sense. More detailed studies can check later on the "truth" of the findings. In any case, the errors in the two data collections are unknown and are probably small when patterns of this nature are being sought.

Two indices are formulated to test the above hypotheses. As in the exploratory studies, an index of the "size" or amount of violence in the system is required to test for the expected periodicities. Secondly, it is felt that the type of war prevalent in the system may be of some value in explaining fluctuations in intensity. Thus variables are included which indicate the relative frequency of the four types of war defined by Professor Wright (see above).

In order to reduce fluctuations caused by random or atypical events when very short time intervals are used, and in order to reduce the necessity for handling many small observations, the 420-year time interval is divided into 84 five-year periods.[6] Any war overlapping an interval for more than one year or totally contained in an interval is included in that interval. Thus, a war extending over 20 years might be counted four or five times (in successive intervals). The purpose in doing this is to include *all violence existing in the system at a given time*, rather than simply outbreaks of war during each period.

The following variables are included as potential indicators of size and type:

Size Indicators

(1) Number of wars occurring within or overlapping the time period.

(2) The number of political groups having combat during the interval.[7]

(3) Number of groups from (2) divided by the number of nations listed in Wright's charts for that year. This normalizing factor is included to see if it might influence the results.

(4) The total number of *participant years* of war in the time interval.

(5) The number of battles fought in the time interval. Wright lists only the total battles for a war. For wars extending over more than one period the number of battles is allocated by the *percent of the war's participant years* which occurred in the given period.

Type Indicators

These indicators are "normalized" for size, since *predominance in the system* is the desired attribute. Thus the type of war variables is

Table 1   Distribution of Wars Over the Five-Year Periods

Number of wars in the observation period	Number of periods with indicated number of wars
0	1
1	9
2	10
3	14
4	12
5	10
6	8
7	7
8	8
9	5
	84

computed as the percent of wars in a time interval that are classified as a given type: (1) percent civil, (2) percent defense of civilization, (3) percent balance of power, and (4) percent imperial.

These items, observed for the 84 time periods, constitute the data for testing the hypotheses. The sample includes 375 wars, and 88 percent of the time periods include at least two wars. Thus, the data as aggregated are sufficiently well distributed to reduce somewhat the fluctuation which might result from occasional atypical events (see Table 1).

Size or Amount of Violence Index

The primary need in this research is for a valid indicator of size or scope of war in the international system during each of the five-year periods. The five measures indicated above are all felt to be manifestations of the general concept *size of violence*. That is, a period with much violence can typically be expected to have many wars, many participants, many battles, etc. However, it also is possible to have few wars but still have much violence during a five-year period if a few large wars occur. Intense violence may also be manifested in many small wars during the period. For these reasons it is felt that no one of these variables is as good an index of "size" as is a combination of them.

An examination of the empirical record indicates that, with the exception of *number of battles,* these variables do correlate quite highly (Table 2). Since Wright often failed to list any battles for non-European conflicts, this variable was deemed of lower quality and is not used in

Table 2   Correlation of Size Variables

	Variables				
	1	2	3	4	5
1. Number of battles	—				
2. Number of belligerents	.36	—			
3. Number of belligerents normalized	.25	.73	—		
4. Number of belligerent years	.35	.85	.69	—	
5. Number of wars	.05	.67	.53	.51	—

Table 3   Index Fit to the Data

	Variables				
	2	3	4	5	6
6. Principal Component	.84	.85	.89	.79	—
7. Construct Mapping	.84	.84	.84	.84	.97

the index. The high correlations indicate that the occurrence of a "few large" or "many small" wars is uncommon and that the number of belligerents is a reasonably good estimator of the others. However, an index can be formulated which is a still better predictor (has higher correlations). Two methods are used to obtain an index of size: (1) a principal component (factor analysis) fit to the data, and (2) a construct-mapping fit. The construct-mapping routine fits a factor between the four retained variables so that each variable is equidistant from that factor (Jones, 1966). Table 3 gives the correlation of these two indices (or factors) with each of the original variables and with one another. The two fits are essentially the same (correlation of .97) and account for almost the same proportion of variance (71+ percent). They are each better predictors than the belligerent variable (67 percent of variance), and the construct-mapping index is arbitrarily chosen for this analysis.

Twenty-five Year Cycle

Figure 3 is a plot of this size-of-war index for the period 1480-1900. Visually it is difficult to note any outstanding trends. An examination of periods 20, 25, and 30 years apart provides little indication of a periodic upswing in violence when the size of conflict variable is divided into periods above and below its mean value (Table 4).

There is some tendency for periods of relatively low violence to occur every 30 years starting in 1490. The hypothesis is in terms of periods of relatively intense violence rather than periods of relative quiet, however.

Table 4   Test for a 25-Year Cycle

		Number of periods when amount-of-war index is					
20-year intervals	Above mean	11	8	12	11		
	Below mean	10	13	9	10		
25-year intervals	Above mean	9	10	7	8	8	
	Below mean	8	7	10	9	8	
30-year intervals	Above mean	7	3	8	8	8	8
	Below mean	7	11	6	6	6	6
		1485	1490	1495	1500	1505	1510

Initial time period

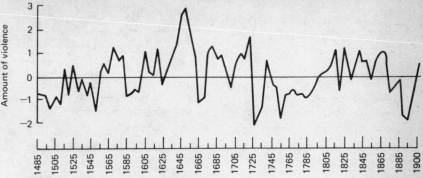

Figure 3. Amount of Violence (Wright's data).

Returning to Figure 3, the fluctuations do appear to be more rapid for the first half of the period than for the second. The two halves, taken separately, might show a trend which is cancelled out when both halves are combined. Such a periodicity could show some drift across a temporal domain of this length. For example, if the periodicity is explained in terms of a generation or a decision-maker "life time" effect, the increase in man's life span would lead to an expectation of a slower fluctuation in the latter part of the sample. Moreover, if such a cycle is considered as probabilistic rather than deterministic, a rare event of two violent periods exceptionally close together (or far apart) could result in a shift in the phase of the cycle, giving a cancelling effect over a period of this length. To test for these possibilities the data are divided into pre-1680 and post-1680 groups.

The data for the period prior to 1680 indicate an upswing in violence about every 20 years, starting with the period ending in 1495 (Table 5). After 1680 there is a history of relatively high violence about every 30 years starting in the time period ending in 1690 (Table 6).

The data are generally supportive of the hypothesis that an upswing in violence occurs about once every generation to a generation

Table 5   Twenty-Year Cycle—Pre-1680

	Number of periods			
Above mean	5	4	9	5
Below mean	6	7	2	6
	1485	1490	1495	1500
Average value of amount-of-war	.07	−.08	.30	.08

Table 6   Thirty-Year Cycle—Post-1680

	Number of periods					
Above mean	5	6	3	1	3	4
Below mean	3	2	4	6	4	3
	1685	1690	1695	1700	1705	1710
Average value of amount-of-war	.04	.58	−.39	−.59	−.03	.09

and a half, if one assumes some change in the life spans making up the "generation."

80- to 120-Year Cycle

While it is necessary to have a relatively short observation time to examine the data for the expected 25-year cycle, it is desirable to average out these fluctuations when examining the data for longer-term trends. Such an averaging makes visual observation easier; that is, short-term fluctuations are masked in order to give a more ordered appearance to the longer fluctuations. The amount-of-war index values are averaged over four periods (20 years) in Figure 4.

A pattern similar to that postulated is exhibited by these data. Throughout the period of the data a consistent rise in the level of violence is exhibited for several years (a minimum of 60 years—a maximum of 120 years). A sharp decline is noted after each rise and the decline is in each case followed by another rise in violence. The dashed

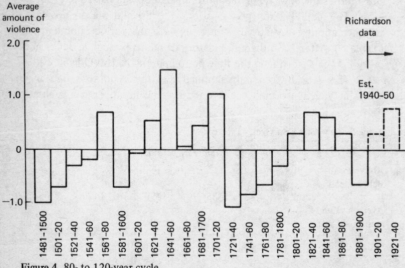

Figure 4. 80- to 120-year cycle.

curve indicates an extrapolation using Richardson's data for the first half of the 20th century.

How significant, statistically, is this visually detectable pattern? That is, could a pattern of this nature result by random chance? The original hypothesis is formulated in nonoperational terms. The prime expected effect is that periods of high violence will be followed by a decrease in violence and that periods of low violence will be followed by an increase in violence. Operationally the hypothesis may be expressed by two statements:

(1) Periods of high violence in the system will be followed by a decrease in the level of violence.

(2) Periods of low systemic violence will be followed by an increase in violence.

These hypotheses imply a system in which conflict (manifested in violence) grows in scale until a reaction against violence per se occurs. This reaction results in lower conflict in the system until conditions permit the growth of new conflict.

The only remaining problem in operationalizing the hypothesis is

Table 7   Levels of Violence and Increase or Decrease in Next Period

	Next Period	
1		
Period with	Increase in violence	Decrease in violence
High Violence*	4	6
Low Violence	10	0

Fisher exact test Significance:≈.025

	Next Period	
2		
Period with	Increase in violence	Decrease in violence
High Violence**	1	5
Low Violence	13	1

Fisher exact test Significance: ≈ .005

* "High" defined as above the *mean* (20-year average of size index).

** "High" defined as more than half a standard deviation above the mean.

to select a level above which violence will be considered high. The mean is one obvious "level" for this purpose. "High violence," on the other hand, intuitively implies a condition somewhat above average, that is, more than just barely above the mean. Two levels of "high" are tested to determine if the results differ depending on how high "high" is.

(1) In the first case, "high" violence is defined as above the *mean* (a 20-year average of the size index). Below the mean is defined as "low" violence.

(2) In the second case, "high" violence is defined as more than half a standard deviation above the mean. "Low" violence is below this value. This test is perhaps more consistent with the original hypothesis. The cushion of half a standard deviation eliminates from the "high" violence category those time periods with *only slightly* greater than average violence.

For both definitions, as we see in Table 7, periods of high violence are usually followed by a decrease in violence and periods of low violence by an increase. The division of high at half a standard deviation above the mean gives a pattern with only two of 20 periods deviating from the expected trend.

Previous research indicated a correlation between high violence and the prevalence of civil conflict (Denton, 1966). Checking on these data it is found that, across the 84 five-year observations, the indicator involving the percent of wars that are civil wars correlates .39 with the

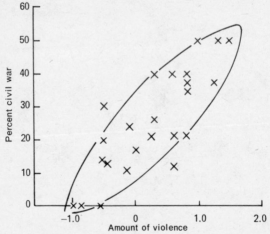

Figure 5. Percent of civil war versus amount of war (20-year averages).

size-of-war indicator. For 84 observations a product moment correlation of .39 (two-tailed test) is significant at about the .01 level. Figure 5 shows a plot of the 20-year average of percent civil war versus the amount-of-war values for the same intervals. The correlation is obviously even higher when the longer time periods are used. Thus, periods in which there is much war are quite consistently periods which exhibit a relatively high frequency of civil wars. In addition to supporting the hypothesis about a long-term cyclical trend in violence, the data also support the postulate that those periods of intense conflict are associated with social change or turmoil, as indicated by a concomitant relatively high frequency of civil conflict.

SUMMARY AND CONCLUSIONS

The primary purpose in this analysis is to test for the existence of two postulated cyclical trends in the historical record of violence. The data generally support the hypotheses. That is, evidence of cyclical patterns exists in the historical data.

Why such patterns and what are the theoretical implications? Some speculation about the "why" was made in the earlier formulation of the hypotheses. Although the data give few direct clues for going beyond the empirical *description* to an *explanation*, it does seem appropriate to discuss possible explanations which might be tested.

Hypothesis 1 suggested an upswing in the level of violence about every 25 years. The data support such a general trend with 20 years providing a "best" fit prior to 1680 and about 30 years thereafter.[8] Others have alluded to a generation effect in decision-making, or a fading into the past of the last war as a permissive condition for future war. Several conditions may be associated with such a periodic increase in violence:

(1) Immediately following an intense war the public remembrance of it is of the horrors, the human suffering, the dislocations. The "horror" remembrance may be reinforced by the memoirs, biographies, novels, or fireside stories written and told by first-hand participants while their own experiences are fresh. At the same time, the glory of defending the fatherland, the adventure, and the grim humor of war are themes which are often expressed. Perhaps, through a defense against the distasteful, the "horror" is suppressed as memories age and as an increasing number of society's members have not had first-hand ex-

posure to violence. On the other hand, the requirement[9] for ensuring willingness to defend the state or society against intruders motivates many to further glorify "dying for one's country." Thus the themes employed in the descriptions of the last great war shift from "horror"-dominant to "glory"-dominant.[10] This shift is then tied to the human life cycle of 20 to 30 years.

Such an hypothesis could be easily tested by examining the frequency of the above themes in the public literature at various times after major conflicts.

(2) Perhaps even more important would be the application of a similar thesis to decision-makers. Subjectively, it seems that major conflicts have created the next generation of decision-makers. Thus, for a number of years after major conflicts, the *system's* decision-makers have had first-hand experience with violence. Decision-makers seldom obtain power before they are 40 to 45 and seldom retain power beyond 65 to 70. A new, "unsullied" group of decision-makers gradually comes into power after the conflict. A generation's time sees almost a complete turnover.

By examining the records of the accession to office of decision-makers, and by comparing the attitudes of "war responsible" decision-makers toward the use of violence with those of the new generation without war responsibility, it would be possible to test such an explanation.

(3) Implicit in the above explanations is an assumption that the opportunities for employing violence are always present. The only condition necessary for violence is the willingness to become engaged. However, Richardson for one has shown that there is an indication of continuity in conflict. In fact, he goes on to explain this continuity in terms of a generation effect (1960, p. 200):

> We may suppose that the generation who had not fought in the earlier war, but who were brought up on tales about its romance, heroism, and about the wickedness of the enemy, became influential 30 to 60 years after the war ended and so delayed the process of forgetting and forgiving [Richardson observed an increase in retaliatory and revenge wars 30 to 60 years after the original conflict].

Thus, a third effect of continuity of conflict and hatred may exacerbate the above trends. A planned research project will involve the investigation of continuity in alliance and conflict patterns.

The generation effect was *a priori* a "reasonable" explanation. The longer-term cycle is, in the minds of the authors, a pattern which is not as easily explained on intuitive grounds. It appears that one promising explanation is offered by an action-reaction process in political philosophy (taken in the broad sense to include the general attitude of the elites[11] toward the "correct" society). Man seems to strive both for a better society and for more security in those desirable aspects of the present society. As previously noted, periods of intense violence are typified without exception by a relatively high frequency of civil conflict. This would, of course, indicate dissatisfaction with the *form of society*. Moreover, periods of intense conflict are consistently followed by periods of low violence with a *very low* frequency of civil violence.

It seems logical to expect that philosophies of change are associated with periods of intense civil violence. The religious controversies of the 16th and 17th centuries, the French and European revolutions of the late 18th and early 19th centuries, and the Communist revolutions of this century are all associated with philosophies of change. A reasonable hypothesis could be made that, in a given society, the relative weight given to the desire for improving society as opposed to *maintaining* society is dependent on the conditions of (1) *rate of change* and (2) discrepancy between the *actual* and the *ideal* society.

Thus, first, conditions of rapid change, associated with high tensions and violence, induce philosophic positions emphasizing the importance of an *orderly* society and philosophies glorifying the *traditional* strengths of the established system. That is, insecurity leads to emphasizing the desire for maintenance of the good aspects of an idealized society of the past.[12]

Second, periods of stability reduce the *experience* and thus the fear of insecurity. Moreover, it seems intuitively true that there is almost always a *discrepancy* between the *real* and *ideal* form of society. It may even be true that periods of social stability (slow change) increase this discrepancy. That is, to use an old adage, "them that has gets." Philosophers, freed from the need to emphasize order in an already orderly society, concentrate on the need for reducing the discrepancy between the *real* and the *ideal*.

The result of these hypotheses is that change tends to be followed by stability, by change again, and so on. The period of the cycle (averaging about a century in length), while it is not intuitively obvious, does not seem unreasonable. Truths die slowly and perhaps the passage of several generations is needed to tilt the scales.

A systematic examination of themes in philosophic positions could provide a test of this hypothesis. These data provide one-half the objective situation, that of violence and social disruption. It might also be worthwhile to determine, as possible, if periods of slow change do increase the discrepancy between real and ideal.

Obviously the above explanations go beyond the data. However, it is believed that they are *consistent* with the patterns exhibited in the history of violence. Moreover, if the explanations are useful models they provide a framework for a theoretical explanation of certain changes in the international (or perhaps more appropriately the inter-political group) system.[13]

In addition to the limitations on our explanations, the assumption that violence reflects the political issues of the day may be questioned by some.[14] Beyond this, the data are not complete (there is some discrepancy between the Wright and Richardson data for the period in which they overlap). The possible effects of improved data are, of course, not known.

REFERENCES

Denton, Frank H., "Some Regularities in International Conflict, 1820-1949," *Background*, 9, 4 (Feb. 1966), 283-96.

Jones, Ronald D., *Construct Mapping*. Kansas City: University of Missouri, mimeographed, June 1966.

Moyal, J. E., "The Distribution of Wars in Time," *Journal of the Royal Statistical Society*, 112 (1949), 446-58.

Richardson, Lewis F., *Statistics of Deadly Quarrels*. Pittsburgh: Boxwood Press, 1960.

Rosecrance, Richard N., *Action and Reaction in World Politics*. Boston: Little, Brown, 1963.

Sabine, George H., *A History of Political Theory*. New York: Holt, Rinehart and Winston, 1961.

Singer, J. David, Melvin Small, and George L. Kraft, *The Frequency, Magnitude, and Severity of International War, 1815-1940*. Ann Arbor: Mental Health Research Institute Preprint 158, University of Michigan, July 1965.

Sorokin, Pitirim, *Social Change and Cultural Dynamics*. Boston: Porter Sargent, 1957.

Wright, Quincy, *A Study of War*. Chicago: University of Chicago Press, 1942.

FOOTNOTES

[1] This index is a weighted composite of the three variables indicated. It is formed by a technique called concept mapping (Jones, 1966). Mathematically the index is similar to a "factor" in factor analysis. The values of this index are standardized. That is, the set of index values has a mean of zero and a variance of one.

[2] Denton (1966). This provides some support for the assumed correlation between international politics and international violence.

[3] " . . . actual independence before or after the war rather than legal status under international law was the criterion used [for including a participant]. . . . Unsuccessful revolutionists, rebels, or insurgents which lacked even *de facto* status, except during the war itself, have not been so listed, and many of the small feudal principalities of the Holy Roman Empire have been ignored. . . . "

[4] See Singer, Small, and Kraft (1965) for a discussion of the merits of these works.

[5] Singer, Small, and Phillips at the University of Michigan, and Denton at the University of Southern California.

[6] This interval length results in four or five observations per period for the shortest cycle hypothesized.

[7] Contrary to Wright's formulation, revolutionary groups are included.

[8] The process of grouping five-year intervals somewhat limits the accuracy of these statements.

[9] Some might go further and say man's inherent evil.

[10] Today's TV programs on World War II seem rather "humor"-dominant.

[11] In the abstract sense, those who influence and/or make decisions.

[12] George Sabine in his *History of Political Theory* explicitly discusses philosophic reactions against violence, though not the reverse. Sabine states that Bodin's "divine right of kings" could be considered as a reaction against the instability and violence of the religious wars (p. 399). He also describes Burke's decision to codify his philosophy as " . . . the beginning of a shift . . . [in] social philosophy from attack to defense and . . . to an emphasis on the value of stability . . . " as the result of the French Revolution (p. 617). In our own time the conservative reaction to instability and change seems to gather greater strength with more change.

[13] In order to (try to) escape the label of "historical determinist" let us offer the following position statement. It is not felt that patterns in history absolutely determine the future. However, it is felt, because of historical learning, ongoing environmental factors, and certain regularities in man's behavioral structure, that he is partially a product of his past. While individually man may enjoy "free will," be "adaptive," etc., as a group he is conservative and not very adaptive, at least in the short run. Even if faced with the "truth" of a needed revision in beliefs he (and a number of his prophets) is perfectly capable of "rationally" showing that

the old ways are best. And who is going to give him the truth? On the other hand, there is obviously a long-term learning process which may change such factors.

[14] The pattern of violence (Figure 4) does seem to reflect known historical trends, however.

8   The Stimulating vs. Cathartic Effects of a Vicarious Aggressive Activity

——SEYMOUR FESHBACH

The present study is concerned with the complex effects of participation in a presumably vicarious aggressive activity upon subsequent aggressive behavior. A number of studies have demonstrated that the expression of aggression—whether directly or in symbolic form—results in a lowering of subsequent aggression (Berkowitz, 1960; Feshbach, 1955; Pepitone & Reichling, 1955; Rosenbaum & de Charms, 1960; Thibaut & Coules, 1952). However, there is also experimental evidence to the effect that aggressive activity has a stimulating effect upon the manifestation of other aggressive acts (Feshbach, 1956; Kenny, 1953); that is, aggression may breed aggression.

Since both possibilities—reduction and stimulation—have been experimentally observed, the pertinent issue then is under what conditions a vicarious aggressive act increases and under what conditions it decreases the probability of subsequent aggressive behavior. One such condition suggested by differences in procedure between the studies that obtained evidence of a cathartic effect and those demonstrating a stimulating effect is the emotional state of the subject at the time the aggressive act is performed; that is, if the subject is angry at the time he engages in the aggressive activity, he can then use the act to satisfy and thereby reduce his hostility.

The general hypothesis is suggested that in order for an activity to have drive reducing properties, components of the drive must be present or evoked during performance of the activity; that is, there must be some functional connection between the vicarious act and the original drive instigating conditions. While it is undoubtedly true that

From the *Journal of Abnormal and Social Psychology*, 1961, *63* (2), 381-385. Reprinted by permission of the American Psychological Association.

the vicissitudes of life will arouse hostilities that cannot be directly discharged, it does not follow that any indirect aggressive act will have the property of reducing hostility that has been evoked under markedly different circumstances. According to the present view, a child's anger toward its mother will not be reduced by an aggressive act toward a doll figure unless its anger toward the mother is aroused when the aggressive act is performed. The evocation of anger may not be a sufficient condition—the doll figure may have to be similar to the mother—but it is probably a necessary condition for drive reduction to take place.

If the subject is not hostile at the time of participating in an aggressive act, his subsequent aggressive behavior will not merely remain unaffected but is very likely to be increased. An increase in aggression following a vicarious aggressive act could result from a number of different processes: a reduction in inhibition or aggression anxiety, reinforcement of aggressive responses, and finally conditioned stimulation of aggressive drive and/or aggressive responses.

On the basis of the foregoing considerations, the following hypotheses are proposed: Participation in a vicarious aggressive act results in a reduction in subsequent aggressive behavior if aggressive drive has been aroused at the time of such participation; if aggressive drive has not been aroused at the time of participation in a vicarious aggressive act, such participation results in an increase in subsequent aggressive behavior.

METHOD

The experimental procedure consisted of arousing a subject's aggressive drive before participation in a vicarious aggressive act or before participation in a neutral act and then obtaining measures of aggression subsequent to these interpolated activities. The variation in level of aggression was accomplished by means of an insult versus noninsult condition and the variation in the interpolated activity consisted of exposure to a fight film versus a neutral film.

Subjects

The subjects were male college student volunteers who were assigned at random to one of the four treatment groups generated by the two experimental variables. One hundred and one subjects were used in the study, with approximately equal numbers in each experimental condition. The subjects were seen in small groups by the experimenter so that nine experimental sessions in all were held, three for the Noninsult Fight Film condition and two sessions for each of the other experimental conditions.

Procedure

Insult versus Noninsult Condition

Subjects assigned to the insult groups were subjected to a number of unwarranted and extremely critical remarks. These comments essentially disparaged the intellectual motivation and the emotional maturity of the students.[1] Previous studies (Feshbach, 1955; Gellerman, 1956) have provided abundant evidence that this technique successfully arouses hostility toward an experimenter. Subjects assigned to the Noninsult condition were given standard test instructions.

Aggressive Film versus Neutral Film Condition

Subjects in the Insult and Noninsult groups then witnessed either a Fight or a Neutral Film. The Fight Film consisted of a film clip of a rather exciting prize fight sequence taken from the motion picture *Body and Soul* while the neutral film depicted the consequence of the spread of rumors in a factory. The duration of each of the films was approximately 10 minutes.

As a rationale for the presentation of the film, the subjects were told before the film was presented that they would be asked to judge the personality of the main character in the film. Following the completion of the film, each subject indicated his impression of the personality of the hero of the film on a questionnaire provided for that purpose.

Dependent Measures of Aggression

All subjects were given a modified word association test which, in a previous study (Gellerman, 1956) had been shown to be sensitive to differences in the arousal of aggression. The test involves the presentation of five aggressive words interspersed among six neutral stimuli as follows: wash, choke, travel, massacre, walk, murder, relax, stab, sleep, torture, listen. The subjects are asked to give in written form a series of associations to each word. The stimuli are presented orally and also visually, the experimenter holding up a 5″ × 8″ card on which the stimulus word is printed. The subject's Aggression score is based on the number of aggressive word associations among the first 10 responses to each of the aggressive stimulus words. The maximum score that can therefore be obtained on this measure is 50.

Table 1   A. Mean Aggressive Word Association Responses Obtained under Each Experimental Condition

Drive (D)	Film (F)	
	Fight (N)	Neutral (N)
Insult	24.5 (25)[a]	28.9 (21)
Noninsult	27.7 (29)	25.3 (25)

B. Summary of Analysis of Variance of Aggressive Word Association Responses

Source	SS	df	MS	F
D	8.93	1	8.93	
F	38.43	1	38.43	
DF	291.59	1	291.59	4.58*
Within	6,111.80	96	63.66	
Total	6,450.75	99		

[a]The word associations of one subject were not scored due to illegibility.
*$p < .05$.

Table 2 Distribution of Aggressive Word Association Responses Falling above and below the Median as a Function of Insult Fight Film and Insult Neutral Film Treatments

Treatment	<27	>27
Insult fight film	17	8
Insult neutral film	10	19

Note.$-\chi^2 = 6.02$; $p < .05$.

Table 3   A Comparison of Mean Scores on the Aggression Questionnaire

	Insult-fight (IF) (N = 26)	Insult-neutral (IN) (N = 29)	Noninsult-fight (NIF) (N = 20[a])	Noninsult-neutral (NIN) (N = 25)
M	14.6	19.5	13.7	15.0
σ	3.72	3.90	2.52	2.95

Note.$-$IF$-$IN $= 4.7$; $p < .01$.
[a]One subject failed to complete questionnaire.

Subsequent to the administration of the word association test, the first experimenter left the room having presumably completed the study. A second experimenter then entered and informed the subjects that the psychology department wished to assess students' opinions of the value of participating in psychological experiments. A questionnaire was then administered dealing with the subjects' attitudes toward the experimenter and with their evaluation of the experiment. The questionnaire which consists of six items, each of which has six alternatives, is described in more detail in a previous study (Feshbach, 1955). It is scored so that the least aggressive choice for a particular item is given a score of 1 and the most aggressive choice, a score of 6.

RESULTS

By hypothesis, it was predicted that the Insult group exposed to the Fight Film would manifest *less* subsequent aggression on each of the two measures of aggression than the Insult group exposed to the Neutral Film while the Noninsult group exposed to the Fight Film would display *more* subsequent aggression than the Noninsult group exposed to the Neutral Film. The word association data bearing upon these predictions are presented in Table 1. The mean differences are in accordance with expectation, the Insult-Fight (IF) Film group responding with fewer aggressive associations than the Insult-Neutral (IN) Film group and the Noninsult-Fight (NIF) Film group responding with more aggressive associations than the Noninsult-Neutral (NIN) Film group. The results of an analysis of variance of the data indicate that the interaction between the Insult and the Film variable is statistically significant. The difference between the IF Film and the IN Film groups falls short of the 5% confidence level, the value of t being 1.9. The difference between the NIF Film and NIN Film groups yields a t value of approximately 1 which is clearly not significant.

The contrast between the IF Film and IN Film groups is more sharply delineated by a simple median split. The chi square for the fourfold table presented in Table 2 is 6.02 which yields a p value of <.02. The word association data, then, indicate that under conditions of anger-arousal, witnessing a fight film results in a lowering of aggression. However, the hypothesized stimulating effect of an aggressive film under nonaroused conditions is not borne out by the data.

The questionnaire data are presented in Table 3. Because of the lack of homogeneity of variance between the IN and NIF Film groups, separate comparisons were made between pertinent groups and, in these

Table 4   Distribution of Aggression Questionnaire Scores Falling above and below the Median as a Function of Insult Fight Film and Insult Neutral Film Treatments

Treatment	<17.5	>17.5
Insult fight film	20	6
Insult neutral film	7	22

Note.$-\chi^2 = 15.1; p < .001.$

comparisons, the variances of the respective distributions are not reliably different. As was the case with the world association data, the IF Film group displays significantly less aggression on the questionnaire than does the IN group. The difference between the Noninsult groups is not in the predicted direction but is small and unreliable.

The difference in subsequent aggressive attitudes between the insulted group exposed to the fight film and the insulted group exposed to the neutral film is further illustrated by a simple median split. The chi square for the fourfold table presented in Table 4 is 15.1, which is significant at less than the .001 level.

DISCUSSION

The experimental results are consistent with the hypothesis that the drive reducing effect of a vicarious aggressive act is dependent upon the aggressive state of the subject at the time of the vicarious aggressive activity. Witnessing the prize fight film resulted in a significant relative decrement in aggression in comparison to witnessing the neutral film only for those subjects in whom aggression had been previously aroused by the insulting comments of the experimenter. The predicted increase in aggression for the noninsulted subjects following exposure to the fight film did not occur, however. Each of these two outcomes warrants further comment.

With regard to the difference between the two Insult groups in subsequent aggression, a possible alternative to a catharsis or drive reduction hypothesis is one that assumes that guilt or revulsion stimulated by the fight film is the primary mechanism responsible for the lowered aggression. Berkowitz (1958, 1960) has strongly argued for such an explanation of a reduction in aggressive behavior following an aggressive act. However, it must be noted that the evidence for a guilt or inhibition process is most indirect and inferential.

With regard to the present study, the guilt alternative is certainly possible, although, for various reasons to be suggested below, not a likely one. If guilt arousal were a ubiquitous process, occurring whenever people are given the opportunity to indulge in aggressive fantasies, then the fight film should similarly have inhibited the aggressive response output of the Noninsult group. The possibility still remains that guilt arousal can account for the aggression reducing effects of fantasy under conditions where aggression has recently been stimulated, as in the Insult condition. As a check on whether the lowered aggression of the IF Film group was due to some inhibitory factor, the word associations were scored for defensiveness. A previous study of the effects of inhibition upon aggressive word associations has shown that when inhibition is experimentally aroused, the number of aggressive responses decreases while the number of defensive responses increases (Gellerman, 1956). While, in the present study, a difference was observed in the number of aggressive associations, the difference between the two Insult groups in the number of defensive associations was negligible and insignificant. The absence of an increment in defensive responses, while not decisive since the experiment cited employed an inhibition procedure more closely resembling fear rather than guilt, is more consistent with a drive reduction rather than guilt explanation of the decrease in aggression following the exposure of the insulted subjects to the Fight Film.

The problem remains of accounting for the failure to obtain the expected increase in aggression in the Noninsult group. One possible reason is the limitation of the questionnaire instrument as a measure of aggression. Although one's preference for or attitude toward another person is frequently used as an index of aggression, as was the case in the present experiment, dislike and aggression are not equivalent dimensions. At the extreme, aversion and aggression are likely to be strongly correlated but within moderate ranges of feeling, the association between dislike and aggression may well be negligible. For this reason, the word association measure is probably a better instrument than the attitude questionnaire for detecting changes in aggression in the noninsulted groups. However, although the relative increment in aggressive associations following exposure of the noninsult group to the Fight Film was in the predicted direction, it was not statistically significant. Whether this failure to obtain evidence of a stimulating effect of a vicarious aggressive activity under relaxed emotional conditions is due to inadequacies in the theoretical analysis or to limitations in the methods utilized cannot be ascertained from the present data.

On the other hand, the data consistently reflect the dependence of the drive reduction effect upon the arousal of aggression at the time the subject is engaging in the vicarious aggressive activity. Presumably vicarious aggressive acts do not willy-nilly serve as outlets for aggressive motivation. This latter process warrants further attention. Aggression is not an ever-present tension system pervading all of an individual's activities. Like other acquired motives, its appearance is very much dependent upon situational factors; and, the more specific the category of objects toward which the aggression is directed, the narrower is both the range of stimuli that can elicit the motivation and the range of situations that can serve as substitute outlets for the aggression.

What would appear to be a relatively simple matter—the effects of a vicarious aggressive activity upon subsequent aggressive behavior—is in actuality a quite complex process. The present study has examined the influence of the drive state of the organism upon this process. Beyond the requirement of replication in a variety of situations, further research is needed to establish the extent to which other variables determine the effects of so-called vicarious aggressive activities and to establish the precise mechanism by which the performance, direct or vicarious, of an aggressive act influences subsequent aggressive behavior.

SUMMARY

Studies of the effects of a presumably vicarious aggressive activity upon subsequent aggressive activity suggest that under certain conditions the activity will tend to increase, and under other conditions decrease, the probability of subsequent aggressive behavior. The purpose of this experiment was to study the effects of one such condition—namely, the emotional state of the subject at the time the vicarious aggressive activity is performed. Specifically, it was proposed that a vicarious aggressive activity results in a *reduction* in subsequent aggressive behavior if the subject is emotionally aroused at the time he is engaging in this activity, but if anger has not been aroused, the activity results in an *increase* in subsequent aggressive behavior. The two independent variables manipulated in the study consisted in an Insult versus Noninsult condition and an Aggressive Film versus Neutral Film condition. One hundred and one college students were assigned at random to the four treatment groups generated by the two experimental variables. The subjects met the experimenter in small groups so that nine experimental sessions in all were held. Subjects assigned to the Noninsult condition were given standard test instructions while subjects in the Insult groups were subjected to a number of unwarranted and extremely critical remarks. The subjects then witnessed either an Aggressive Film or a Neutral Film. The former consisted of a film clip depicting a prize fight sequence while the latter depicted the consequences of the spread of rumors in a factory. They were then administered a word association test and under the guise of a departmental assessment of the value of students' serving as experimental subjects, a second experimenter administered a questionnaire dealing with the subjects' attitudes toward the first experimenter and with their evaluation of the experiment. The degree of aggression manifested on the attitude questionnaire and the number of aggressive responses on the word association test constituted the dependent measures.

A significant interaction in the predicted direction was obtained for the Word Association measure—the Insult-Aggressive Film group responding with fewer aggressive associations than the IN Film group, and the Noninsult-Aggressive Film group responding with more aggressive associations than the NIN Film group. A similar significant difference between the two Insult groups was found on the attitude questionnaire, but the difference between the two Noninsult groups on this measure was unreliable and was not in the predicted direction.

The results were interpreted as being consistent with a drive reduction theory, although an inhibitory process (guilt arousal) cannot be excluded by the evidence at hand. The dependence of the *aggression* reducing effect of exposure to a film depicting violent activity upon the prior or simultaneous arousal of aggressive drive was stressed.

REFERENCES

Berkowitz, L. The expression and reduction of hostility. *Psychol. Bull.*, 1958, 55, 257-283.

Berkowitz, L. Some factors effecting the reduction of overt hostility. *J. abnorm. soc. Psychol.*, 1960, 60, 14-22.

Feshbach, S. The drive-reducing function of fantasy behavior. *J. abnorm. soc. Psychol.*, 1955, 50, 3-11.

Feshbach, S. The catharsis hypothesis and some consequences of interaction with aggressive and neutral play objects. *J. Pers.*, 1956, 24, 449-462.

Gellerman, S. The effects of experimentally induced aggression and inhibition on word association response sequences. Unpublished doctoral dissertation, University of Pennsylvania, 1956.

Kenny, D. T. An experimental test of the catharsis theory of aggression. Unpublished doctoral dissertation, University of Michigan, 1953.

Pepitone, A., & Reichling, G. Group cohesiveness and the expression of hostility. *Hum. Relat.*, 1955, 3, 327-337.

Rosenbaum, M. E., & de Charms, R. Direct and vicarious reduction of hostility. *J. abnorm. soc. Psychol.*, 1960, 60, 105-111.

Thibaut, J. W., & Coules, J. The role of communication in the reduction of interpersonal hostility. *J. abnorm. soc. Psychol.*, 1952, 47, 770-777.

FOOTNOTE

[1] The author wishes to express his gratitude to Abraham Wolf for his competence and courage in carrying out this phase of the experiment.

9 Weapons as Aggression Eliciting Stimuli

—*LEONARD BERKOWITZ*
ANTHONY LEPAGE

Human behavior is often goal directed, guided by strategies and influenced by ego defenses and strivings for cognitive consistency. There clearly are situations, however, in which these purposive considerations are relatively unimportant regulators of action. Habitual behavior patterns become dominant on these occasions, and the person responds relatively automatically to the stimuli impinging upon him. Any really complete psychological system must deal with these stimulus-elicited, impulsive reactions as well as with more complex behavior patterns. More than this, we should also be able to specify the conditions under which the various behavior determinants increase or decrease in importance.

From the *Journal of Personality and Social Psychology*, 1967, 7, 202-207. Reprinted by permission of the American Psychological Association. The present experiment was conducted by Anthony LePage under Leonard Berkowitz' supervision as part of a research program sponsored by Grant G-23988 from the National Science Foundation to the senior author.

The senior author has long contended that many aggressive actions are controlled by the stimulus properties of the available targets rather than by anticipations of ends that might be served (Berkowitz, 1962, 1964, 1965). Perhaps because strong emotion results in an increased utilization of only the central cues in the immediate situation (Easterbrook, 1959; Walters & Parke, 1964), anger arousal can lead to impulsive aggressive responses which, for a short time at least, may be relatively free of cognitively mediated inhibitions against aggression, or for that matter, purposes and strategic considerations.[1] This impulsive action is not necessarily pushed out by the anger, however. Berkowitz has suggested that appropriate cues must be present in the situation if aggressive responses are actually to occur. While there is still considerable uncertainty as to just what characteristics define aggressive cue properties, the association of a stimulus with aggression evidently can enhance the aggressive cue value of this stimulus. But whatever its exact genesis, the cue (which may be either in the external environment or represented internally) presumably elicits the aggressive response. Anger (or any other conjectured aggressive "drive") increases the person's reactivity to the cue, possibly energizes the response, and may lower the likelihood of competing reactions, but is not necessary for the production of aggressive behavior.[2]

A variety of observations can be cited in support of this reasoning (cf. Berkowitz, 1965). Thus, the senior author has proposed that some of the effects of observed violence can readily be understood in terms of stimulus-elicited aggression. According to several Wisconsin experiments, observed aggression is particularly likely to produce strong attacks against anger instigators who are associated with the victim of the witnessed violence (Berkowitz & Geen, 1966, 1967; Geen & Berkowitz, 1966). The frustrater's association with the observed victim presumably enhances his cue value for aggression, causing him to evoke stronger attacks from the person who is ready to act aggressively.

More direct evidence for the present formulation can be found in a study conducted by Loew (1965). His subjects, in being required to learn a concept, either aggressive or neutral words, spoke either 20 aggressive or 20 neutral words aloud. Following this "learning task," each subject was to give a peer in an adjacent room an electric shock whenever this person made a mistake in his learning problem. Allowed to vary the intensity of the shocks they administered over a 10-point continuum, the subjects who had uttered the aggressive words gave shocks of significantly greater intensity than did the subjects who had spoken the neutral words. The aggressive words had evidently evoked implicit aggressive responses from the subjects, even though they had not been angered beforehand, which then led to the stronger attacks upon the target person in the next room when he supposedly made errors.

Cultural learning shared by many members of a society can also associate external objects with aggression and thus affect the objects' aggressive cue value. Weapons are a prime example. For many men (and probably women as well) in our society, these objects are closely associated with aggression. Assuming that the weapons do not produce inhibitions that are stronger than the evoked aggressive reactions (as would be the case, e.g., if the weapons were labeled as morally "bad"), the presence of the aggressive objects should generally lead to more intense attacks upon an available target than would occur in the presence of a neutral object.

The present experiment was designed to test this latter hypothesis. At one level, of course, the findings contribute to the current debate as to the desirability of restricting sales of firearms. Many arguments have been raised for such a restriction. Thus, according to recent statistics, Texas communities having virtually no prohibitions against firearms have a much higher homicide rate than other American cities possessing stringent firearm regulations, and J. Edgar Hoover has maintained in *Time* magazine that the availability of firearms is an important factor in murders (Anonymous, 1966). The experiment reported here seeks to determine how this influence may come about. The availability of weapons obviously makes it easier for a person who wants to commit murder to do so. But, in addition, we ask whether weapons can serve as aggression-eliciting stimuli, causing an angered individual to display stronger violence than he would have shown in the absence of such weapons. Social significance aside, and at a more general theoretical level, this research also attempts to demonstrate that situational stimuli can exert "automatic" control over socially relevant human actions.

METHOD

Subjects

The subjects were 100 male undergraduates enrolled in the introductory psychology course at the University of Wisconsin who volunteered for the experiment (without knowing its nature) in order to earn points counting toward their final grade. Thirty-nine other subjects had also been run, but were discarded because they suspected the experimenter's confederate (21), reported receiving fewer electric shocks than was actually given them (7), had not attended to information given them about the procedure (9), or were run while there was equipment malfunctioning (2).

Procedure

General Design

Seven experimental conditions were established, six organized in a 2 × 3 factorial design, with the seventh group serving essentially as a control. Of the men in the factorial design, half were made to be angry with the confederate, while the other subjects received a friendlier treatment from him. All of the subjects were then given an opportunity to administer electric shocks to the confederate, but for two-thirds of the men there were weapons lying on the table near the shock apparatus. Half of these people were informed the weapons belonged to the confederate in order to test the hypothesis that aggressive stimuli which also were associated with the anger instigator would evoke the strongest aggressive reaction from the subjects. The other people seeing the weapons were told the weapons had been left there by a previous experimenter. There was nothing on the table except the shock key when the last third of the subjects in both the angered and nonangered conditions gave the shocks. Finally, the seventh group consisted of angered men who gave shocks with two badminton racquets and shuttlecocks lying near the shock key. This condition sought to determine whether the present of *any* object near the shock aparatus would reduce inhibitions against aggression, even if the object were not connected with aggressive behavior.

Experimental Manipulations

When each subject arrived in the laboratory, he was informed that two men were required for the experiment and that they would have to wait for the second subject to appear. After a 5-minute wait, the experimenter, acting annoyed, indicated that they had to begin because of his other commitments. He said he would have to look around outside to see if he could find another person who might serve as a substitute for the missing subject. In a few minutes the experimenter returned with the confederate. Depending upon the condition, this person was introduced as either a psychology student who had been about to sign up for another experiment or as a student who had been running another study.

The subject and confederate were told the experiment was a

study of physiological reactions to stress. The stress would be created by mild electric shocks, and the subjects could withdraw, the experimenter said, if they objected to these shocks. (No subjects left.) Each person would have to solve a problem knowing that his performance would be evaluated by his partner. The "evaluations" would be in the form of electric shocks, with one shock signifying a very good rating and 10 shocks meaning the performance was judged as very bad. The men were then told what their problems were. The subject's task was to list ideas a publicity agent might employ in order to better a popular singer's record sales and public image. The other person (the confederate) had to think of things a used-car dealer might do in order to increase sales. The two were given 5 minutes to write their answers, and the papers were then collected by the experimenter who supposedly would exchange them.

Following this, the two were placed in separate rooms, supposedly so that they would not influence each other's galvanic skin response (GSR) reactions. The shock electrodes were placed on the subject's right forearm, and GSR electrodes were attached to fingers on his left hand, with wires trailing from the electrodes to the next room. The subject was told he would be the first to receive electric shocks as the evaluation of his problem solution. The experimenter left the subject's room saying he was going to turn on the GSR apparatus, went to the room containing the shock machine and the waiting confederate, and only then looked at the schedule indicating whether the subject was to be angered or not. He informed the confederate how many shocks the subject was to receive, and 30 seconds later the subject was given seven shocks (angered condition) or one shock (nonangered group). The experimenter then went back to the subject, while the confederate quickly arranged the table holding the shock key in the manner appropriate for the subject's condition. Upon entering the subject's room, the experimenter asked him how many shocks he had received and provided the subject with a brief questionnaire on which he was to rate his mood. As soon as this was complete, the subject was taken to the room holding the shock machine. Here the experimenter told the subject it was his turn to evaulate his partner's work. For one group in both the angered and nonangered conditions the shock key was alone on the table (nonobject groups). For two other groups in each of these angered and nonangered conditions, however, a 12-gauge shotgun and a .38-caliber revolver were lying on the table near the key (aggressive-weapon conditions). One group in both the angered and nonangered conditions was informed the weapons belonged to the subject's partner. The subjects given this treatment had been told earlier that their partner was a student who had been conducting an experiment.[3] They now were reminded of this, and the experimenter said the weapons were being used in some way by this person in his research (associated-weapons condition); the guns were to be disregarded. The other men were told simply the weapons "belong to someone else" who "must have been doing an experiment in here" (unassociated-weapons group), and they too were asked to disregard the guns. For the last treatment, one group of angered men found two badminton racquets and shuttlecocks lying on the table near the shock key, and these people were also told the equipment belonged to someone else (badminton-racquets group).

Immediately after this information was provided, the experimenter showed the subject what was supposedly his partner's answer to his assigned problem. The subject was reminded that he was to give the partner shocks as his evaluation and was informed that this was the last time shocks would be administered in the study. A second copy of the mood questionnaire was then completed by the subject after he had delivered the shocks. Following this, the subject was asked a number of oral questions about the experiment, including what, if any, suspicions he had. (No doubts were voiced about the presence of the weapons.) At the conclusion of this interview the experiment was explained, and the subject was asked not to talk about the study.

Dependent Variables

As in nearly all the experiments conducted in the senior author's program, the number of shocks given by the subjects serves as the primary aggression measure. However, we also report here findings obtained with the total duration of each subject's shocks, recorded in thousandths of a minute. Attention is also given to each subject's rating of his mood, first immediately after receiving the partner's evaluation, and again immediately after administering shocks to the partner. These ratings were made on a series of 10 13-point bipolar scales with an adjective at each end, such as "calm-tense" and "angry-not angry."

RESULTS

Effectiveness of Arousal Treatment

Analyses of variance of the responses to each of the mood scales following the receipt of the partner's evaluation indicate the prior-shock treatment succeeded in creating differences in anger arousal. The subjects getting seven shocks rated themselves as being significantly angrier than

Table 1   Analysis of Variance Results for Number of Shocks Given by Subjects in Factorial Design

Source	df	MS	F
No. shocks received (A)	1	182.04	104.62*
Weapons association (B)	2	1.90	1.09
A X B	2	8.73	5.02*
Error	84	1.74	

*$p < .01$

Table 2   Mean Number of Shocks Given in Each Condition

	Shocks received	
Condition	1	7
Associated weapons	2.60a	6.07d
Unassociated weapons	2.20a	5.67cd
No object	3.07a	4.67bc
Badminton racquets	–	4.60b

Note.—Cells having a common subscript are not significantly different at the .05 level by Duncan multiple-range test. There were 10 subjects in the seven-shocks-received-badminton racquet group and 15 subjects in each of the other conditions.

the subjects receiving only one shock ($F = 20.65$, $p < .01$). There were no reliable differences among the groups within any one arousal level. Interestingly enough, the only other mood scale to yield a significant effect was the scale "sad-happy." The aroused-seven-shocks men reported a significantly stronger felt sadness than the men getting one shock ($F = 4.63$, $p < .05$).

Aggression toward Partner

A preliminary analysis of variance of the shock data for the six groups in the 3 X 2 factorial design yielded the findings shown in Table 1. As is indicated by the significant interaction, the presence of the weapons significantly affected the number of shocks given by the subject when the subject had received seven shocks. A Duncan multiple-range test was then made of the differences among the seven conditions' means, using the error variance from a seven-group one-way analysis of variance in the error term. The mean number of shocks administered in each experimental condition and the Duncan test results are given in Table 2. The hypothesis guiding the present study receives good support. The

Table 3   Mean Total Duration of Shocks Given in each Condition

| Condition | Shocks received | |
	1	7
Associated weapons	17.93_c	46.93_a
Unassociated weapons	17.33_c	39.47_{ab}
No object	24.47_{bc}	34.80_{ab}
Badminton racquets	–	34.90_{ab}

Note.—The duration scores are in thousandths of a minute. Cells having a common subscript are not significantly different at the .05 level by Duncan multiple-range test. There were 10 subjects in the seven-shocks-received-badminton racquet group and 15 subjects in each of the other conditions.

strongly provoked men delivered more frequent electrical attacks upon their tormentor in the presence of a weapon than when nonaggressive objects (the badminton racquets and shuttlecocks) were present or when only the shock key was on the table. The angered subjects gave the greatest number of shocks in the presence of the weapons associated with the anger instigator, as predicted, but this group was not reliably different from the angered-unassociated-weapons conditions. Both of these groups expressing aggression in the presence of weapons were significantly more aggressive than the angered-neutral-object condition, but only the associated-weapons condition differed significantly from the angered-no-object group.

Some support for the present reasoning is also provided by the shock-duration data summarized in Table 3. (We might note here, before beginning, that the results with duration scores—and this has been a consistent finding in the present research program—are less clear-cut than the findings with number of shocks given.) The results indicate that the presence of weapons resulted in a decreased number of attacks upon the partner, although not significantly so, when the subjects had received only one shock beforehand. The condition differences are in the opposite direction, however, for the men given the stronger provocation. Consequently, even though there are no reliable differences among the groups in this angered condition, the angered men administering shocks in the presence of weapons gave significantly longer shocks than the nonangered men also giving shocks with guns lying on the table. The angered-neutral-object and angered-no-object groups, on the other hand, did not differ from the nonangered-no-object condition.

Mood Changes

Analyses of convariance were conducted on each of the mood scales, with the mood ratings made immediately after the subjects received their partners' evaluation held constant in order to determine if there were condition differences in mood changes following the giving of shocks to the partner. Duncan range tests of the adjusted condition means yielded negative results, suggesting that the attacks on the partner did not produce any systematic condition differences. In the case of the felt anger ratings, there were very high correlations between the ratings given before and after the shock administration, with the Pearson rs ranging from .89 in the angered-unassociated-weapons group to .99 in each of the three unangered conditions. The subjects could have felt constrained to repeat their initial responses.

DISCUSSION

Common sense, as well as a good deal of personality theorizing, both influenced to some extent by an egocentric view of human behavior as being caused almost exclusively by motives within the individual, generally neglect the type of weapons effect demonstrated in the present study. If a person holding a gun fires it, we are told either that he wanted to do so (consciously or unconsciously) or that he pulled the trigger "accidentally." The findings summarized here suggest yet another possibility: The presence of the weapon might have elicited an intense aggressive reaction from the person with the gun, assuming his inhibitions against aggression were relatively weak at the moment. Indeed, it is altogether conceivable that many hostile acts which supposedly stem from unconscious motivation really arise because of the operation of aggressive cues. Not realizing how these situational stimuli might elicit aggressive behavior, and not detecting the presence of these cues, the observer tends to locate the source of the action in some conjectured underlying, perhaps repressed, motive. Similarly, if he is a Skinnerian rather than a dynamically oriented clinician, he might also neglect the operation of aggression-eliciting stimuli by invoking the concept of operant behavior, and thus sidestep the issue altogether. The sources of the hostile action, for him, too, rest within the individual, with the behavior only steered or permitted by discriminative stimuli.

Alternative explanations must be ruled out, however, before the present thesis can be regarded as confirmed. One obvious possibility is that the subjects in the weapons condition reacted to the demand characteristics of the situation as they saw them and exhibited the kind of behavior they thought was required of them. ("These guns on the table mean I'm supposed to be aggressive, so I'll give many shocks.") Several considerations appear to negate this explanation. First, there are the subjects' own verbal reports. None of the subjects voiced any suspicions of the weapons and, furthermore, when they were queried generally denied that the weapons had any effect on them. But even those subjects who did express any doubts about the experiment typically acted like the other subjects. Thus, the eight nonangered-weapons subjects who had been rejected gave only 2.50 shocks on the average, while the 18 angered-no-object or neutral-object men who had been discarded had a mean of 4.50 shocks. The 12 angered-weapons subjects who had been rejected, by contrast, delivered an average of 5.83 shocks to their partner. These latter people were evidently also influenced by the presence of weapons.

Setting all this aside, moreover, it is not altogether certain from the notion of demand characteristics that only the angered subjects would be inclined to act in conformity with the experimenter's supposed demands. The nonangered men in the weapons group did not display a heightened number of attacks on their partner. Would this have been predicted beforehand by researchers interested in demand characteristics? The last finding raises one final observation. Recent unpublished research by Allen and Bragg indicates that awareness of the experimenter's purpose does not necessarily result in an increased display of the behavior the experimenter supposedly desires. Dealing with one kind of socially disapproved action (conformity), Allen and Bragg demonstrated that high levels of experimentally induced awareness of the experimenter's interest generally produced a decreased level of the relevant behavior. Thus, if the subjects in our study had known the experimenter was interested in observing their *aggressive* behavior, they might well have given less, rather than more, shocks, since giving shocks is also socially disapproved. This type of phenomenon was also not observed in the weapons conditions.

Nevertheless, any one experiment cannot possibly definitely exclude all of the alternative explanations. Scientific hypotheses are only probability statements, and further research is needed to heighten the likelihood that the present reasoning is correct.

REFERENCES

Anonymous. A gun-toting nation. *Time*, August 12, 1966.

Berkowitz, L. *Aggression: A Social Psychological Analysis*. New York: McGraw-Hill, 1962.

Berkowitz, L. Aggressive cues in aggressive behavior and hostility catharsis. *Psychological Review*, 1964, 71, 104-122.

Berkowitz, L. The concept of aggressive drive: Some additional considerations. In L. Berkowitz (Ed.), *Advances in Experimental Social Psychology*. Vol. 2. New York: Academic Press, 1965. Pp. 301-329.

Berkowitz, L., & Geen, R. G. Film violence and the cue properties of available targets. *Journal of Personality and Social Psychology*, 1966, 3, 525-530.

Berkowitz, L., & Geen, R. G. Stimulus qualities of the target of aggression: A further study. *Journal of Personality and Social Psychology*, 1967, 5, 364-368.

Buss, A. *The Psychology of Aggression*. New York: Wiley, 1961.

Easterbrook, J. A. The effect of emotion on cue utilization and the organization of behavior. *Psychological Review*, 1959, 66, 183-201.

Geen, R. G., & Berkowitz, L. Name-mediated aggressive cue properties. *Journal of Personality*, 1966, 34, 456-465.

Loew, C. A. Acquisition of a hostile attitude and its relationship to aggressive behavior. Unpublished doctoral dissertation, State University of Iowa, 1965.

Walters, R. H., & Parke, R. D. Social motivation, dependency, and susceptibility to social influence. In L. Berkowitz (Ed.), *Advances in Experimental Social Psychology*. Vol. 1. New York: Academic Press, 1964. Pp. 231-276.

FOOTNOTES

[1] Cognitive processes can play a part even in impulsive behavior, most notably by influencing the stimulus qualities (or meaning) of the objects in the situation. As only one illustration, in several experiments by the senior author (cf. Berkowitz, 1965) the name applied to the available target person affected the magnitude of the attacks directed against this individual by angered subjects.

[2] Buss (1961) has advanced a somewhat similar conception of the functioning of anger.

[3] This information evidently was the major source of suspicion; some of the subjects doubted that a student running an experiment would be used as a subject in another study, even if he were only an undergraduate. This information was provided only in the associated-weapons conditions, in order to connect the guns with the partner, and, consequently, this ground for suspicion was not present in the unassociated-weapons groups.

10 Pain as a Cause of Aggression

—ROGER ULRICH

Aggressive behavior has often been noted among both domestic and wild animals (Utsurikawa, 1917; Zuckerman, 1932; Davis, 1935; Hall and Klein, 1942; Yerkes, 1943). Many situations seem to occasion aggressive behavior. Animals have been observed fighting for the possession of a mate, territory, or position in the group. Or, the aggressive response may involve food or responding to attack from predators. Man himself, although his aggressive behavior is controlled and complicated by "civilizing" factors, would probably occupy a prominent position on a list of the more aggressive animals.

Although observations of animals living in "natural" situations are often helpful, and speculations about the causes of aggression in humans are sometimes interesting, better understanding of the factors which cause and maintain aggressive behavior can come from experimentation under the controlled conditions provided by the laboratory. Concepts and techniques useful in the analysis of many kinds of behavior under a wide variety of environmental conditions have been developed by experimental psychologists, most notably Skinner (1938) and his students.

In the research described in this paper, these techniques and concepts have been used in the experimental analysis of aggression as a response to painful stimuli. The term, pain, is used synonymously with aversive stimuli, which are, in turn, defined as stimuli which the organism will work to eliminate from its environment.

Many observers have noted the relationship between pain and aggressive behavior. For example, as soon as young mice have teeth,

From the *American Zoologist*. 1966, *6*, 643-662. Reprinted by permission of the American Society of Zoologists. The research on which this paper is based was conducted first at the Behavior Research Laboratory at Anna State Hospital, Anna, Illinois, then at the Psychology Laboratories, Illinois Wesleyan University, and more recently at Western Michigan University. This research has received support from the National Science Foundation, the National Institute of Mental Health, the Illinois Academy of Science, the Illinois Psychiatric Research and Training Authority, and the Western Michigan University Research Fund.

they try to bite anything which pinches their tails (Scott, 1946). Often the tactile stimulation caused by one mouse's rough grooming of another will start a fight. The painful stimulation received by a peaceful male mouse when attacked will also produce fighting (Scott and Fredericson, 1951). Early investigators (O'Kelly and Steckel, 1939; Daniel, 1943; Richter, 1950) noted that electric shock delivered to a pair of rats through a grid floor (footshock) will produce fighting.

This investigator in collaboration with N. H. Azrin first encountered pain-elicited aggression in an attempt to replicate an earlier experiment (Miller, 1948) in which fighting between rats was reinforced by termination of shock. We found that reinforcement was not necessary—the rats would fight when the shock was first presented. The series of experiments described below followed from this attempted replication.

A BASIC EXPERIMENT IN PAIN-ELICITED AGGRESSION

In this section the procedure which formed the basis for our experiments in pain-elicited aggression is described in detail. Most of the conditions—except for the ones which were deliberately varied—were held more or less constant during subsequent investigations of the effects of different variables on the phenomenon.

The Experimental Animals

The experimental animals used in initial experiments (Ulrich and Azrin, 1962) in pain-elicited aggression were male laboratory rats of the Holtzman Sprague-Dawley strain. They were approximately 100 days old and weighed around 300 g. The animals were maintained in separate living cages with adequate food and water.

Sprague-Dawley rats are very docile, a necessary characteristic if the animal is to be used for any period of time. Fighting in more aggressive animals often results in casualties before the experiment is terminated. Use of a docile animal also eliminates possible complications due to genetically determined tendencies to fight.

Apparatus

The animals were placed in an experimental compartment measuring 12 in. by 9 in. by 8 in. Two sides were metal, the other two clear plastic. The floor consisted of steel rods, 3/32 in. in diameter and spaced 0.5 in. apart. A shielded bulb at the top of the compartment provided illumination, and a speaker produced a "White" masking noise. An exhaust fan provided additional masking noise, as well as ventilation. The temperature was maintained at about 75° F.

The presentation of stimuli was programmed by electrical apparatus located in a room separate from the experimental chamber. A cumulative recorder (Skinner, 1938), counters, and timers recorded the frequency of the responses. Shock was delivered through the grid floor by an Applegate constant current stimulator. A shock scrambler provided a changing pattern of polarities across the grids. Scrambled polarities prevented the animals from escaping shock by standing only on grids of like polarity.

Definition and Elicitation of the Response

The fighting response was defined as a striking or biting movement by either animal while standing on its hind legs in the sterotyped fighting posture. Only responses separated from one another by approximately 1 sec were recorded. Observers pressed microswitches to register each response on the recording equipment. Agreement between observers on the number of fighting responses was very close—within 5%.

Before delivery of the first shock, neither animal emitted any fighting responses. However, after delivery of a 2 ma shock of 0.5 sec duration, the subjects would assume and maintain the fighting posture and would produce several rapid fighting movements. The latency of this response was typically very short (approximately 0.2 sec) and the duration of the striking movements was usually less than 1 sec. The appearance of the response seemed to require that the animals were fairly close to and facing one another.

Behavior, such as this, which is elicited by a preceding stimulus with no prior conditioning, is termed unconditioned respondent, or reflexive behavior (Skinner, 1938).

SOME FACTORS INFLUENCING PAIN-ELICITED AGGRESSION

Once pain-elicited aggression was established as respondent, or reflexive, behavior, the effects on the phenomenon of various parameters and variables were studied.

Variations in the Aversive Stimulus and in Other Environmental Conditions

Although characteristics of the organism influence the occurrence of pain-elicited aggression, the characteristics of the aversive stimulus and

Table 1. Examples of the Consistency of Fighting Elicited by Shock from Three Pairs of Subjects During Two Sessions at Each of the Different Shock Frequencies. (The consistency of the fighting reflex is expressed as the percentage of shocks that resulted in a fighting response.)

Frequency of shocks (shocks/min)	Consistency of fighting reflex (responses) (shocks)		
	Pair 1	Pair 2	Pair 3
0.1	0.33	0.66	0.66
0.6	0.61	0.55	0.61
2.0	0.83	0.58	0.58
6.0	0.83	0.94	0.77
20.0	0.92	0.91	0.82
38.0	0.85	0.89	0.93

of the general environment exert by far a more profound influence on the response. A series of studies of the effects of environment on aggression is described below.

Frequency of Shock Presentation

The 2 ma shocks were delivered to three pairs of rats at frequencies of 0.1, 0.6, 2, 20, and 38 shocks per min (Ulrich and Azrin, 1962). Each of these frequencies was administered in a random order during each of three different 10-min sessions. A 24-hr interval was usually allowed between sessions.

Table 1 shows the rate of fighting (response/shocks) for each of the three pairs as a function of frequency of shock presentation. The rate of fighting for each pair of subjects increased from a mean of 0.55 at a frequency of 0.1 shock/min to a mean of 0.89 at a frequency of 38 shocks/min. The increase in fighting at higher frequencies seemed to be related to the positions assumed by the animals in the intervals between shocks. Since the fighting response seems to require that the subjects be near to and facing one another, the probability of response decreased during lower frequencies when the animals had time between shocks to come out of the fighting posture and move about the chamber. At higher frequency, the animals usually remained in the fighting posture.

The relationship between fighting and frequency reversed at very high frequency. In another study (Ulrich and Azrin, 1962) with two pairs of rats, discrete shocks were presented so frequently as to be virtually continuous. Although fighting occasionally occurred, much of the rats' behavior seemed to be directed toward escape from the experi-

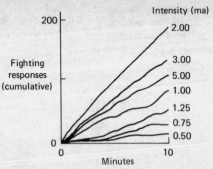

Figure 1. Typical cumulative records of the fighting responses elicited from one pair of rats at various intensities of foot-shock.

mental chamber. Similar behavior has also been noted when naive subjects are first shocked. However, in the case of naive animals given non-continuous shock, fighting soon replaced the escape behavior.

Intensity of the Shock

Shocks of various intensities were delivered to three pairs of rats at a fixed frequency of 20 shocks/min (Ulrich and Azrin, 1962). Each intensity was presented for periods of at least 10 min. The different intensities were presented in a varied sequence, and several 10-min periods were given at each intensity.

Figure 1 shows the response of a typical pair of rats receiving shocks of various intensities. Increasing the shock intensity from 0 to 2 ma produced an increase in the rate of fighting. At still higher intensities of 3 to 5 ma, however, the response rate was somewhat reduced. Tedeschi, et al. (1959) also found shocks of 2 to 3 ma best for eliciting aggression between mice.

Visual observation indicated that, at lower intensities, the fighting response was less vigorous, shorter in duration, and had longer latency. Indeed, at lower intensities the definition of a movement as a fighting response was often arbitrary, while at higher intensities the attack movement was unmistakable. Also, at lower intensities the rats' physical position had a great influence on the likelihood that a response would appear. If the rats' feet were making good contact across several of the floor grids and the rats were also facing one another, a fighting response was likely to result.

The decrease in frequency of fighting at intensities of 3 and 5 ma appeared to be partly a consequence of the debilitating effects of the shock. Prolonged exposure to shocks of 5 ma often resulted in paralysis

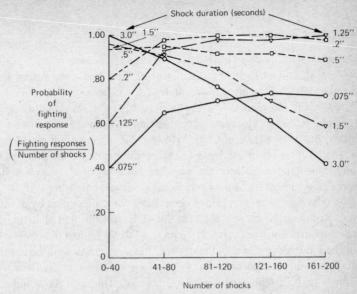

Figure 2. For each duration of shock, the probability of fighting (responses/shocks) is presented as a function of successive blocks of 40 shocks. The curves are based on the mean scores for six pairs of rats.

of one or both of these subjects. Higher intensities also generated escape behavior which competed with the fighting response.

Duration of the Shock

Two hundred shocks of various durations were given to six pairs of rats each day for 12 days (Azrin, Ulrich, Hutchinson, and Norman, 1964). A single duration was used for each day (each duration was used for two experimental sessions), and the shocks were separated by an interval of 3 sec.

Figure 2 shows the results of these sessions. The vertical axis shows the probability of fighting. The horizontal axis breaks the 200-shock sequence up into blocks of 40. Initially, shocks of greater duration elicited a higher rate of fighting. However, by the end of the 200-shock sequence, the effect was somewhat reversed, two of the shocks of higher duration (3.0 and 1.5 sec) eliciting the lowest rates of fighting. Also, by the end of the sequence, durations of 0.2 and 0.125 sec, which initially had elicited rates of 0.40 and 0.60 responses/shock were eliciting a response rate of nearly 1.00. Again, escape-like movements, such as jumping about, seemed to interfere with the fighting behavior as the delivery of shocks of long duration increased.

Heat, Cold, and Noise as Elicitors of Aggression

Attempts were made to elicit aggression by exposing the rats to other aversive stimuli (Ulrich and Azrin, 1962). The rats were placed in a chamber with a thin metal floor that could be heated. If the rats were placed in the chamber before the floor was heated, they would display signs of discomfort (jumping about, licking their feet) as the floor grew hotter, but they would not fight. However, if the rats were placed on a preheated floor, their attempts to escape the heat were interrupted by stereotyped fighting. Once the rats had been exposed to a preheated floor, gradual heating of the floor would elicit some fighting.

The rats were also placed on a metal floor precooled by dry ice. No fighting behavior was elicited, but it is possible that the stimulus was not sufficiently aversive. No pain was felt by human observers who touched the floor for less than 2 sec. Since the rats were constantly moving about, it is possible that no one foot remained on the floor long enough to be affected aversively by the cold. The subjects would also occasionally escape the cold by lying on their backs.

Presentation of intense noise was also not effective in producing fighting. Noise of 135 db enclosing a band from 200-1500 cps was delivered in durations ranging from bursts of less than 1 sec to sustained periods of more than 1 min.

Extinction as an Elicitor of Aggression

Azrin, Hutchinson and Hake (1966) investigated the relation between removal of positive reinforcement, or extinction, to aggression. A pigeon's key-pecking was conditioned by making food available after every peck. After the response was well established, an experimental session was conducted in which, after the bird emitted 20 pecks and received 20 reinforcements, no further reinforcements were delivered.

When no other animal was present in the chamber, the bird emitted the flurry of responses typical of performance just after the onset of extinction. However, when another pigeon was located nearby, the final burst of responding did not occur, and the bird would instead rush over to the other pigeon and attack its head. A stuffed pigeon would also be attacked by a bird placed on extinction.

Although extinction does not involve any specific aversive stimulus, both experimental and casual observation have suggested that animals find it aversive. Ferster (1957) has found that withdrawal of positive reinforcement will serve as punishment for pigeons. When placed on extinction, animals will usually show signs of what has commonly been called "frustration." Pigeons will ruffle up their feathers and coo and run about the chamber. The fact that extinction elicits aggression, at least in pigeons, supports the notion that extinction is aversive.

Intra-cranial Stimulation as an Elicitor of Aggression

Much research has been done relating aggression to electrical stimulation of certain areas of the brain. An early experiment by Ranson (1934) showed that stimulation of the hypothalamus of a cat resulted in almost exact reproduction of the cat's normal rage behavior. Masserman (1950, 1964) repeated Ranson's experiments and further concluded that the rage behavior continues as long as the electric current is present. Hess (1949, 1954; Hess and Akert, 1955) also reported eliciting rage behavior in a cat by stimulating the hypothalamus. According to their observations, however, if the hypothalamus was stimulated long enough or with an unusually strong current, the response would persist beyond termination of the current.

The importance of the exact anatomical site of hypothalamic stimulation has been stressed by Wasman and Flynn (1962) who found that the probability of attack or rage was related to the site of stimulation. Egger and Flynn (1961) demonstrated a more complicated relation when they showed that attack induced by stimulation of the hypothalamus can be suppressed by simultaneous stimulation of the amygdala. A more recent study has shown that electrical stimulation of the hypothalamus of the cat will elicit attack and stalking responses toward both animals and inanimate objects.

Similar results have been found in chickens (von Holst and von Saint Paul, 1962). Stimulation of one region of the brain elicits the characteristic attack that one hen makes against another hen of lower rank. This response, although its characteristics vary from object to object, will be directed at alive or stuffed hens or, if the stimulation is increased slightly, against a human hand and against various other species.

Other Environmental Effects

We have seen that aversive stimuli will elicit stereotyped aggression. In this section we shall examine the effects of two other environmental variables—the size of the chamber and the length of the session—on the rate at which aggression is elicited by these aversive stimuli.

Size of the Chamber

Calhoun (1950) and others (Crew and Mirskaia, 1931) have observed that overcrowded rats fight more often than rats living at lower popula-

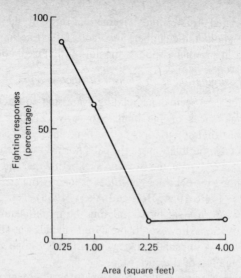

Figure 3. Rate of elicitation of fighting responses (responses × 100/shocks) from a pair of rats by foot-shock in a square chamber of constant height and variable floor area.

tion densities. Indeed, in the present experiments, size of chamber was found to have a marked effect on the rate at which pain-elicited aggression occurred (Ulrich and Azrin, 1962). As mentioned above, the chamber ordinarily used in studies of pain-elicited aggression is approximately 12 × 9 × 8 in. The chamber used in studying the effects of the size of the chamber had an adjustable floor area although the height was held constant at 17 in. In this study, a pair of rats was given 2 ma shocks at the rate of 20 shocks per min for 10-min sessions.

Figure 3 shows the relationship between floor space and rate of fighting. With a very small space of 6 × 6 in., the rate of fighting was very high—about 90%. As the size of the floor increased, fighting decreased until a minimal rate of 2% was reached on a floor measuring 24 × 24 in.

This increased rate of fighting when space is limited is probably due to the effect of proximity (mentioned above) on fighting. When rats are near or facing one another, they are much more likely to fight. Since the small experimental chambers kept the subjects close together, they kept the fighting rate high.

Length of the Session

The fighting response seems to be very resistant to reflex fatigue. Shocks were presented to a pair of rats at a rate of 40/min for 7.5 hr

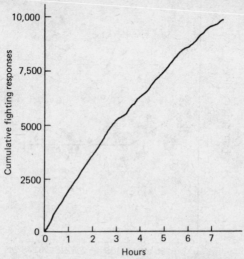

Figure 4. Cumulative record of fighting responses elicited from a pair of rats during a 7.5-hr period of frequent (every 1.5 sec) shock presentation.

(Ulrich and Azrin, 1962). During the first hour (2400 shocks) 82% of the shocks elicited fighting. After the third hour (7200 shocks) 70% of the shocks still elicited fighting. Only during the last 1.5 hr, after nearly 15,000 shocks had been delivered, did the rate of elicitation drop to 40%. At the end of 7.5 hr, and after approximately 10,000 fighting responses had been elicited, the rats were still responding to the shock (Fig. 4).

Characteristics of the Experimental Animals

Various modifications were made in the types of animals used as subjects. They are described briefly below.

Sex

Male rats have been reported to fight more often than female rats in "natural" situations (Beeman, 1947; Scott and Fredericson, 1951). To investigate the influence of the sex of the subjects upon pain-elicited aggression, a pair of female rats and a pair consisting of one male and one female rat were each administered 2-ma shocks at a frequency of 20 shocks/min (Ulrich and Azrin, 1962). The usual rate of fighting was not influenced by the sex of the subjects. Indeed, sexual behavior which would normally have occurred between the male and female rats was completely displaced by the elicited fighting response.

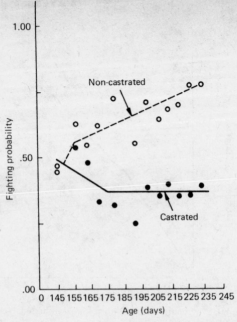

Figure 5. Fighting probability (responses/shocks) in castrated and non-castrated adult rats during 12 successive weekly sessions.

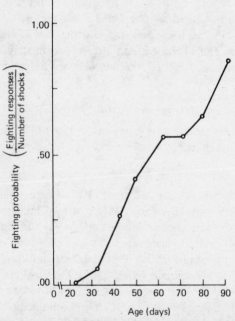

Figure 6. Fighting probability (responses/shocks) in eight pairs of rats, aged 24, 33, 43, 53, 63, 73, 83, and 93 days.

Castration

The effect of castration both before and after puberty and the accompanying reduction of androgens (Hutchinson, Ulrich, and Azrin, 1965) was studied. Several investigators (Beeman, 1947; Kislak and Beach, 1955) have noted that castration and hormonal therapy reduced the aggressiveness of rats. Rats castrated before maturity and those castrated after maturity fought about half as frequently as non-castrated rats. However, in the rats castrated after maturity, the effect took several weeks to develop. Although they fought at the same rate as the non-castrated rats, eventually the rate declined until they fought only half as often (Fig. 5).

Age

The effect of age on pain-elicited fighting is probably related to the effect of hormonal level which occurred above with castration. Ten pairs of rats, each pair of a different age ranging from 24 to 93 days, were exposed to 100 shocks (Hutchinson, Ulrich, and Azrin, 1965). The subjects had had no prior experience with shock and had been raised in community cages. As is shown in Figure 6, fighting increased monotonically with age.

Sensory Impairment

The orientation of the subjects seemed to have a great effect on the probability that the subjects would fight. When the animals were facing one another, shock was much more likely to result in an aggressive response than when the rats were turned away from one another. This observation suggested that the sight of another rat was an additional stimulus in eliciting fighting. In other species, such as the Siamese fighting fish (Thompson, 1965), visual stimuli are sufficient to elicit fighting.

In order to determine the effect of visual stimuli on pain-elicited aggression in rats, the rate of fighting was recorded for four pairs of rats both with and without hoods (Flory, Ulrich, and Wolff, 1965). The rats fought much less frequently when hooded. During 10-min sessions without hoods, from 130 to 195 responses occurred, while sessions with hoods produced from 45 to 112 responses.

Since the hoods entirely covered the rats' heads, it was possible that the decrease in fighting was also due to the impairment of senses other than sight. In order to determine more reliably the effect of blindness on rate of fighting, the eyes of one pair of rats were removed. Before the operation, the rats fought at a mean rate of 162 responses per session for 10 sessions. The mean rate for 10 sessions after the removal of their eyes was 118. Thus the rats fought less after having their eyes removed, but the decrease was not as great as when hooded. The vibrissae of the pair of blinded rats were then cut as close to the skin as possible. In 10 sessions with both eyes and vibrissae removed, the mean number of responses was 94. Thus, the rate of reponse with both vibrissae and eyes removed was close to the rate of response when wearing hoods.

Strain

The basic procedure described above was repeated using various strains of laboratory rats (Ulrich and Azrin, 1962). Long-Evans hooded, Wistar, General Biological hooded, and Charles River Sprague-Dawley rats were used as well as the Holtzman Sprague-Dawley which originally served as subjects. All strains except the Wistar produced response rates similar to that of the Holtzman Sprague-Dawley, responding to at least 70% of the shocks. The Wistar rats responded to less than 50% of the shocks. The 2-ma shock appeared to have a debilitating effect on the Wistar rats similar to the effect of a 5-ma shock on Holtzman Sprague-Dawley rats. In fact, four of the Wistar rats died after exposure to the shocks. Since shocks of high intensity tend to generate competing escape behavior in the Holtzman Sprague-Dawley rats, it is not surprising that the same effect was observed at lower intensities in the more sensitive Wistar strain.

Species

Although differences between strains of rats were negligible, differences in the occurrence of pain-elicited aggression were found between species. Hamsters (Ulrich and Azrin, 1962), mice (Tedeschi, et al., 1959), cats (Ulrich, Wolff, and Azrin, 1964) and monkeys (Azrin, Hutchinson, and Hake, 1963) have all displayed pain-elicited aggression.

However, guinea pigs have never responded to shock with fighting behavior (Ulrich and Azrin, 1962). The intensity and frequency of the shock were varied, and the animals were deprived of food for periods up to 72 hr, but they still did not exhibit a fighting response.

Interspecies Fighting

When a Holtzman Sprague-Dawley rat was paired with a hamster, both animals fought in response to shock. When a rat was paired with a guinea pig, however, the rat attacked the guinea pig's head, but even when attacked by the rat, the guinea pig did not fight (Ulrich and Azrin, 1962).

When a cat was paired with a rat, the cat would attack the rat. Because the cat's attacks would often seriously injure the rat, it was difficult to measure the frequency of the cat's attacks. However, it was apparent that rats attacked cats much less frequently than vice versa.

Previous Social Experience Between the Subjects

Seward (1945) has reported that nonreflexive fighting behavior appears to be influenced by previous familiarity between the animals. In order to determine the effect of familiarity on respondent aggression, 12 pairs of animals which had been housed together in pairs for several weeks, were submitted to the experimental procedure (Ulrich and Azrin, 1962). This prior experience between members of the experimental pairs did not influence the rate of the fighting response.

Number of Rats in the Experimental Chamber

When 2, 3, 4, 6, or 8 rats were placed in the same chamber and shocked, the stereotyped fighting response also occurred. Often two or more animals would "gang up" in aggressing against a single rat.

Delivery of Shock to Only One Subject of a Pair

In order to study the effect of shocking only one of a pair of rats, it was necessary to deliver the shock through electrodes implanted beneath a fold of skin (Ulrich and Azrin, 1962). A harness and swivel arrangement allowed the rat complete freedom of movement.

When the rat was alone in the experimental chamber, a shock of 2 ma for 0.5 sec delivered through the electrode resulted only in a spasmodic movement of the animal. However, when the shock was delivered in the presence of another rat, the stimulated rat would usually assume the stereotyped fighting posture and attack the unstimulated animal. After being attacked, the unstimulated rat would often assume the stereotyped fighting posture and would retaliate. Fighting was less frequent and less violent, however, when only one animal received stimulation.

Inanimate Objects

When an insulated doll was being placed in the experimental chamber while a rat was being shocked, the rat did not attack the doll (Ulrich and Azrin, 1962). Nor would a rat attack a conducting doll or a recently deceased rat. Dolls moved rapidly about the cage also failed to occasion fighting. Only when a dead rat was moved about the cage on a stick was a fighting response elicited.

In contrast, pigeons will attack stuffed pigeons (Levi, 1957;

Smith and Hosking, 1955). Also, Azrin, Hake, and Hutchinson (1965) have found that monkeys will attack tennis balls when aversively stimulated by a shock or blow on the tail.

Indeed the experiments with the monkey and the tennis ball have provided a method for checking on the accuracy of a human observer. In the study by Azrin, et al. (1965) the tennis ball was attached to a micro-switch which automatically recorded each blow delivered to it. In this way recording of the aggressive response could be accomplished automatically as well as through the human observers. A high correlation was apparent between the automatic device and recordings by the human observer.

Social Isolation

We have seen that the rate of pain-elicited aggression increases with the age of the subject. In order to discriminate the influence of the rats' social histories on this increase and on rate generally, rats raised in isolation from the time of weaning were tested. Three groups of five pairs of rats each were tested at 30, 60, and 90 days of age, respectively (Hutchinson, Ulrich, and Azrin, 1965). The pair tested at 30 days of age did not fight. Those 60 days old had a fighting rate of 0.38 response/shock, or 38%. Those 90 days old fought in response to 48% of the shock. Compared to those shown in Table 1 these rates are low. The higher rates produced by the socially-raised animals suggest that some kind of conditioning in the rats' social environment operates to strengthen the aggressive response. Since fighting can be conditioned according to both operant and respondent principles (Reynolds, Catania, and Skinner, 1963; Ulrich, Johnston, Richardson, and Wolff, 1963; Stachnik, Ulrich, and Mabry, 1966; Vernon and Ulrich, 1966), either operant or respondent conditioning—or a combination of the two—may be increasing the frequency of the aggressive response in socially-experienced rats. However, the fact that subjects with no history of social interaction did fight indicates that pain-elicited aggression has an instinctive, or unconditioned, basis as well.

Successive Elicitation

Since the animals raised in groups fought at a higher rate than those with no social history, an attempt was made to determine the effect of a long social history of pain-elicited aggression upon the animals' fighting rate (Hutchinson, Ulrich, and Azrin, 1965). To test this, five pairs of rats were tested over an 80-day period, ranging from the time when they were 21 days old to the time when they were 100 days old. Each

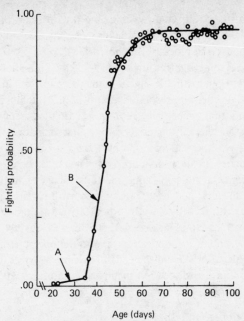

Figure 7. Fighting probability (responses/shocks) in five pairs of socially isolated rats tested daily from 21 to 100 days of age. Tests were not conducted on days 26-36 (point A) or on days 40-42 (point B).

rat was kept in isolation when not in the experimental chamber. There were two gaps in the testing due to failure of the apparatus, but except for these lapses, the animals were given a variable number of shocks every day. The results are shown in Figure 7. At 100 days of age, the rats were fighting at a rate of 93%. This rate was about the highest ever observed in our laboratory. Casual observation indicated that the animals fought more viciously as well, often cutting and bruising each other severely. Rats shocked every day from 21 to 90 days of age also attained the same high rate. Their increase in fighting also occurred at the same time as the increase at 37 days shown for the other rats.

CONDITIONED AGGRESSION

Generally speaking, conditioning involves increasing the frequency of occurrence of a response. Respondent behavior can be conditioned by pairing a neutral stimulus with the unconditioned stimulus which elicits the response (Pavlov, 1927). When conditioning has occurred, the neutral stimulus alone becomes able to elicit the response. Operant behavior can be conditioned by following it with a reinforcer—a stimulus whose presentation following a response increases the probability that the response will reoccur (Skinner, 1938).

The fact that fighting increased with the animals' social experience suggested that stimuli in the animals' social environment were conditioning the fighting response. Also Ulrich, Hutchinson, and Azrin (1965) found that when rats kept in isolation, except during daily experiments on elicited aggression, were placed in the chamber with naive rats, any movement or noise by the naive animal would elicit the fighting response in the experienced rat. The movements of another rat in the experimental situation appeared to become a conditioned elicitor of fighting, through consistent association with shock. Similar results were reported by O'Kelly and Steckle (1939). Support has been given these observations, since aggressive behavior has, in fact, been conditioned in the laboratory, both according to the operant and the respondent paradigms. Both are discussed briefly below, operant aggression being relevant to this discussion mainly as it interacts with respondent aggression.

Respondent Conditioning of Aggression

In recent work by Vernon and Ulrich (1966), conditioned responses were obtained after 2000 pairings of a conditioned stimulus with an unconditioned stimulus. Three different intensities were used (2.0, 2.5, and 3.0 ma) in each of three different pairs of animals. The conditioned responses were elicited by the buzzer at rates relatively comparable to the rates obtained for the different intensities in the unconditioned elicitations. Shocks of 2 ma elicit the highest rates of unconditioned responding, and the rats conditioned to the 2-ma shocks also responded at the highest rates to the conditioned stimulus. These results support the principle that the stronger the unconditioned stimulus is as an elicitor the stronger its conditioned stimuli will be.

Operant Conditioning of Aggression

As noted above, the rate of aggressive behavior can be increased by following it with reinforcers. Miller (1948) reported the successful conditioning of operant aggression in rats by reinforcing successive approximations to the aggressive response with termination of shock. However, as we have also noted above, the ability of shock to elicit the aggressive behavior which Miller was trying to shape may have complicated his results. Reynolds, Catania, and Skinner (1963) have successfully conditioned operant aggression in pigeons, and operant aggression was conditioned in rats using water as an unconditioned reinforcer by Ulrich, Johnston, Richardson, and Wolff (1963). In the latter experiment animals were first paired with control animals which had not been deprived of water. Then, the experimental animal's successive approximations to attacks on the control animal were reinforced. Fighting behavior was successfully conditioned in the experimental animals. The control animals, however, seldom fought even when attacked.

When two experimental animals were paired, rate of fighting in each increased, although the reinforcement procedure was not changed. In fact, occasionally the animals would not stop fighting to obtain the water. A pattern of dominance also emerged, one experimental rat becoming more aggressive and the other more submissive until the submissive rat's behavior began to resemble that of the control rats.

It seemed that, in animals which had acquired the fighting response through operant conditioning, further fighting could be elicited by the attack of another animal. Reynolds, Catania, and Skinner (1963) also reported that operant aggression in pigeons seemed to be complicated by respondent aggression resulting from the presence of another bird. In Reynolds' experiment, the respondent aggression triggered by the operant aggression maintained itself after reinforcement ceased. However, Ulrich, et al. (1963) found in rats that, after a slight increase in aggressive behavior following the onset of extinction, all fighting ceased. This slight increase has long been noted in the performances of animals beginning the extinction procedure on such responses as bar pressing and key pecking. It is possible that this flurry of responses coincident with extinction may be a form of aggression produced by the aversive aspects of extinction.

These results, of course, suggest some of the complex effects that occur when aggression is produced and maintained by both operant and respondent stimuli. It appears that the interaction results in an increase in aggressive behavior, rather than an inhibition of one type of aggression by another. On some occasions, the development of operant aggression seems to be prerequisite to the elicitation of respondent aggression. It is also likely that the operant reinforcement of respondent aggressive behavior will increase its rate and strength.

Some Interesting Interactions Between Operant Behavior and Respondent Aggression

Another interesting interaction between operant behavior and respondent aggression is provided by experiments using the opportunity to complete an attack against another animal as positive reinforcement for an operant response (Azrin, Hutchinson, and McLaughlin, 1965). As previously mentioned, monkeys will attack tennis balls following the presentation of an aversive stimulus. In addition, it was found that, after being shocked, monkeys will emit an operant chain-pulling response when it results in the presentation of the tennis ball. The only function of the chain-pull is to present the tennis ball; it has no relation to termination of shock.

It has also been pointed out above that the extinction procedure will act as an aversive stimulus eliciting a fighting response in pigeons. Azrin (1964) found that a pigeon placed on extinction will peck a key which causes another bird to appear, which is then attacked.

Still another interesting relationship has been explored by Delgado (1963). Using an established monkey colony, he found that a subordinate monkey will repeatedly press a lever which stimulates the caudate nucleus of the dominant monkey, thereby inhibiting his aggressive behavior. Thus, the inhibition of the dominant monkey's aggression serves as reinforcement for the lever-pressing response.

PAIN-ELICITED AGGRESSION AND AVERSIVE CONTROL

One effective way of controlling behavior is through the use of aversive stimuli. Three paradigms of aversive control are commonly used: escape, avoidance, and punishment. In escape, an aversive response is presented, and an operant response is required for its removal. This procedure, in which a response is reinforced by removal of an aversive stimulus, is also known as negative reinforcement. In avoidance, the animal, after the presentation of a discriminative stimulus (which may simply be a time lapse), emits an operant response which causes the aversive stimulus not to be presented. In punishment, an aversive stimulus is presented after a response. The effects of punishment are usually studied by establishing and maintaining an operant response by following the response with positive reinforcement, and, then, by punishing that same response.

The interaction between punishment and pain-elicited aggression is an important area, since punishment is so widely used in the control of behavior. However, although investigations of this interaction are currently being conducted in our laboratories, no definitive treatment of this area has yet been made. Therefore, the following discussion will be confined to the effects of pain-elicited aggression on escape and avoidance.

Rats have been known to escape or avoid shock by standing on grids of the same polarity in an experimental chamber which has not been equipped with a scrambler. They also seem to avoid the aversive stimulus of a heated floor by shifting feet so that no one foot remains on the floor long enough to feel an appreciable amount of heat. When heated floors are used as aversive stimuli, rats will also lick their feet to cool them—a possible escape or avoidance response. In addition, rats exposed to cold floors will sometimes lie on their backs to escape the cold. These are instances in which the organisms, in spite of the presence of an aversive stimulus, make adaptive responses that lessen or eliminate the effects of the stimulus.

Yet, many of the responses made in the presence of aversive stimuli are nonadaptive. Reflexive fighting is a prime example, since the fighting behavior does nothing to decrease the impact of the stimulus. In fact, it is possible that fighting increases the aversiveness of the shocks.

When alone in the experimental chamber, animals soon develop adaptive responses under both escape and avoidance conditions. However, when two animals are present in the chamber during escape or avoidance conditioning, the tendency of the aversive stimuli to elicit aggression interferes with the seemingly more adaptive escape or avoidance behavior.

Pain-Elicited Aggression and Avoidance

The interference of pain-elicited responses can be seen in the case of avoidance conditioning. One study (Ulrich, Stachnik, Brierton, and Mabry, 1965) used a Sidman (1953) avoidance schedule which presented discrete shocks of 1 ma at a rate of 40 shocks/min. The animals could avoid these shocks by pressing a bar which would instate a 20-sec, shock-free period. Thus, theoretically, by pressing a bar at a rate slightly higher than 3 presses/min, the animal could avoid shock altogether.

The experiment was undertaken in two phases. Under the first, eight of the 20, 120-day-old, naive, male, Sprague-Dawley rats used in the experiment were divided into three pairs and two single animals. They were run daily for 18, 4-hr avoidance sessions. None of the subjects had prior experience with the experimental procedure.

In the second set of observations, the three pairs used in the first procedure were broken up into six single animals, and the two rats trained singly in avoidance behavior were paired with two naive rats.

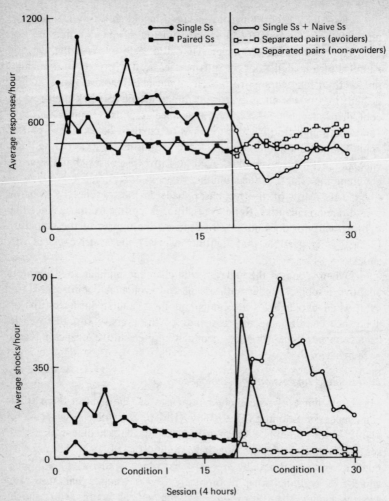

Figure 8. Avoidance responses (top graph) and shocks received (bottom graph) by ten rats. The left-hand portion (condition I) of the top graph shows the mean number of responses per hour for two rats run singly (closed circles) and for six rats run in pairs (closed squares). Heavy horizontal lines show the mean number of responses per hour for eight sessions. The right-hand portion (condition II) of the top graph shows the mean number of responses per hour for the subjects formerly run singly, when paired with naive subjects (open circles) and for the formerly paired subjects when separated and run singly (open squares). The avoidance behavior of the good avoiders from the separated pairs is indicated by a hatched line. The avoidance behavior of the nonavoider from the separated pairs is designated by a solid line. The bottom graph shows mean number of shocks received per hour by the same groups of rats under the same conditions.

Although both the single and the paired subjects in the first procedure acquired avoidance behavior, the single subjects consistently performed better than the paired. They emitted more avoidance responses and received less shocks (see Condition I in Fig. 8).

Initially, both the paired and single subjects responded to the shocks by jumping around the chamber and occasionally landing on the bar. However, the paired rats would often assume the fighting posture and produce a relatively large number of aggressive responses. Naturally, this fighting behavior interfered with the jumping about and accidental bar-presses. Since the number of accidental bar-presses was reduced, the paired animals were slower in acquiring the avoidant bar-pressing response. However, the 1-ma shocks interfered less with the avoidance behavior during the acquisition stage than they did later, since rate of respondent fighting increased as the session progressed.

As the avoidance behavior developed, one animal of each pair would do most of the avoiding. While the avoider was pressing the bar, the nonavoider would often continue to attack him. The avoider would also attempt to attack while pressing the bar. These attack movements would interfere with the avoidant bar-presses, disrupting the avoidance performance of the animal and increasing the number of shocks received.

In the second part of the experiment (Condition II, of Fig. 8) when the pairs were separated, the number of bar-presses emitted by both the avoiders and the nonavoiders remained at the level emitted by the pair (when the avoider was doing most of the responding). However, at the beginning of the procedure, the avoider received fewer shocks than the nonavoider. Although the nonavoiders pressed the bar as often as the avoiders, it actually took them longer to acquire effective avoidance responding than it takes single naive animals.

When a naive subject was paired with a rat which had been trained singly in avoidance responding, fighting behavior comparable to that which occurred with naive pairs was observed (Fig. 8). In both cases, bar-pressing was at first almost entirely displaced by fighting. Even though successful avoidance increased as the animals became more experienced, it never reached the level attained by single animals.

In addition to the procedures described above, five pairs of naive rats were given avoidance training using shocks of 2 ma. Although some avoidance responding was acquired, it was never maintained at the level observed for 1-ma shocks. In fact, only one out of the five pairs was able to continue past the seventh 4-hr session. Fighting occurred at a higher rate and appeared to be more violent. Since the avoider of each pair was able to defend himself only imperfectly from the attacks of the nonavoider, while also attempting to avoid shock, the more violent attacks elicited by the 2-ma shocks soon incapacitated the avoider. It is, of course, obvious from these results that pain-elicited aggression is not, as some people claim, simply a defensive response.

Another study of avoidance behavior and pain-elicited aggression has been reported by Ulrich and Craine (1964). Two rats with a long history of stable, single, avoidance performances were paired, one with a rubber dummy and one with a naive rat. The avoidance procedure was similar to that described above (shock intensity was 1 ma), except that an auditory signal was presented 2 sec before the onset of the shock.

No appreciable difference occurred in the performance of the animal paired with the dummy. The avoidance behavior of the rat paired with the naive animal, however, was initially almost entirely displaced by fighting. Nevertheless, by the end of the experiment, fighting had decreased to a very low level. Yet, in spite of the decrease in fighting, avoidance behavior never returned to the level maintained by the rat when alone.

Pain-Elicited Aggression and Escape

When exposed to high-intensity shocks or to long, continuous shocks, rats have been observed making futile attempts to escape. We have also seen that when two animals are placed in the same chamber, adaptive avoidance behavior is disrupted by pain-elicited aggression. Additional investigations were made of the interaction between pain-elicited aggression and escape behavior.

In one study (Ulrich and Craine, 1964) any behavior which was *not* the stereotyped fighting response was designated as the escape response. After paired rats had been in the experimental chamber for 60 sec, continuous shock was presented until a non-fighting response occurred in both animals. When fighting stopped, the shock was immediately terminated. After a 60-sec lapse, the procedure was repeated, and each session consisted of 15 repetitions. Figure 9 gives the results. The duration of the fighting behavior during the shock actually increased as the sessions progressed.

These findings were particularly surprising in view of the usual effect of removing shock contingent upon a response. In this case since nonaggression was the reinforced response it would have been likely that it would increase.

One possible explanation for the fact that the punished response

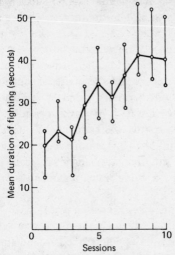

Figure 9. Mean number of seconds (dark circles) spent fighting by two pairs of rats during ten sessions when nonfighting was reinforced by termination of shock. Open circles indicate the range.

(i.e., fighting) increased rather than the reinforced response (i.e., non-fighting) can be found in the nature of the required escape response. Most escape responses (key-pecks, bar-presses, etc.) are specific. However, in the study mentioned above, the behavior, "not fighting," was unspecific. Since reinforcement has been shown to be more effective when following a specific than when following a general response, the unspecific nature of nonfighting may have lessened the reinforcing effect of the removal of shock.

Studies of avoidance behavior mentioned above lend support to this explanation. In these investigations a specific response (bar-pressing) was used, and, although fighting did reduce the number of avoidance responses, avoidance also reduced fighting.

Another investigation (Ulrich, 1965) using cooperative bar-presses and discrete shocks, has helped clarify this point. Two rats were placed in an experimental chamber with two response bars and with a transparent plexiglass partition dividing it in half separating them. The rats were first trained individually in escape behavior. A rat was placed in its half of the chamber, and every one sec a shock of 0.5 sec duration was delivered to the animal. The rat could escape from these shocks by emitting a bar-press, which would instate a 20-sec, shock-free period. At .

the end of the shock-free period another shock was delivered, and the rat was then able to instate another shock-free period by emitting another bar-press.

After the subjects were emitting a stable rate of individual escape behavior, they were placed together in the chamber with the partition separating them. A cooperative response was then required. Only when both animals pressed their bars within 15 sec of one another would a shock-free period be instated. After this period of training in cooperative escape, a 2-hr session was conducted with the partition removed. This session was followed by another 2-hr session with each rat in a separate chamber requiring only individual escape behavior. This alteration of cooperative and individual sessions was continued throughout the experiment.

As long as the partition was in place, the cooperative behavior of the subjects occurred at much the same rate as their individual behavior. In both cases, responding varied around a mean rate of approximately 6.7 responses/min. This rate was also maintained during the individual sessions. However, in the sessions when the partition was removed and a cooperative response was required, responding deteriorated to a mean rate of 2.7 responses/min. In all cases but one, the subjects spent over 50% of the 2-hr session in the shock interval. On several occasions fighting was so severe that the session had to be discontinued.

In order to control for effects of alternating individual sessions with cooperative, partition-removed sessions, additional rats were trained first singly, then in cooperative responding. Then, after the partition was removed, no sessions requiring individual escape were interspersed between the cooperative sessions. Results were similar to those described above. However, in this case, although responding decreased and the number of shocks received increased, the amount of fighting behavior varied greatly both from session to session and from pair to pair. Animals frequently had to be withdrawn from the sessions because of injuries sustained in fighting. Even the few animals that maintained a stable rate of escape never completely discontinued fighting. Both fighting and escape responses were often delivered while the animals were in the stereotyped fighting posture. Thus, even in the case of a discrete response and stimulus, reflexive fighting interfered with the adaptive escape behavior.

Although the effects of escape, avoidance, and punishment are well known for animals individually placed on these schedules, the effects of these aversive controls when a social situation is involved are just beginning to be recognized. Indeed, research in areas such as those related above may completely revise our notions of the effectiveness of aversive control in social situations.

AGGRESSION IN HUMANS

While interspecies differences in pain-elicited aggression have been found, it is probably fair to say that interspecies similarities are stronger. We have seen that pain-elicited aggression has not been found in guinea pigs and that rats will not attack inanimate objects, although pigeons and monkeys will. On the other hand, the fact that pain-elicited aggression has been observed to occur and to follow similar laws in such widely ranging species as rats, pigeons, and monkeys suggests that the interspecies generality of the phenomenon may be far-reaching. Thus, it is entirely possible that pain may be a source of human aggression.

Casual observation seems to support this possibility. Humans frequently strike out in anger at animals or children who have hurt or annoyed them. Inanimate objects also seem to be objects of attack, as when someone stubs his toe and then kicks the offending object.

If it is true that pain-elicited aggression occurs in humans, then casual observation suggests that it is also better controlled in humans by learned responses. Thus, past punishment for striking animals or other humans may curtail the aggressive response. We may also learn that aggressive responses toward inanimate objects are generally non-adaptive.

These speculations are supported by recent research done by Judy Elbert and the author.[1] Four, 10-yr-old school children were used as subjects. Three (subjects I, III, and IV) were male. Each subject was given the task of stacking 10 bottle stoppers into 2 stacks of 5 each. When the subjects completed this task successfully, a dime was delivered as reinforcement. At the beginning of the experiment the task was explained to each subject. In addition, the subjects were told that another, hypothetical subject in another room was engaged in the same task. The hypothetical subject supposedly had in his room a button which he could push, causing the top of the subject's table to vibrate, upsetting his stack and depriving him of reinforcement. The actual subject also had a button which he could press, supposedly to shake the hypothetical subject's table. Presses on this button were registered on a cumulative recorder and counters in an adjoining room.

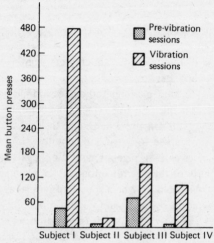

Figure 10. Average number of button-presses per session for each of four subjects during periods before and after vibrations of the table were instituted.

After the subject's stacking behavior was well-established, the experimenters who observed the subject behind a one-way mirror, would occasionally introduce vibrations of the table.

Figure 10 shows the responses on the button for each subject both before and after the vibrations started. In all cases, aggressive responses increased after the vibrations were introduced. The wide variation from subject to subject in button-pressing rate both before and after the vibrations was interesting. Casual observation indicated that subject II, who showed very low rates of button-pressing, was most aggressive toward inanimate objects in the environment. Also, some of the children indicated, in interviews conducted after the session, that they had not pushed the button as much as they would have liked, since they had been taught that such behavior was wrong.

Studies such as this show that the objective study of complex human aggression is possible. Naturally, the moral and practical difficulties in studying aggression in humans are tremendous. Yet, as our knowledge of aggression in lower animals progresses and as more and more feasible methods of studying aggression in humans are developed, a clear picture of the variables which initiate, maintain, and eliminate aggression in human beings should emerge. Hopefully, a good understanding of human aggression will lead to its control, and to a subsequent reduction in the aversive and maladaptive occurrences which often accompany aggressive responses.

REFERENCES

Azrin, N. H. 1964. Aggression. Paper read at Amer. Psychol. Assoc., Los Angeles.

Azrin, N. H., D. F. Hake, and R. R. Hutchinson. 1965. Elicitation of aggression by a physical blow. *J. Exptl. Anal. Behav. 8*:55-57.

Azrin, N. H., R. R. Hutchinson, and D. Hake. 1963. Pain-induced fighting in the squirrel monkey. *J. Exptl. Anal. Behav. 6*:620-621.

Azrin, N. H., R. R. Hutchinson, and D. F. Hake. 1966. Extinction induced aggression. *J. Exptl. Anal. Behav. 9*:191-204.

Azrin, N. H., R. Ulrich, R. R. Hutchinson, and D. G. Norman. 1964. Effect of shock-duration on shock-induced fighting. *J. Exptl. Anal. Behav. 7*:9-11.

Azrin, N. H., R. R. Hutchinson, and R. McLaughlin. 1965. The opportunity for aggression as an operant reinforcer during aversive stimulation. *J. Exptl. Anal. Behav. 8*:171-180.

Beeman, E. A. 1947. The effect of male hormone on aggressive behavior in mice. *Physiol. Zool. 20*:373-405.

Calhoun, J. B. 1950. The study of wild animals under controlled conditions. *Ann. New York Acad. Sci. 51*:1113-1122.

Crew, F. A. E., and L. Mirskaia. 1931. The effects of density on adult mouse population. *Biol. Generalis 7*:239-250.

Daniel, W. J. 1943. An experimental note on the O'Kelly-Steckle reaction. *J. Comp. Psychol. 35*:267-268.

Davis, F. C. 1935. The measurement of aggressive behavior in laboratory rats. *J. Genet. Psychol. 43*:213-217.

Delgado, J. 1963. Cerebral heterostimulation in a monkey colony. *Science 141*:161-163.

Egger, M. K., and J. P. Flynn. 1961. Amygdaloid suppression of hypothalamically elicited attack behavior. *Science 136*:43-44.

Ferster, C. B. 1957. Withdrawal of positive reinforcement as punishment. *Science 126*:509.

Flory, R. K., R. Ulrich, and P. C. Wolff. 1965. The effects of visual impairment on aggressive behavior. *Psychol. Record 15*:185-190.

Hall, C. S., and S. J. Klein. 1942. Individual differences in aggressiveness in rats. *J. Comp. Psychol. 33*:371-383.

Hess, W. R. 1949. *Das Zwischenbirn. Syndrome, Lokalization, Functionen.* Schwabe, Basil.

Hess, W. R. 1954. *Diencephalon. Autonomic and Extrapyramidal Functions.* Grune and Stratton, New York.

Hess, W. R., and K. Akert. 1955. Experimental data on role of hypothalamus in mechanisms of emotional behavior. *Arch. Neurol. Psychiat. 73*:127-129.

Hutchinson, R. R., R. Ulrich, and N. H. Azrin. 1965. Effects of age and related factors on the pain-aggression reaction. *J. Comp. Physiol. Psychol. 59*:365-369.

Kislak, J. W., and F. A. Beach. 1955. Inhibition of aggressiveness by ovarian hormones. *Endocrinol. 56*:684-692.

Levi, W. M. 1957. *The Pigeon.* Levi, Sumter, S. C.

Masserman, J. H. 1950. A biodynamic psychoanalytic approach to the problems of feeling and emotion. In M. L. Reymert [Ed.], *Feelings and Emotions.* McGraw-Hill, New York.

Masserman, J. H. 1964. *Behavior and Neurosis.* University of Chicago Press, Chicago.

Miller, N. E. 1948. Theory and experiment relating psychoanalytic displacement to stimulus-response generalization. *J. Abnorm. Soc. Psychol. 43*:155-178.

O'Kelly, L. W., and L. C. Steckle. 1939. A note on long-enduring emotional responses in the rat. *J. Psychol. 8*:125-131.

Pavlov, I. P. 1927. *Conditioned Reflexes.* Oxford University Press, New York.

Ranson, S. W. 1934. The hypothalamus: its significance for visceral innervation and emotional expression. *Trans. Coll. Physicians,* Philadelphia, p. 222-242.

Reynolds, G. S., A. C. Catania, and B. F. Skinner. 1963. Conditioned and unconditioned aggression in pigeons. *J. Exptl. Anal. Behav. 1*:73-75.

Richter, C. P. 1950. Domestication of the Norway rat and its implications for the problem of stress. In Harold G. Wolff, et al., [Ed.], *Life Stress and Bodily Disease.* Williams and Wilkins, Baltimore.

Scott, J. P. 1946. Incomplete adjustment caused by frustration of untrained fighting mice. *J. Comp. Psychol. 39*:379-390.

Scott, J. P., and E. Fredericson. 1951. The causes of fighting in mice and rats. *Physiol. Zool. 24*:273-309.

Seward, J. P. 1945. Aggressive behavior in the rat. I. General characteristics, age and sex differences; II. An attempt to establish a dominance hierarchy; III. The role of frustration; IV. Submission as determined by conditioning, extinction and disuse. *J. Comp. Psychol.* *38*:175-197.

Sidman, M. 1953. Avoidance conditioning with brief shock and no exteroceptive warning signal. *Science 118*:157-158.

Skinner, B. F. 1938. *The Behavior of Organisms.* Appleton-Century-Crofts, New York.

Smith, S., and E. Hosking. 1955. *Birds Fighting.* Farber and Farber, London.

Stachnik, T. J., R. Ulrich, and J. H. Mabry. 1966. Reinforcement of aggression through intra-cranial stimulation. *Psychom. Sci. 5:* 101-102.

Tedeschi, R. E., D. H. Tedeschi, A. Mucha, L. Cook, P. A. Mattis, and E. J. Fellows. 1959. Effects of various central acting drugs on fighting behavior of mice. *J. Pharmacol. Exptl. Therapeut. 125*:28-34.

Thompson, T., and T. Strum. 1965. Classical conditioning of aggressive display in Siamese fighting fish. *J. Exptl. Anal. Behav. 8*:397-403.

Ulrich, R. 1965. Disruptive effects of aggression. Paper read at Amer. Psychol. Assoc., Chicago, 1965.

Ulrich, R., and N. H. Azrin. 1962. Reflexive fighting in response to aversive stimulation. *J. Exptl. Anal. Behav. 5*:511-520.

Ulrich, R., and W. H. Craine. 1964. Behavior: Persistence of shock-induced aggression. *Science 143*:971-973.

Ulrich, R., R. R. Hutchinson, and N. H. Azrin. 1965. Pain-elicited aggression. *Psychol. Record 15*:111-126.

Ulrich, R., M. Johnston, J. Richardson, and P. C. Wolff. 1963. The operant conditioning of fighting behavior in rats. *Psychol. Record 13*:465-470.

Ulrich, R., T. J. Stachnik, G. R. Brierton, and J. H. Mabry. 1965. Fighting and avoidance in response to aversive stimulation. *Behaviour 26*:124-129.

Ulrich, R., P. C. Wolff, and N. H. Azrin. 1964. Shock as an elicitor of intra- and inter-species fighting behavior. *Animal Behav. 12*:14-15.

Utsurikawa, N. 1917. Temperamental differences between out-bred and inbred strains of the albino rat. *Animal Behav. 7*:11-129.

Vernon, W., and R. Ulrich. 1966. Classical conditioning of pain-elicited aggression. *Science 152*:668-669.

Von Holst, E., and U. von Saint Paul. 1962. Electrically controlled behavior. *Scientific Am. 3*:50-59.

Wasman, M., and J. Flynn. 1962. Direct attack elicited from hypothalamus. *Arch. Neurol. 6*:60-67.

Yerkes, R. M. 1943. *Chimpanzees, A Laboratory Colony.* Yale University Press, New Haven.

Zuckerman, S. 1932. *The Social Life of Monkeys and Apes.* Harcourt-Brace, New York.

FOOTNOTE

[1] Elbert, Judy, and Roger Ulrich. An experimental analysis of frustration-produced aggression in children. (In preparation.)

11   Studies on the Basic Factors in Animal Fighting: VII. Inter-Species Coexistence in Mammals

—ZING YANG KUO

A. PROBLEM

The purpose of this study is twofold: (*a*) to prevent the formation of certain anti-social habits in the dog and cat such as prey and hunting and killing of smaller animals, antagonism between dogs and cats, and fighting between dogs and cats, and fighting between and domination over members of the same species. And (*b*) to extend or transfer certain sociable behavior patterns to other species. Our ultimate goal was, of course, to establish a peaceful and tranquil mammalian society not only without fighting, but also with the dominance orders or social ranks eliminated. Throughout this investigation the developmental or ontogenic methods were used.

From *The Journal of Genetic Psychology*, 97, 1960, 211-225. Reprinted by permission of the Journal Press, Provincetown, Mass.

Table 1   Species and Number of the New Born Animals Reared Together in the Same Pen

Species / Pen No.	Cats (C) and cockatoos (Co) / Pen 9	Cats (C) and canaries (Ca) / Pen 8	Dogs (D) and canaries (Ca) / Pen 7	Dogs (D) and cockatoos (Co) / Pen 6	Cats (C) and rabbits (Rb) / Pen 5	Cats (C) and rats (R) / Pen 4	Dogs (D) and rabbits (Rb) / Pen 3	Dogs (D) and rats (R) / Pen 2	Dogs (D) and cats (C) / Pen 1
I	1C 5Co	1C 5Ca	1D 5Ca	1D 5Co	1C 5Rb	1C 5R	1D 5Rb	1D 5R	1D 5C
II	1Co 5C	1Ca 5C	1Ca 5D	1Co 5D	1Rb 5C	1R 5C	1Rb 5D	1R 5D	1C 5D
III	1C 5Co	1C 5Ca	1D 5Ca	1D 5Co	1C 5Rb	1C 5R	1D 5Rb	1D 5R	1D 5C
IV	5Co 5C	5Ca 5C	1Ca 5D	1Co 5D	1Rb 5C	1R 5C	1Rb 5D	1R 5D	1C 5D
V	Ten dogs and ten cats reared in isolation								

GROUPS

183

B. THE EXPERIMENTS

The subjects employed in these experiments were newly born puppies, kittens, albino rats, and rabbits. In addition, two species of very young birds, canaries, and cockatoos were included in this research. The subjects were divided into five main groups, each group was subdivided into nine pens, each pen consisting of one animal of one species and five of another species. Thus, in Group 1 Pen 1, one dog and five cats were reared together—see Table 1. The newborn animals were put to live together in pens according to the plans of the study as soon as they were old enough to be hand-fed with cow's milk. The sexes of each group were, whenever possible, equally divided. Of the five groups, the first two were used as experimental subjects, the next two groups as controls, and Group V for the study of the effects of isolation on cats and dogs; other species were not involved in this group. Each pen in which six animals of two species lived together was placed in a wire net enclosure $18' \times 18' \times 10'$ (high). The detailed arrangements of the four groups are presented in Table 1. In the table, the letters $D, C, R,$ etc. stand for dogs, cats, rats, etc. respectively. The figures preceding the letters indicate the number of animals reared together in a pen; thus, 1D 5C means one newborn dog reared together with five newborn cats.

From our studies on the fighting of birds and our preliminary observations on the social behavior of various mammals we had come to the conclusion that competition over food, and over sex, playing activities, hostility towards strangers, and living in isolation were among the major environmental factors leading to the development of such antisocial behavior patterns as mentioned above. In this study, therefore, special emphasis was placed on the control of these factors during development. In the control groups (III and IV) no attempt was made to interfere with the spontaneous development of eating, and play activities of the subjects; these control animals were not allowed to see any strange animals until they were 10 months old. In Group V (dogs and cats reared in isolation) each subject was tested for its reactions to strange animals after they were 10 months old (dogs, cats, rats, rabbits, guinea pigs, canaries and cockatoos as well as several other species of birds and mammals).

1. Socializing Eating Habits

The object of this part of the investigation was to prevent competition over food by socializing the eating habits of the animals in Groups I and

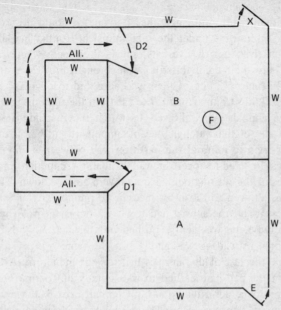

Figure 1. Ground plan of feeding platform. E—entrance to the observation compartment; A—observation compartment; B—eating compartment; F—food dish; D1—glass door leading from A to alley way; A11—alley way leading from D1 to D2; D2—glass door from alley way to eating compartment; X—exit from eating compartment; W—glass walls.

II. Only dogs, cats, and cockatoos were used. The cockatoos were included in this study because they could be brought up to become very fond of bread soaked in milk which they could pick up from the food dish for dogs and cats. The procedure is briefly as follows:

When the dogs, the kittens, and cockatoos were about two months old, they were daily brought to eat by turn in the feeding platform, the detailed ground plan of which is shown in Figure 1. All the partition walls and doors were made of plate glass. The walls were moveable so that the two compartments, *A* and *B*, as well as the alley way could be widened or narrowed according to the sizes of the animals. There were several sets of doors which could be changed in accordance with the sizes of the two compartments and the alley way. Each door was fixed with a spring hinge and could be opened by pulling the string attached to it. When the string was released, the door closed immediately. Generally, Compartment *A* was so set as to have sufficient room for six animals to move and turn about. The alley way was set to allow only one animal to walk through at a time so that if a second animal attempted to overtake the first one, it could do so only by squeezing or crawling under the first animal in front. When all the six animals (or five dogs or five cats only in those pens where rabbits or canaries were their coinhabitants) were brought into Compartment A through Entrance E, one of them was admitted through D1, the alley way, and D2 into Compartment B to eat from the food dish (F) and the other five animals were allowed to watch the first admitted animal eating. If any of the watchers became impatient and began to bark or meow, or tried to jump at the partition wall between A and B, a black curtain was lowered to cover the wall until all the impatient movements and noises in the waiting compartment were quieted down. The animal eating was allowed to finish his meal in 10 minutes. After 10 minutes the bell rang, and the animal had to leave Compartment B by the exit (X). Otherwise, he was gently pushed out through X. This done, another animal in waiting was allowed in the same manner to take his meal while the rest of the animals had to wait their turn. After all the animals in the pen had learned to wait quietly in Compartment A for their turns to be admitted to eat in Compartment B, the next step in the training was to open Door D1 and let the waiting animals in Compartment A file one by one into the alley way while one was already eating in Compartment B. At first, the experimenter had to push them one by one into the alley way when the door D1 was opened. But later on all learned to do this without help as soon as D1 was opened. While lining up in the alley way, if one of the animals attempted to overtake another either by squeezing through or by crawling under, the overtaker was immediately brought back to the end of the waiting line.

After all the subjects had learned to line up in the alley way instead of Compartment A, each waiting for its turn to be admitted to Compartment B, door D2 was kept open. But if the animal in front of the waiting line rushed into Compartment B before the animal in eating had finished his meal and left the place, the door was immediately closed and the violator was taken up and replaced at the end of the waiting line. Through a number of trials, all the animals in the pen learned not to enter D2, though it was open, until Compartment B was clear of any animal.

One month after every animal in the pen had mastered "the dining car etiquette" (a nickname we used in the laboratory because the process resembled waiting in the dining car in a crowded American train) they were tested for their reactions to a food situation (not in the feeding platform) in which the other animals had either to wait beside the one eating, or to push him away in order to occupy the food dish which did not allow more than one animal to eat at the same time. Each subject was given three tests.

Since it is not our purpose in this report to analyze the learning process and due to space limitation, the records of the animals of these three species in learning the "dining car etiquette" will not be presented here. Suffice it to state that all the cockatoos, cats, and dogs in Groups I and II had eventually become "socialized" in their eating habits, and that the dogs were the quickest learners, the cats next, the cockatoos the slowest (three to 10 trials for the dogs, eight to 25 trials for the cats, 21 to 29 trials for the cockatoos).

In the three tests in a situation outside the feeding platform all the animals showed that the eating habits, that is, to wait, not to push, if the food dish were occupied, and not to compete with another while trying to reach the food dish which was just unoccupied, were preserved even in a new environmental setting, although 56 per cent of the dogs, 42 per cent of the cats, and 33 per cent of the cockatoos often came close to the food dish and attempted to sniff at it; but neither did the animal in eating object to it, nor did the sniffer make any move to join him in eating.

In the control groups (III and IV), no attempt was made to prevent fighting or domination in eating. The feeding plate was large enough for all the six animals in the pen to eat together. The phenomenon of food competition ending in domination without fighting or as a result of one or at most two fights was observed in the two groups: (66 per cent for dogs, 68 per cent for cats, 48 per cent for rats, and 72 per cent for cockatoos). A dominant dog occupying the food plate or having a piece of food in its mouth would growl to warn off the other animals. A dominant cat would snarl or hiss. A dominant rat would bite another rat. The dominant cockatoo would peck or chase other birds away. The non-dominant ones (dogs, cats, as well as cockatoos) learned to quietly steal a piece of food and run to eat in a corner. A non-dominant cockatoo would steal with its beak one piece of food and put it to one of its feet to hold it, and with the free beak, steal another piece before it went away to eat the stolen food. All this happened when the dogs were about three to four weeks old, cats about four to eight weeks old, rats and cockatoos about two to three months old. All such behavioral phenomena were observed only during feeding time. At other times all the animals living in the same pen were friendly to each other, playing together, and even fondled or were attached to one another.

It would seem that there is a striking contrast in the behavioral patterns relative to eating habits between the experimental and control groups.

2. Play

In the experimental groups, play of the animals was closely watched every day. The experimenter, with a water spray in hand (the kind commonly used in ironing clothes) would quickly spray water on the faces of the animals as soon as fighting started during play. In the chow, puppies of three to four weeks would start to fight during play, or to bite at the handler, but with one to three sprays of water, fighting or biting would not occur again. We also employed water spray on a puppy carrying something in his mouth, who would growl or bite when another animal or the handler tried to snatch it from his mouth. However, in the case of play between cats, between cats and dogs, between dogs and cockatoos, and between cockatoos and cats, no fighting was ever observed. In the case of rabbits and canaries there are no acts which can be interpreted as play activities, except the preliminary and final sex acts.

In the two control groups, fighting between dogs during play was not stopped by the experimenter. Such fighting lasted often only for a few seconds rarely resulting in bodily injury, but might end with the establishment of the dominance relationship in favor of the stronger puppy.

In play activities there is one significant fact which must be pointed out here. There was no species barrier in play. The same kind of play patterns of a dog was manifested in the play either between a dog and a cat or between a dog and a cockatoo. Five of the puppies in all the first four groups even attempted at sham coitus movements on the cats or cockatoos or rabbits. The kittens displayed cat-like play patterns with the cockatoos or with dogs. The cockatoos in playing with cats or dogs would nod its head, ruffle its crest, flutter its wings, touching the floor with its beak, attempting billing movements with a dog or a cat, or gently touch the fact or neck of the dog or cat with its beak, making tender sounds in the meantime.

In inter-species social life while there is no species barrier in play as well as in attachment or fondling activities, nevertheless, there is still species preference. It will be noted that in the first four groups of animals in this investigation there are six animals of two different species living in one pen. Thus, in Pen 1 in Group I there were one puppy and five kittens living together; in Pen 9 Group I one cat to five cockatoos; in Pen 6 Group II five dogs to one cockatoo, etc. (see Table 1). Now, in the pen with one cockatoo and five dogs, all the attachment, fondling activities, and play of the cockatoo were built around the five puppies. When all the puppies were taken out from the pen, the cockatoo became restless, making loud noises until one of the five puppies was returned to the pen. Even when a strange well-tamed cockatoo was put into this pen, instead of returning a puppy mate, the restless cockatoo adopted an indifferent attitude towards the stranger (or even hostile if the restless cockatoo belonged to Pen 6 Group IV instead of Pen 6 Group II). On the other hand, the attitude of the five puppies in the same pen towards this single avian pen mate can be described as "friendly, occasional playmate, but not indispensable." They did not play with the bird as much as they would among themselves. If four puppies were taken from pen leaving one of the five with the cockatoo, the lone little dog would become as restless and in great distress as if the avian pen mate were not there. The reversal of the same phenomena was observed in the case in which there were five cockatoos to one puppy in the same pen. In this case, the pup became attached almost solely to the cockatoos, acted as if they were his sole companions and play mates while indifferent to other animals including dogs. On the other hand, the cockatoos reacted to this lone dog with a tolerant and indifferent attitude. When two or four cockatoos were paired, their love making activities were confined to their sex mate, becoming indifferent not only to the lone dog, but also to the other cockatoos. And this applies to the relationship between kittens and cockatoos and between kittens and puppies throughout this investigation. In other words, when one young animal grows up together with mates of its species as well as a different species, it is more attached to the mates of its own species than to the mates of another species whereas when one lone young animal of one species is reared together with one or more young of a different species, after the critical time (two to three months), its attachment is in most cases entirely fixed with the mates not belonging to its own species so that when it meets a stranger of its own species its attitude is either indifferent or hostile depending on its past experience with strange animals.

3. Sex and Fighting

From the two previous sections we have noted that by socializing eating habits and by experimental intervention of fighting resulting from play

Table 2

Species of visitors Types of reaction to visitors Group and animals	No.	*Indifferent No.*	*Friendly No.*	*Hostile No.*	*Attack No.*	*Indifferent No.*	*Friendly No.*	*Hostile No.*	*Attack No.*
		Cats				*Dogs*			
Group I									
Dogs	5	2	3	0	0	0	5	0	0
Cats	4	1	3	0	0	0	0	0	0
Group II									
Dogs	25	4	21	0	0	5	20	0	0
Cats	21	8	13	0	0	19	2	0	0
Group III									
Dogs	5	1	0	2	2	0	0	3	2
Cats	4	1	0	3	0	3	0	1	0
Group IV									
Dogs	25	2	4	12	7	2	2	8	13
Cats	21	2	2	16	1	4	0	19	2
Group V									
Dogs	10	0	0	6	4	0	1	4	5
Cats	10	1	1	6	2	2	0	6	2
		Rabbits				*Rats*			
Group I									
Dogs	5	4	1	0	0	3	2	0	0
Cats	4	4	0	0	0	4	0	0	0
Group II									
Dogs	25	5	0	0	0	10	15	0	0
Cats	21	20	1	0	0	18	3	0	0
Group III									
Dogs	5	2	0	0	3	2	0	0	3
Cats	4	3	0	0	1	2	0	2	0
Group IV									
Dogs	25	13	3	0	10	13	3	0	2
Cats	21	4	0	18	3	3	2	8	12
Group V									
Dogs	10	1	1	0	8	1	1	0	8
Cats	10	2	0	8	0	2	0	0	8

Species of visitors / Types of reaction to visitors / Group and animals	No.	Indifferent No.	Friendly No.	Hostile No.	Attack No.	Indifferent No.	Friendly No.	Hostile No.	Attack No.
		Guinea Pigs				*Canaries*			
Group I									
Dogs	5	2	3	0	0	4	1	0	0
Cats	4	4	0	0	0	4	0	0	0
Group II									
Dogs	25	14	11	0	0	13	12	0	0
Cats	21	21	0	0	0	19	2	0	0
Group III									
Dogs	5	1	0	0	4	4	1	0	0
Cats	4	3	0	1	0	4	0	0	0
Group IV									
Dogs	25	9	0	0	16	14	11	0	0
Cats	21	8	0	8	9	16	3	0	6
Group V									
Dogs	10	2	2	0	6	4	1	0	5
Cats	10	4	0	0	6	5	0	0	5
		Sparrows				*Parrots*			
Group I									
Dogs	5	5	0	0	0	1	4	0	0
Cats	4	4	0	0	0	4	0	0	0
Group II									
Dogs	25	13	12	0	0	8	17	0	0
Cats	21	19	2	0	0	20	1	0	0
Group III									
Dogs	5	5	0	0	0	4	1	0	0
Cats	4	3	0	0	1	4	0	0	0
Group IV									
Dogs	25	9	0	0	16	14	2	4	5
Cats	21	8	0	0	17	8	0	0	17
Group V									
Dogs	10	3	0	0	7	2	3	0	5
Cats	10	3	0	0	7	4	0	6	0

activities, our experimental animals of Groups I and II had become practically "domination free" and "fighting free" and that although fighting and domination in eating and in play were observed in the two control groups, they were quite mild and were rarely observed except during feeding time and during play. There was no antagonism between cats and dogs or between cats and rats or any other species used in this investigation. This is true not only of the experimental animals, but also of the control goups. But, now, when the animals in our investigation reached sex maturity, we were confronted with a much more complicated problem. In the first place, we had not succeeded in devising a satisfactory method of procedure, as we had with regard to eating behavior, to socialize sex behavior of the subjects. Secondly, from our past experience previous to this study, we had already known that in a sexually inexperienced but momentarily sexually aroused male chow, there was nothing we could do to stop him from his fierce struggle to reach the female in heat except by a very strong electric shock or an injection of a very heavy dose of estrogen to counteract the presumably momentary hormonal unbalance neither of which would serve our purpose in this type of investigation. Furthermore, fighting over sex between two male cats is frequently violent, and very much more so between two male chows. Such dog fights often ended in a serious injury to both or even death. In order not to upset the peaceful animal societies without or with very little social domination or fighting which we had built for more than 10 months, we decided to suspend our investigation on this aspect of behavior in dogs and cats. Whenever the first signs of the estrous cycle in a female dog or cat were noticed, the animal was immediately removed from the pen and kept in isolation until the receptive period was over (except in Groups I and III in which there was only one dog or one cat in a pen, and Group V in which every one of the 10 dogs and 10 cats lived in isolation). (In other species in this study, fighting as a result of competition over sex had not been observed.)

C. PREY ON SMALL ANIMALS, HUNTING, REACTION TO STRANGERS, AND THE EFFECTS OF ISOLATION

Kittens and puppies in Group V naturally were not allowed to see any strange animal of whatever species during the period of isolation. From the tenth month on, each was tested bi-weekly for its reaction to strangers including dogs, cats, rats, rabbits, guinea pigs, a number of varieties of parrots, canaries, and sparrows. Similarly, the control animals (Groups III, IV) had not seen any strange animal until they were 10 months old, and, thenceforth, each was given a bi-weekly test with the animals of those species used to test Group V. On the other hand, those animals in Groups I and II, from the start of their collective life in the pens were allowed to spend 10 minutes every day with one strange animal of the various species used to test animals in Groups III, IV, and V. These testing animals had been trained to be very tame, having no fear or hostility to strange animals of any species, birds, or mammals but in most cases were very friendly to the animals in the pen they were introduced to visit. The reactions of the kittens and dogs of the five groups to the strange visitors are summed up in Table 2 (the reactions of other species were not presented here as our primary interest in this investigation was kittens and puppies. The reason why other species such as rats, rabbits, canaries, and cockatoos were also used was because of the problem of prey on small animals, and hunting in dogs and cats.)

A glance at the table will reveal a number of important points: (*a*) Those cats and dogs which received a strange visitor every day from the very start of the collective life in the pen showed neither hostile nor attacking reaction to the strange animals of any species during the testing period. Gone were their preying and hunting tendencies. Their reaction to strange animals was either indifferent or friendly. Gone, too, was the antagonism or hostile attitude between dog and cats. (*b*) But it is not the case with the cats and dogs in Groups III and IV with which no strange animals were allowed to pay visits. Upon tests, their reaction was by and large hostile or attacking. There were, however, a small number of cases in these two groups in which an indifferent or even friendly attitude was displayed by some of the dogs and cats. Our records show that these dogs or cats were the subjects which were brought up in the same pen with the animals which belong to the same species as the testing animals used as strange visitors to which the dogs and cats in question had displayed an indifferent or friendly attitude instead of a hostile or attacking reaction. (*c*) When we come to the dogs and cats brought up in isolation in the first 10 months, their reactions to strangers are predominantly hostile or attacking although there are a few cases in which the dogs or cats showed an attitude of indifference or a mild degree of friendliness. Our records show 10 cases in which small birds and rats were killed and eaten by the isolated cats, while the dogs chased and killed rats, guinea pigs as well as rabbits and birds. (*d*)

The antagonism between cats and dogs, between two dogs and between two cats is very strong as an inspection of the reaction of the animals in Group V to visiting cats and dogs (Table 2) will show.

1. Suppression of Aggression

From Table 2 it will be seen that there are 22 dogs in Groups III, IV, and V which would rush to attack strange dogs whenever they were introduced into their enclosure. We purposely selected as visitors to these enclosures, dogs of extreme timidity who will invariably run to escape whenever the resident or residents of the enclosure rushed out and, when an escape was blocked, they would lie down on their backs and let the attackers attack without resistance.

Now, as a last step in this investigation, we made use of 20 (all chows) of these 22 aggressive dogs (the other two, Boston terriers) as experimental subjects on suppression of aggression. The suppressor was a medium sized, short-haired chow with a broad muzzle and with a bodily make-up excellent for a fighter. When we employed this chow as the suppressor for the 20 chow aggressors, it was about three years old. For all the three years of its life this dog never initiated a fight, but when attacked, never failed to make a counter-attack, and never lost a single combat. In any environmental setting, familiar or unfamiliar, whenever he met one or more strange dogs or whenever one or more dogs rushed to challenge him, his first and invariable reaction was to stand straight with his tail tightly bent forward to his back, making neither motion nor noise. Such a reaction pattern was in a majority of cases sufficient to send the challengers away without a fight. However, if he was attacked he always fought so well that unless he was ordered to stop by his trainer, the fight was finished almost always with a very serious injury, or death to his opponent. We made this three-year-old chow pay visits to those 20 chows by turn. In case of some mild injury or fatigue, the suppressor was rested until full recovery. Results: 13 of them (65%) were sent away peacefully by his firm and steady stand, the other seven got him into battles which were ended with three cases of severe injury, two cases of light injury, and two cases of death. After the wounds were healed, all these dogs were given four more tests with the suppressor as the visitor and with other new visiting dogs. Results: of the 13 which did not have any fight with the suppressor, six tried to keep in some distance from him, the others got near him but soon left him alone. All the 13 continued to challenge or threaten to attack other visiting dogs. All the injured dogs had the typical reactions of an underdog in the presence of the suppressor. All those which continued to have hostile or attacking reactions to the other dog visitors did so with a milder degree than before.

It must be noted in passing that in Groups III and IV when a top dominant dog or cat was removed from the pen the next dominant animal moved up to take its place as the top ranking animal. The hierarchy or rank order, remained constant until one of the dominant animals was removed.

D. SUMMARY AND CONCLUSION

In a number of experiments it has been demonstrated: (*a*) That in such anti-social behavioral patterns in dogs and cats as have been popularly known as prey on small animals, hunting, mutual antagonism between cats and dogs, aggressive or hostile reactions between two dogs, and hostility to or attacking strangers could be prevented from appearance if appropriate environmental situations were in force during development. (*b*) That sociable behavior patterns in mammals such as attachment to one another, fondling acts and courting could be brought to cross the species barrier and even extended from mammals to birds and vice-versa. This was accomplished also by the developmental or ontogenic methods. (*c*) That by adequate control of environmental conditions during development, eating habits in dogs and cats and even cockatoos could be socialized to resemble human etiquette. (*d*) That during development by elimination of competitions over food, by prevention of fighting during play, and by avoidance of competition over sex, a mammalian society could be immunized against not only fighting but also what has been supposed to be a normal phenomenon in vertebrate animals, namely, the so-called dominance-submission relationship. (*e*) That in mammals as well as in fish and birds, isolation during development has been found to be a major factor in developing hostile reaction to other animals, which reaction is a step leading to fighting. (*f*) That in dogs (cats, also, the data concerning this are not presented in this paper) "standing firm" instead of running or submission has a 65 per cent chance to stop or avoid an aggressive attack. This confirms our findings in fish and birds. (*g*) And that in dogs, too, we confirm the conclusions based on the results of our investigation on fish and birds than an aggressor could be suppressed or subdued by a stronger or more powerful aggressor. But the suppression is never complete nor permanent, and rather costly as such a suppression may lead to severe injuries or even death. Here again, in mammals as in other classes of vertebrates, when the most aggressive animal is removed from the group, another moves up to take its place and establishes itself as the top aggressor. It appears, then, that suppression of an aggressor or removal of a dominant animal from the group will not end social domination of fighting in an animal society.

Our observation on the behavior of animals towards strangers raised some doubt about the validity of the concept of defense of territory in animals. We are rather inclined to think that the so-called "territorial defence" is, after all, merely a fancy name for the reaction patterns to strangers flavored with anthropomorphism and the 19th century Darwinism. Further and more systemic experimental explorations are necessary to decide this issue.

The results of our investigations on animal fighting have some bearing on the concept of species-specific behavior.

As has been repeatedly pointed out in the previous reports of this series,* fixation of behavioral patterns (habituation) during development, plays an important role as a determinant of behavior. Its significance can not be overemphasized. It is an essential controlling factor not only in social domination and fighting but also in the development of behavior in general.

In conclusion, let us restate that although social domination and fighting are very common phenomena in animal societies, and although our studies have demonstrated that once such behavior patterns are formed and once the social hierarchy is established it is extremely difficult, if not impossible, to stamp them out, there is no reason to accept the fatalistic view that such behavioral patterns are "necessary evils" in social life. Our hope for a peaceful and tranquil society without fighting and social domination lies in the prevention of the development of such behavior patterns. Our findings have demonstrated that by appropriate control of environmental conditions during development we can immunize against certain types of anti-social behavior patterns in animals. We seem to be able to accomplish such an immunization without resorting to such drastic measures as depriving the animals of the essential nutritional factors such as thiamin or castration of the gonads. Our experimental attack on the various problems of the elimination of social domination and fighting is just beginning and further and more extensive investigations are required in order to work out a detailed and systematic program for the prophylaxis of aggression in animal society. However, the developmental or ontogenic approach seems to have encouraged us to hope that the day may not be so far off when it will be just as feasible to immunize against social domination and fighting as immunization against smallpox or poliomyelitis.

FOOTNOTE

* Kuo published seven reports in all during the year 1960. This report is the seventh in the series. See *Journal of Genetic Psychology*, vols. 96 and 97. THE EDITORS

12   Ghetto Sentiments on Violence

—HARLAN HAHN

Although the riots that have erupted in American cities failed to yield many needed structural and policy changes, they have inspired increased attempts to understand the conditions that spawned the disorders. Immediately after the violence of 1967, for example, political leaders responded to the need for an explanation of the events by securing the appointment of a presidential commission. The extensive soul-searching and research produced by this commission in documenting the vast gaps between white and black America, as well as the abject poverty and despair in urban slums, may have been the most concrete—and one of the potentially important—results of the riot. While some of the findings of this study were reported in a document prepared by the so-called Kerner commission on civil disorders, an even more intensive analysis has been provided by a series of investigations begun under the auspices of the commission.

One of the most useful means of exploring alternative theories regarding the riots as well as the current mood of the ghetto is through the analysis of public opinion. A critial part of the research ordered by the President's committee, therefore, is a survey of public attitudes conducted by Angus Campbell and Howard Schuman in 15 major U.S. cities. The results of this survey are of far-ranging significance not only because they represent one of the few reports on this subject but also

Reprinted by permission from *Science and Society*, *33* (No. 2), Spring, 1969.

because they constitute perhaps the most comprehensive available examination of the sentiments of black Americans in urban areas of the North. The findings of the study may be crucial for estimating the possible outbreak of future violence and for probing the causes of prior disorders. As a result, both the degree and intensity of opinions expressed in the survey became unusually important.

Although a principal purpose of the research conducted by Campbell and Schuman was "to provide reliable information concerning the prevalence and distribution of the attitudes and understanding of racial problems held by the general Negro and white population,"[1] the survey also was designed to fulfill the broad mandate of the Kerner commission to provide an interpretation for the outbreak of rioting. In the initial description of the survey findings, this attempt to investigate the sources of urban unrest seemed to be a primary objective. A substantial portion of the report was devoted to an examination of support for violence in black ghettos.

The survey reported by Campbell and Schuman was based on a random probability sample of 2,582 whites and 2,814 black residents in 15 major cities—Baltimore, Boston, Chicago, Cincinnati, Cleveland, Detroit, Gary, Milwaukee, Newark, Brooklyn, Philadelphia, Pittsburgh, San Francisco, St. Louis, and Washington, D.C. Only five of the cities had experienced what the Kerner commission termed "major" disorders by the time the survey was completed in 1968.[2]

Most of the respondents in this survey, which provides much of the available evidence on attitudes regarding civil disorders, did not live in cities that had undergone riots. It is somewhat puzzling that urban blacks who had not experienced riots were used as a basis for assessing sources of support for violence nearly a year after most disturbances had occurred. Perhaps the fundamental issue of this investigation was, why have black residents of the ghetto engaged in rioting? To answer this question, the opinions of ghetto residents who have been involved in a major riot also might be appropriately explored. As a result, particular importance must be assigned to the need for a comparative examination of sentiments about violence among black Americans living in areas that have erupted in riots as well as among the residents of cities that have been relatively calm.

To complement data from the Campbell-Schuman research, this study will explore the results of a modified quota sample survey of 37 whites and 270 black residents of the Twelfth Street neighborhood of Detroit conducted by the author shortly after the riots of 1967. Fortunately, the two surveys included numerous questions on violence that were sufficiently similar to permit meaningful comparisons. All the respondents in the latter survey, however, had participated—either directly or emotionally—in one of the worst disorders in urban history.

Although the time during which surveys are conducted may influence the results, the research completed shortly after the Detroit riot may have had the advantage of eliciting responses when the issue of violence was especially salient. As time elapsed and memories dimmed, opinions of rioting may have been modified or obscured by other events.

Unfortunately, the initial presentation of data from the Campbell-Schuman survey was restricted to "what are commonly called 'marginals,' showing the distribution of the responses of the Negro and white samples to the various questions of the interview."[3] No attempt was made to distinguish the opinions of the residents of cities that had exploded in riots and those living elsewhere or to examine separately the attitudes of respondents who had personal experience with violence. As a result, the descriptions of the Detroit survey also will be limited to the total proportion of persons who selected the available alternatives to the questions. Only the responses from the black samples in both surveys will be included in this study. Although the questions from the two sets of interviews were not identical, they did seem to be comparable and to encompass most of the major issues concerning violence.

The major conclusion of the Campbell-Schuman study in fifteen cities was stated as follows: "The most apparent fact that emerges from the data . . . is that the Negro mass is far less revolutionary in its outlook than its more militant spokesmen. . . . [T] he changes they have in mind are essentially conservative in nature."[4] This survey of the attitudes of black citizens in the North found that only a small segment of the population was ready to resort to violence, few wanted to overthrow the existing social and political system and only a fraction sought a direct encounter with white authorities. The study promoted the impression not only that black Americans were willing to use traditional methods of achieving their goals but also that they desired to avoid conflict and confrontation in the process.

This generalization might be questioned on the basis of results from the survey conducted in the riot-torn Twelfth Street neighborhood of Detroit. Perhaps the principal finding of that study was the widespread local recognition of the effectiveness and power of violence as a vehicle for accomplishing rapid social and political change. Residents of the Twelfth Street area displayed a greater ideological commitment to violence, more willingness to participate in civil disorders, increased militancy, and higher expectations of basic social changes than black Americans who lived in other urban ghettos.

Perhaps the most direct test of personal support of violence in the Campbell-Schuman survey was a question that asked respondents to evaluate "the best way for Negroes to try to gain their rights." Only 15 percent selected violence rather than demonstrations or legal action.[5] A similar question about the fastest way "for Negroes to get what they want" in the Detroit survey revealed that twice as many persons in the Twelfth Street neighborhood preferred violence rather than peaceful protests or legal action.

The tendency of black citizens who had not been exposed to major disorder to seek to avoid violence also was reflected in attitudes toward riot participation. The Campbell-Schuman study referred directly to the Detroit riots in asking respondents if they would join, try to stop, or stay away from a similar disturbance in their neighborhoods.[6] While only 8 percent stated that they would participate in the riot, Campbell and Schuman pointed out that this figure was comparable to the proportion of self-reported riot participants uncovered by other studies in Watts and Detroit.[7] The conclusions of those investigations, therefore, have created the impression that participation in serious disorders has been limited to a somewhat small and unrepresentative segment of the community.

The actual proportion of local residents who have participated in major riots, however, has been difficult to measure. As Campbell and Schuman noted, the percentage of "don't know" responses on the question concerning riot participation was relatively high, indicating perhaps "either genuine uncertainty or an understandable reluctance to speak frankly on this particular subject."[8] In addition, direct questions about riot participation may be successful only in identifying a particular type of riot participant—Campbell and Schuman observed that "surveys of actual riot participation may also tend to represent and describe most adequately the more self-conscious and probably ideological riot participants, but underrepresent others who may join for reasons they are less willing to discuss with an interviewer."[9] The two basic weaknesses in the estimates of riot participation admitted by Campbell and Schuman, therefore, have cast some doubts on existing information about this subject.

The results of the Detroit survey conducted soon after the 1967 riots indicated that participation in the disturbances may have been more extensive than has been previously estimated. In this survey, questions probing self-reported participation were not included in the interviews to protect respondents from possible self-incrimination. Nonetheless, a high percentage of the residents of the Twelfth Street neighborhood were engaged or involved in the disturbances. Sixty-one percent, for example, reported that they "got out to see what was happening" during the riot. Most of those persons were either spectators or participants in the disturbances—or possibly both. The proportion of persons who might be drawn into active involvement once a major disorder has started probably had not been adequately determined by prior research.

In addition, residents of the Twelfth Street area were asked to guess the proportions of their neighbors who participated in the riots. Since people often tend to project acts on others that they are reluctant to admit themselves, responses to this question may reflect accurate estimates of riot participation. Although one-fourth of the residents declined to conjecture about this issue, nearly four-fifths placed actual participation at 10 percent or more of their neighborhood, and nearly one-third reported it at 25 percent or more. The indirect evidence offered by persons who had experience with a riot, therefore, revealed higher participation than self-reported data.

In order to determine how ghetto residents would react to incidents that precipitate riots, both surveys included questions about hypothetical events that might foment violence. Respondents in the Campbell-Schuman survey were questioned about methods of changing a possible situation that involved a "white storekeeper in a Negro neighborhood" who "refuses to hire any Negro clerks." The vast majority (73 percent) initially suggested attempts to boycott the store or other formally approved methods of redress such as petitions, appeals to government agencies, or nonviolent demonstrations. Only two percent proposed direct violence, and another two percent limited their reactions to veiled threats.[10]

In many respects, however, the hypothetical incident involving the white storeowner may not have been a useful test of predispositions about violence. None of the major riots in U.S. cities were sparked by a controversy over a white storeowner; most of them were triggered by an encounter with the police. As the Kerner Commission noted, "Almost invariably the incident that ignites disorder arises from police action."[11]   A hypothetical event focusing on hostile contact with the police, therefore, may provide a more accurate index of the proportion of residents who would respond by engaging in behavior that might induce violence.

In the Twelfth Street area survey in Detroit, respondents were asked how they would react to "a policeman beating up someone from the neighborhood." Only 44 percent said they would respond in the officially sanctioned manner of filing a complaint with public officials. Just 3 percent claimed that they would ignore the beating, and 15 percent were uncertain about their probable reactions to the incident. Twenty-five percent stated they would engage in potentially provocative conduct by "going to other people for help," and ten percent claimed that they would "wait around to see what happens." Not all of the people who reported they would find local help probably would participate in violence, but the proportion seeking assistance from official sources was much lower in the Detroit survey after the riots than in the Campbell-Schuman study.

Resentment of the police may have been particularly strong in the Twelfth Street area of Detroit. While a plurality of the respondents in the Campbell-Schuman survey did not think that the police used unnecessary force, insulting language, or harassment techniques in their neighborhoods,[12] 80 percent of the Detroit residents believed that police treatment there was unequal and unfair. Only 17 percent of the Twelfth Street residents felt that the law was basically fair to all people, and less than 5 percent agreed that "laws are enforced equally." Police conduct during the riot apparently intensified this negative image. Seventy-six percent of all respondents stated that they had "heard stories that some policemen were involved in taking things or burning stores." While concrete evidence of criminal conduct by the police was difficult to obtain, there was little doubt that such rumors were widely circulated in the neighborhood after the riot.

The grievances of Twelfth Street residents also were reflected in increased militancy. In response to a question about whether black people should "integrate with whites or try to get along without whites," support for black separation in the Detroit ghetto was recorded at 22 percent. On the other hand, only 9 percent of the respondents in the Campbell-Schuman study agreed that "Negroes should have nothing to do with whites if they can help it."[13] On this issue, black residents who had undergone a riot were much more radical than their counterparts who had no personal experience with violence.

In part, the reduced militancy of the general black sample may have overshadowed special local conditions. Although Campbell and Schuman did not ask any direct questions about the outbreak of riots, they repeatedly referred to the disorders as "spontaneous."[14] This statement was not confirmed by the residents of Detroit who had immediate contact with a riot. Fifty-seven percent felt that the riot "had to happen sooner or later," only 27 percent believed that it could have been avoided, and 15 percent were undecided about the issue. Unlike other observers, black citizens who had experienced a riot perceived the violence as inevitable or unavoidable.

In addition, most black persons who were relatively untouched by violence did not view the riots as planned or organized in any major degree. In response to a question about whether "the large disturbances such as those in Detroit and Newark were planned in advance," only 18 percent of the people in the Campbell-Schuman survey expressed the belief that the riots were generally planned, a third thought there was some planning, and another third believed there was no planning at all.[15]

The Detroit survey, however, disclosed that 37 percent stated flatly that "the people who started the trouble were mainly organized," while slightly less than half felt that "they were mainly acting on their own." In addition, nearly one-third of the Detroit respondents reported that both the arsonists and snipers in the riot, respectively, "knew each other" rather than "acting on their own." While the replies raised some difficulties of interpreting precisely what was meant by organization, there was little doubt that the inhabitants of the Twelfth Street area considered the riot as purposeful and directed action that might yield beneficial results.

In large measure, the differences in the results of the two surveys seemed to be related to the political goals of the two populations studied. Many black Americans have regarded the riots as a natural product of centuries of frustration and as one available means of protest or securing recognition for their needs and aspirations. Persons who lived in an area that exploded in violence probably had higher levels of frustration and demanded more fundamental social and political change than the residents of untroubled neighborhoods. As Campbell and Schuman noted, "The word 'protest' is a key one."[16] Instead of considering the riots as a kind of protest many inhabitants of the Twelfth Street area seemed to imply that they regarded the disorders a form of rebellion, or a confrontation with white authorities.

In fact, respondents in the Detroit survey were somewhat less likely to view the riots as a method of protest than persons interviewed in the Campbell-Schuman Study. More than 59 percent of those in the Campbell-Schuman survey perceived the riots as a protest rather than as "a way of looting," and an additional 28 percent saw it partly as a protest.[17] On the other hand, 53 percent of the Detroit sample felt that the main purpose of the people who started the riot was "calling attention to their needs," 38 percent thought that they were "just taking things and causing trouble," and 8 percent expressed no opinion. As a method of protest, people who had direct experience with violence seemed somewhat less sympathetic with the actions of rioters than black citizens who resided elsewhere.

Persons who had lived through a riot did not perceive the disturbances solely as a means of protest, but they seemed to have other objectives. At least two other goals were evident in the interviews with members of the Twelfth Street community. Initially, many local people were motivated by a desire for vengeance. More than 64 percent said that most of the stores burned during the riot were owned by whites, and only 28 percent guessed that "it didn't make much difference." In addition, 54 percent reported that the stores were burned because the owners deserved it rather than for other reasons.

Another incentive that may have inspired those who engaged in the riot was the wish for what might be termed "instant social mobility." For many, the disorders represented a rare opportunity to short-circuit the frustrations of middle-class strivings by acquiring prized consumer goods sooner than ordinarily would be possible. Although 71 percent of the Twelfth Street residents said the looters took "anything they could get their hands on" rather than "things they needed but could not afford," 55 percent also stated that they took things for themselves rather than to sell to others. Only 30 percent claimed that the goods were stolen primarily for resale through illegal channels. Ghetto residents who had not been involved in a riot probably lacked some of the immediate objectives that were prevalent among those who had experienced violence.

The principal differences between the Detroit sample and the respondents in the Campbell-Schuman survey were reflected in their expectations about government officials. The tendency to see the riots as protests or attempts to obtain recognition from white authorities also was associated with increased anticipations of government efforts to aid the ghettos. In the Campbell-Schuman study, 47 percent felt that their local mayor was "trying as hard as he could" to solve the problems of the city. More than one-third ascribed the same motives to federal and state governments.[18]

On the other hand, residents of a riot-torn community who re-garded the riot as part of a rebellion or revolution voiced little con-fidence in public officials or government programs. Only 23 percent of the Detroit respondents stated that the way Mayor Jerome Cavanaugh of Detroit handled the trouble was "very good"; 36 percent called it "not good" or "poor." Forty-three percent evaluated the way that Governor George Romney of Michigan handled the disorders as "not good" or "poor"; only 13 percent said it was "very good." Perhaps even more significant was the fact that 22 percent stated they would endorse a movement led by local conservatives to remove Mayor Cavanaugh from office. Apparently, many residents were sufficiently alienated from the incumbent and allegedly liberal mayor to hope new leadership would be more responsive to their needs and aspirations.

Moreover, 78 percent of the Detroit respondents said that before the riot the government was merely trying "to keep things quiet," and only 18 percent believed that government leaders "really were inter-ested in solving the problem that the Negro faces in this city." The apparent hope that the disorders would produce fundamental changes in government structures and policies yielded growing political op-timism after the violence had ended. The proportion of Twelfth Street residents who believed that government really would be interested in solving problems after the riot increased to 33 percent.

Black respondents in the Campbell-Schuman survey who viewed the riots as a protest or simply another attempt to secure a response from officialdom seemed to be grateful for and desirous of existing policies to aid the ghettos. The federal anti-poverty program, for example, was known by 91 percent of the people and 75 percent be-lieved that it was doing a "good" or a "fair" job.[19] On the other hand, 41 percent of the Detroiters disapproved of the local anti-poverty pro-gram, only 36 percent approved, and 23 percent were undecided. Their negative judgments were not based on a lack of information; more than 85 percent were familiar with anti-poverty projects. Persons who had experienced a riot were much more critical of existing government programs than those who had no direct contact with violence.

The residents of the Twelfth Street ghetto had more revolu-tionary expectations than black citizens who lived elsewhere. In com-parison with the Campbell-Schuman survey, more than twice as many people in the Detroit study believed that the riots would help "what most civil rights groups are trying to accomplish" or "the cause of Negro rights."[20] As a young waitress exploded, "Black people have not been given their rights. White people don't understand nothing but violence." Persons who had direct experience with violence were much more likely to stress the positive consequences of a riot than the bulk of black Americans who had not been involved in a major disorder. In addition, there was some evidence from Detroit that the riot was viewed as a class rather than as an exclusively racial struggle. Almost 80 percent of the black respondents—as well as most whites—felt that Negroes and whites in the riot were "fighting on the same side" rather than "against each other."

The specific objectives of the Twelfth Street neighborhood seemed to be expressed in the hope that present political structures would be recast to reflect a stronger sense of community and to increase the opportunities for political participation. After the riot a larger proportion of the residents believed that they would receive "more attention from city officials," that they would have "more to say about what should be done in this neighborhood," and that they would possess "more power." The residents of a riot-torn area desired —and anticipated—that the scope of public activities would be expanded and that they would be granted increased involvement in political decisions.

The same patterns were evident in the responses to similar questions in the two surveys about actions that should be taken to avoid further violence. While black residents in other cities felt that obtaining better jobs was the most important means of averting violence, the Detroit respondents felt that elimination of the entire system of segregation and racism was the most pressing problem. An end to discrimination was the second most important matter in the Campbell-Schuman study, and increased discussions with the white leaders ranked third. [21] Negotiations with whites was the least significant issue to the Detroit respondents, however, while improvements in the police and expanded social services and publicly supported programs were rated high on the list. Most of the black residents of the relatively untroubled cities seemed to regard civil disorders as a means of protesting long-standing grievances. On the other hand, those who had undergone a riot seemed to regard it as a form of rebellion or at least a direct confrontation.

The comparison of surveys in a riot-torn neighborhood and in other major cities has indicated that persons who experienced violence were substantially more militant than most ghetto residents. For many people in the Twelfth Street area, violence may have been a radicalizing experience; but the sentiments and conditions that promoted the riot probably existed before the trouble started.

Public officials should not misjudge the sentiments of black Americans by focusing on the general distribution of opinions and by neglecting the attitudes of special pockets of the society. While some ghetto residents may endorse conservative political goals, other segments of the population seem to support radical hopes and objectives.

FOOTNOTES

[1] Angus Campbell and Howard Schuman, "Racial Attitudes in Fifteen American Cities," in *Supplemental Studies for the National Advisory Commission on Civil Disorders* (Washington, 1968), p. 11.

[2] *Report of the National Advisory Commission on Civil Disorders* (New York, 1968), pp. 112-13, 158-59.

[3] Campbell and Schuman, op. cit., p. 12.

[4] Ibid., p. 61.

[5] Ibid., p. 52.

[6] Ibid.

[7] Raymond J. Murphy and James M. Watson, *The Structure of Discontent* (Los Angeles, 1967); Nathan S. Caplan and Jeffery M. Paige, "A Study of Ghetto Rioters," *Scientific American*, CCXIX (August 1968), pp. 15-21.

[8] Campbell and Schuman, op. cit., p. 52.

[9] Ibid., p. 53.

[10] Ibid., pp. 51-52.

[11] *Report of the National Advisory Commission on Civil Disorders*, op. cit., p. 206.

[12] Campbell and Schuman, op. cit., pp. 42-43.

[13] Ibid., p. 16.

[14] Ibid., pp. 8, 49, 62.

[15] Ibid., p. 48.

[16] Ibid., p. 62.

[17] Ibid., p. 47.

[18] Ibid., p. 41.

[19] Ibid., pp. 41-42.

[20] Ibid., pp. 48-49.

[21] Ibid., p. 48.

PART 4   Modification and Control:

Alternatives to Aggression, Hostility, and Violence

War

Not bad, but miserable
Drenched in gray sadness
Lonely grief handed out to all.
    Sarah Mason, Age 10, U.S.A.

Behavioral scientists are devoting an increasing amount of their time to the discovery of workable solutions to our many social problems. It is apparent that both the public and the scientific community regard human aggression as one of the more serious of these problems. Some techniques and suggestions for the control of aggression appear in this last section, and we are encouraged by finding so many papers devoted to this end. Some of these techniques are in use, others remain largely untried. Whether we adopt one or many of these methods on a large scale remains to be decided by the people. But it is obvious that information about prospective techniques and programs must be made available for public debate. This is particularly true of those suggestions that require a great deal of public supervision, for some people may suspect a hint of 1984 in the measures.

    We will distinguish two approaches in the search for a means to eliminate aggression and warfare as problems. Both stem from the controversy over origins or causes and therefore they diverge somewhat. The first assumes that aggression in its natural form is biologically useful and necessary. It is simply its form and mode of expression which have become increasingly suspect. That we argue and lose our tempers or even that we fight is not a sufficient problem in itself according to this view. Rather it is that we hold grudges, that we resort to plot and subterfuge, and use weapons on each other which creates the danger. If one takes a hard look at the nonhuman animal world, aggression does seem to work in a positive way but not always according to the principles of equality of opportunity. Although generally only the physically strong breed at high rates, this animal world is significantly more orderly and peaceful than the human. Lions, for example, for all their ferocity as hunters, seldom if ever inflict serious injury on their own kind. This is not to say that they do not have their disputes, but they have an elaborate behavioral way of handling them (see Lorenz, Reading 3), as do most animal species. In such ways survival of the species is guaranteed. Unfortunately human society has no such guarantees and, tragically, our demise is likely to mean the end of all life on this planet.

Those who postulate the necessity of aggression in man suggest that we need a healthy and safe outlet for our impulses. This, once again, assumes the presence of a specific energy for aggression that must be allowed to dissipate in a harmless manner. If we have an aggressive urge, they say, why not channel it in useful directions? But we must ask whether we really need this aggressive energy—if, indeed, it is there at all? Similarly, the relationship between aggression and competition must be assessed. We may be confusing actual and symbolic levels of behavior when we equate these two terms.

The second approach argues for the total elimination of aggression in our society. It is obvious that this position has its meritorious as well as its questionable sides. For example, disputes can be and frequently are settled without resorting to violence, and this is the preferred method. There have also been individuals who were able to conduct their lives in a nonviolent fashion and successfully achieve eminence and respect. But would it be wise to eliminate aggression totally from our behavioral repertoire? This is a very important value question as well as a pragmatic one. Young people greatly fear the prospects of dehumanization. In fact, we all suffer some amount of dehumanization each day of our lives. But the idea that human emotions could be controlled in some fashion is abhorrent to our society, even though emotion is likely to be at the root of our problems with violence. Such problems gain importance when we recognize recent research findings which suggest that most violent crimes occur among people who know each other. So it is necessary to decide whether measures are needed to protect society from the violence of the crowd, or whether our right to assemble in a violent crowd is perhaps in greater need of protection. Is it then safe to create a nonviolent society? Can we insure that change will come without revolution? Do we have the means to achieve a just society without the use of power? But after asking such questions let us pose a more fundamental one. In examining the history of violence (particularly warfare) can we ever really be sure that the solution arrived at through war was ultimately more beneficial than a solution arrived at through peaceful negotiation? We have the same problem we have been emphasizing throughout our discussion. Without a control group, which is impossible when dealing with such large phenomena as nations, we cannot be certain that violence has ever really been useful. In other words we cannot say with any certainty that violence is a better way to achieve just goals than other methods. There is simply no empirical evidence to justify such a statement.

One technique that is effective in the control of aggression needs further comment here. The technique of behavior modification shows great promise as a method of change in humans, but it also has its problems. Some people have labeled its techniques as inhumane and dangerous. This criticism probably stems from the fear that behavior modification is some mysterious force which in the hands of evil scientists could control the world and enslave mankind. Although it is true that the necessity of using punishment on occasion makes some of the therapy seem unpleasant, the ends frequently justify the means; that is, the behaviors ultimately changed are frequently so destructive that intervention with the most powerful of methods becomes necessary in order to restore to the individual some hope of improvement. Lest we give the wrong impression, however, let us hasten to add that behavior modification is not *all* punishment. In fact, it is important to note that the systematic use of reward is far more common. Many hospitals now operate reward communities, known as "token economy units," which attempt to shape appropriate behavior by giving rewards for proper conduct. This system is not unlike the so-called "real world" in that it requires some amount of conformity to established norms of behavior. We think that it is a promising technique. The reader may assess its desirability as a system by reading B. F. Skinner's novel *Walden II* for a detailed treatment of what a community could be like under a system of positive control. In a somewhat different light, Aldous Huxley has some interesting views on both reward and punishment in *Brave New World Revisited.* Both of these literary masterpieces consider some of the issues that we have been discussing.

In the final analysis, it would seem that we are left with value decisions. It is not enough simply to know which systems work, for we must have the wisdom to foresee the consequences of our decisions. For now, we must be content to debate the efficacy of our proposals, but many more difficult decisions await us and we must be prepared to make them. As he studies the papers that follow, the reader should imagine himself in the role of policy maker. Indeed, as a voting citizen, he is just that. He can evaluate these suggested courses of action not only for their scientifit merit but, more importantly, for their prospective utility in the quest for a more peaceful world.

REFERENCES

Huxley, Aldous, *Brave New World* and *Brave New World Revisited*. New York: Harper and Row, 1960.

Skinner, B. F., *Walden Two*. New York: Macmillan, 1948.

SUGGESTED READING

Horsburgh, H. J. N., *Non-Violence and Aggression*, London: Oxford University Press, 1968.

An excellent discussion of Ghandi's ethics of nonviolence. Interesting chapters on the relationship of nonviolence to the concept of defense are included.

Kaufmann, H., *Aggression and Altruism*, New York: Holt, Rinehart and Winston, 1970.

The author emphasizes human social behavior in his discussion of these two concepts. Instinct is examined at the beginning of the book and later chapters suggest means of dealing with the problem of human aggression and war.

Menninger, K., *The Crime of Punishment*, New York: Viking, 1968.

In this volume the author, a distinguished psychiatrist, evaluates the United States penal system and makes some interesting suggestions for change.

Rapoport, A., *Fights, Games and Debates*, Ann Arbor: University of Michigan Press, 1960.

Rapoport, an expert in gaming theory, explores the nature of conflict with an emphasis on recent research in gaming.

Schwebel, M. (Ed.), *Behavioral Science and Human Survival*, Palo Alto, Calif.: Science and Behavior Books, 1965.

These papers were selected from the 1963 meeting of the American Orthopsychiatric Association. The scientists who contributed addressed themselves to the problems surrounding the maintenance of peace.

Toch, H., *Violent Men*, Chicago: Aldine Publishing Company, 1969.

A sociological study of violent men in the context of their environment. Toch focuses on the violent "event" and provides interviews with the subjects of his study. Contained within are suggestions for the rehabilitation of violent men.

Wells, D., *The War Myth*, New York: Pegasus, 1967.

The author presents his ideas on the rationalization behind war making in Western culture. A chapter on the causes of war gives heavy emphasis to the ideas of philosophers and some recent research by psychiatrists. Some very valuable information regarding the war-related research following both world wars is included.

13   The Reduction of Intergroup Hostility: Research Problems and Hypotheses

—IRVING L. JANIS and DANIEL KATZ

I. INTRODUCTION

It is a startling fact that almost the last area to be investigated empirically by the social sciences is the area of the constructive forces in human nature and society which make for the reduction of intergroup conflict. Research attention has focused upon the destructive tendencies in human conduct—upon antisocial action, hostility, distorted perceptions, irrational fears, authoritarian personality structure, and a wide variety of psychopathological disorders. Concern with positive forces has not reached the point of stimulating systematic investigation, perhaps because of the antinormative position of present-day scientists,

Reprinted by permission from the *Journal of Conflict Resolution*, 3 (No. 1), 1959, 85-100.

most of whom feel inclined to leave such matters to philosophers and social reformers. Yet an adequate social science must study the social norms and ethical principles by which men live.

Naess has suggested that an analysis of ethical principles with respect to the psychological processes they implicate and the social conditions which maximize their effectiveness may be the most important next step for social science, from the point of view both of improving its theoretical adequacy and of contributing to the problem of social survival. His systematization of the ethical code of Mahatma Gandhi includes an explicit statement of testable hypotheses (7). Most of these hypotheses take the form of predicting that certain types of social action will have the long-run effect of achieving the humanistic aims of a non-violent political movement while, at the same time, reducing the probability of hostile attacks from rival groups. Similar hypotheses can be extracted from the writings of John Dewey, William James, and other philosophers who have emphasized that the means one employs in a social struggle determine the ends that will ultimately be achieved. Additional hypotheses that may warrant reformulation and investigation probably can be extracted from writings on ethics by other modern philosophers such as B. Croce, L. T. Hobhouse, G. E. Moore, J. Royce, B. Russell, and A. E. Taylor.

Our main purpose in this paper is to examine some of the new and promising areas of research in the field of social psychology that are suggested by various ethical propositions concerning methods of reducing intergroup hostility and enhancing mutual adherence to a shared set of ethical norms. First, we shall call attention to some of the key variables that might be investigated and the types of research method that might be employed. Then we shall formulate a series of sample hypotheses that are offered for their suggestive value, illustrating some of the basic theoretical issues in contemporary psychology to which a systematic research program on intergroup conflict could contribute a great deal of pertinent evidence.

II. SOME KEY VARIABLES

A major set of problems requiring both theoretical analysis and rigorous empirical investigation is that of evaluating the social and psychological consequences of the positive ethical means employed by any social movement, organization, or group to achieve socially desirable goals in its struggle against rival groups. A large-scale program of research would be needed to determine under what conditions the various ethical means (independent variables) have the intended effects (dependent variables).

Examples of Independent Variables

An excellent source of various ethical procedures that are illustrative of the means to be investigated is the analysis of the Gandhian ethical system prepared by Naess (7). Most of the normative propositions and hypotheses which specify the forms of conduct that will achieve the ultimate ethical goals can be restated in terms of means-consequence relationships. As examples, we have selected eight norms, all of which are here formulated as procedures or policies of social struggle which are means for attaining the various humanitarian ends. Although loosely defined at present, these means can be readily translated into operational terms and investigated as independent variables in systematic research studies:

1. *Refraining* from any form of verbal or overt *violence* toward members of the rival group

2. Openly *admitting* to the rival group one's plans and intentions, including the considerations that determine the tactics one is employing in the current struggle as well as one's longer-range strategic objectives

3. *Refraining* from any action that will have the effect of *humiliating* the rival group

4. Making visible *sacrifices* for one's cause

5. Maintaining a consistent and persistent set of *positive activities* which are explicit (though partial) realizations of the group's objectives

6. Attempting to initiate direct personal *interaction* with members of the rival group, oriented toward engaging in *friendly verbal discussions* with them concerning the fundamental issues involved in the social struggle

7. Adopting a consistent attitude of *trust* toward the rival group and taking overt actions which demonstrate that one is, in fact, willing to act upon this attitude

8. Attempting to achieve a high degree of *empathy* with respect to the motives, affects, expectations, and attitudes of members of the rival group

Besides the foregoing list, many additional examples of positive means could be culled from Gandhi's ethical code (4), from Dewey's

Human Nature and Conduct (2), and from other ethical writings which also contain propositions concerning the positive and negative social consequences of using alternative ethical procedures.

Dependent Variables

The effectiveness of the positive means can be assessed in relation to the following outcomes, which constitute the dependent variables to be investigated:

1. A reduction in the incidence and intensity of acts of violence
2. An increase in the willingness of the rival group to engage in arbitration and to overcome the obstacles that interfere with peaceful settlement of disputes
3. Favorable attitude changes among members of the rival group toward the group behaving according to ethical principles
4. Greater motivation on the part of group members to continue working toward the attainment of humanitarian and social welfare goals
5. Greater success of the group in achieving its specific humanitarian objectives
6. Favorable attitude changes among members of the group in the direction of greater commitment to peaceful settlement of disputes with all rival groups
7. Favorable attitude changes among spectators of the struggle (i.e., people who are unaffiliated with either of the contending groups) in the direction of being more attracted to the group using positive ethical means, placing greater reliance in their public communications, and thereby becoming more influenced in the direction of accepting their policies and objectives

In general, the predictions would be that the positive means such as those listed here would, singly or in combination, lead to favorable outcomes as specified by the seven dependent variables. But, in addition to these global predictions, a number of much more refined hypotheses would need to be tested in order to determine the intervening processes which mediate the predicted effects. In the course of investigating the social and psychological consequences of any one of the various means, it will probably turn out that there are a number of different component factors involved that must be separated and investigated as independent variables. For example, the policy of openly admitting one's intentions and plans to a rival group might give rise to three quite separate effects.

1. Revealing material that is ordinarily kept secret may influence the rivals' attitude concerning the *moral status* of the acting group (e.g., they may become suspicious that something more important is being kept secret, or they may become much more respectful of the sincerity of the group).

2. Revealing tactical plans that will handicap the acting group may influence the rivals' attitudes concerning the *strength* of the acting group (e.g., admission of one's plans may be perceived as signs of weakness and ineptness in conducting the struggle or as signs of an exceptionally powerful movement that is capable of being successful without resorting to secrecy).

3. Predicting in advance the deprivations that will be inflicted upon the rivals may have the effect of increasing or decreasing the magnitude of frustration and the intensity of the aggressive impulses aroused when the deprivations subsequently materialized.

Thus investigating positive ethical means may lead to the discovery of a number of different mediating processes, some of which may tie in with broad sectors of theory and research in the human sciences.

Implicit in the foregoing discussion is the expectation that objective evaluations of the consequences of the positive ethical means will include careful investigation of the *unfavorable* outcomes as well as the favorable ones. Obviously, the research would have to be carried out in such a way as to detect readily any instance in which the outcome was the reverse of that specified in the foregoing list of favorable outcomes. In this connection it will be necessary to specify a number of additional dependent variables, representing other types of adverse outcomes. For example, a certain type of positive ethical means may prove to be extremely frustrating to the members who are committed to using it and incline some of them to become defensively *apathetic* and to *disaffiliate* themselves from the group. In some cases the intrapersonal conflicts engendered by prolonged suppression of aggressive impulses might conceivably engender a marked increase in *anxiety* or other *symptoms of emotional tension*. In the long run, consistent adherence to certain of the positive means might result in a marked change in the composition of the membership, with a preponderance of masochistic and other *deviant personalities* being attracted to it.

To detect such unfavorable consequences, the research investigator would need to be alert to any indications of unintended effects that arise in the course of carrying out empirical investigations. Comparisons of instances of favorable outcomes with those of unfavorable outcomes should provide valuable evidence concerning the conditions under which the use of various positive means does and does not lead to the intended effects.

Conditioning Factors

One major set of conditions determining favorable versus unfavorable outcomes has to do with the *combination* of positive means that are employed by the group. For example, admission of one's own plans and refraining from violence may be interpreted as weakness and perceived as relatively ineffective unless accompanied by visible sacrifices for one's own cause and a program of persistent, clear activity demonstrating the group's objectives. Moreover, the use of one means, such as refraining from violence, may strengthen the commitment to the group goal, and this intervening psychological change may facilitate the effective execution of other means, such as making visible sacrifices for one's cause. Thus it will be necessary to study the independent variables in combination and in interaction as well as singly.

The nature of the group struggle is another conditioning factor in the operation of these variables. At least three dimensions of group struggle must be taken into account.

The first dimension is the degree of conflict of interest relative to the community of interest between competing groups. It is generally assumed that non-violent means and positive ethical practices are more applicable to factions within the same institution, since they have so much in common, than to rival nations, where the conflict of interest is high. Nevertheless, it is conceivable that the suicidal character of modern methods of violent group conflict has made this distinction less important, since the common interest in survival has become increasingly clear. In any case, it may be possible to discover auxiliary means of making common interests salient to rival nations and thereby increasing the chances of success for limiting international clashes to non-violent conflicts.

A second dimension concerns the psychological closeness of the group conflict to the people involved. The dynamics of enmity between close personal associates and distant peoples may be different. The distance between competing nations makes their struggle less intense on a personalized basis than that between rival factions in the same political party. On the other hand, the more remote, the fewer the reality checks and hence the easier it is for autistic perception, projected fantasies, and hostile distortions to play their role.

A third related dimension has to do with the degree of institutionalization of the channels, or means, of conducting group and national competition and conflict. Violent means of resolving personal and group conflicts may be a direct reflection of the personal aggression of the protagonists, as in frontier community violence, which is an anticipation of legal institutions. But more commonly at the group level, practices have become institutionalized so that there is no one-to-one correspondence between the warlike actions of a nation and the warlike character of its people (5). Most wars are probably fought not because the great majority want to fight but because they accept the legitimacy of the process which has led them into war. All these considerations suggest the need for taking into account the nature of the group struggle in studying the effectiveness of ethical forms of social action. In a final section some aspects of the institutionalization of aggression will be discussed.

Closely related to research on limiting conditions is another field of investigation comprising the study of psychological and social conditions which facilitate the willingness of group members to use the positive ethical principles referred to as the "independent variables." In other words, it is also necessary as part of a systematic program of research to consider the use of the positive means as *dependent* variables and to find out the predisposing factors which enable individuals and groups to limit themselves to positive ethical policies in their struggles with opposing groups.

III. METHODS AND TECHNIQUES

We envisage three phases for the development, refinement, and testing of hypotheses about the peaceful resolution of inter-group conflicts. They need not constitute a discrete temporal sequence, since there is much to be gained from an overlap in the timing of the phases.

The first phase would consist of the use of existing data at two levels: documentary evidence and primary-source data. The former would call for comparative case studies of *historical instances* of social and political struggles in which the given action policies were and were not employed—e.g., studies of various radical, pacifist, religious, and nationalist movements whose social effects can be appraised from available documentary evidence. Primary-source data could be drawn from *interview and questionnaire studies* bearing on industrial conflict situations and factional disputes within social movements, military organizations, political parties, and schools. Of particular relevance would be data on the correlates of different demands and practices on the part of supervisors, union officials, military officers, political leaders, and teachers. This stage of the investigation would furnish some preliminary testing of hypotheses but would serve mainly for the more precise formulation of significant variables and their interrelationships.

The second phase would consist of field studies of *current* and *developing instances* of social and political struggles in which the given action policies are and are not being employed—e.g., collecting systematic interview data in United States southern communities where Negro organizations are attempting to bring about desegregation. The emphasis here would be upon specifying the relevant types of data in advance, whereas in the first phase the studies would be limited by the data which happen to be available. Again this stage could contribute both to the testing and to the reformulation of hypotheses.

The final or experimental phase would consist of field and laboratory experiments. The second phase gives better control over the collection of relevant data than the first, but adequate control of the operative variables requires the use of experimental techniques.

Field experiments, which involve the use of controlled experimental techniques in natural settings, have the advantage that the necessary controls can be taken into account in advance of the investigation. They also have the merit of dealing with the full power of social variables as they occur in a real community setting. Such experiments could be devised, for example, in connection with the program of a social or political group in which alternative action policies are carried out in equivalent towns. (E.g., the co-operation might be obtained of a research-minded national organization which is currently engaging in a social or political struggle within many different communities throughout the country. Certain local chapters in one designated set of communities might be asked to use a given action policy, whereas other chapters in an equivalent set of communities might be asked to use a contrasting action policy. The effects could be ascertained by interviewing representative samples within the two sets of communities and by using behavioral indexes such as incidence of overt violence on the part of rival groups, increases or decreases in membership of the competing groups, etc.)

Laboratory experiments of the type employed in current research on group dynamics could investigate some of the variables of interest in contrived settings, but the manipulations would be relatively weak. The advantage of this method would be the possibility of isolating single variables and varying their strength fairly precisely, although within limited ranges. The most efficient use of this method would probably be to deal with very specific questions which might arise from field experiments about the properties of a given variable.

IV. SAMPLE HYPOTHESES AND PROBLEMS

This section will be devoted largely to presenting a series of hypotheses concerning the psychological processes which mediate the anticipated favorable and unfavorable effects of using various violent and nonviolent procedures in intergroup conflicts. We shall present (1) some general propositions concerning the influence of instrumental actions on group goals and the role of leadership in using means consistent with the goals; (2) some of the major psychological changes that might account for the "corrupting" effects of using violent means; (3) a number of additional explanatory hypotheses bearing on the converse process—the "constructive" effects of abstaining from violence; and (4) hypotheses concerning the attitude changes produced by positive ethical means which involve consistently treating the members of opposing groups as potential allies.

A final section will consider the problem of the consequences of the institutionalization of violence.

Influence of Instrumental Actions on Group Goals

That individuals and groups can be involved in antisocial practices in the interests of desirable social goals and still maintain these goals in relatively pure fashion is a doctrine for which there is little psychological support. Once people act in a certain manner, they tend to develop beliefs and attitudes to make that behavior part of their value system. Thus psychologists have long talked about mechanisms becoming drives or instrumental activities becoming functionally autonomous (1). An important factor in the doctrine that the end justifies the means is the separation this imposes in fact between means and ends. John Dewey and other writers have emphasized that an expedient means chosen without regard for the goal sought will not be an intrinsic part of an integrated pattern of means-end activity. It becomes increasingly difficult for the person himself, as well as those who observe his actions, to identify the goal which he is seeking from the instrumental means he employs. When an individual devotes his major energies to using expedient means, he will tend to see the justification of his behavior not in what he actually does every day but in the great goal which lies somewhere beyond. And, of course, it is relatively easy to justify one's morality by goals which are remote and which permit little reality testing. Concrete everyday activities, however, do not permit easy rationalization when they have to be considered on their own merits. It may be just as important, therefore, for a group to tie its ethical standards to means as to ends, since the means can be checked and observed more readily than the goals.

The central point of what has just been said is that repeated behavior of an antisocial character, though originally in the interests of altruistic social goals, will probably lead to the abandonment of those goals as directing forces for the individual. This proposition applies to the leaders as well as the followers within any group or organization.

Persons in positions of leadership, of course, play the major role in proposing and executing the ethical policies that are used in any social struggle and in inducing the rest of the membership to adopt them. The leaders of groups with humanitarian goals may be able to execute certain of their functions more effectively if they adopt expedient means on an opportunistic basis. But, in the long run, opportunistic leaders will probably be less effective in moving their followers toward achieving the ultimate objectives of their organization than leaders who insist upon using means that are perceived by the members as being consistent with humanitarian goals. This principle has been recognized by those political and social movements which attempt to maintain a fictitious divorce between their ideology and their opportunistic methods by assigning different people to the two functions. Such groups sometimes try to keep their ethical ideology "pure" by not invoking it for every opportunistic measure.

There are at least four different considerations which make it likely that the long-run losses will offset the short-run gains whenever the leader of an altruistic movement indorses expedient means that are not consistent with the group's ultimate objectives. (1) If leaders justify bad means for good ends, it will create perceptual ambiguity for their followers. Many followers are not steeped in the ideology of the movement, and it is difficult for them to distinguish in many instances the means from the end. (2) They will have less confidence in the sincerity of a leader who is not prepared to sacrifice for the cause. His espousing of expediency may be interpreted by the members as indicating that he is taking the easy way and is not sufficiently devoted himself to take the harder route to his objectives. (3) To restore confidence, the leader is likely to resort to aggressive behavior toward his opponents, to impute to them an exaggerated evil intent, and even to advocate violence toward them. (4) Even a single opportunistic practice by a leader sets a precedent and makes subsequent opportunism easier for the leader and his followers to accept. Since the principle has already been compromised once, further compromise will do little additional harm.

Thus, on the one hand, confidence in the leader's sincerity is likely to be undermined by his use of opportunistic methods, and, on the other hand, the goals of the group become obscured for the members whenever their leaders succeeed in inducing them to accept expedient means which are obviously inconsistent with the group's objectives. When the expedient means involve the use of violence against opposing groups, these tendencies will tend to be accentuated. We turn now to some additional hypotheses which specify the psychological changes that occur within any participant who engages in hostile actions against people who are opposing the program or ideology of his group.

"Corrupting" Effects of Using Violent Means

Why and under what conditions would the use of violent means be expected to have extremely adverse effects on the individual participants in a social movement or organization? More specifically, what psychological changes within each participant might account for the following two consequences of the use of a violent means for the alleged purpose of attaining socially desirable goals: (*a*) an increase in the probability that such means will be used again in the future when similar, and perhaps even less demanding, occasions arise, and (*b*) a decrease in the probability that the group will work toward the achievement of socially desirable goals (i.e., violent means "corrupt" the ends)?

One obvious answer might be that a violent means will tend to corrupt the ends because it promotes counteraggression on the part of the group's opponents, and this creates a need to use more and more violence, ending up by engaging all the energies of the group in a violent struggle with the rivals instead of enabling positive actions to be taken toward the attainment of the long-run social goals. But even when we set aside the possibility of evoking counteraggression, there are at least three other psychological processes that may come into play, any one of which could have the effect of "corrupting" the members of a group that participates in the use of violent means:

1. Even when the violent means is socially sanctioned, the users may react with some degree of guilt (as a consequence of earlier moral training or as a consequence of generalization from non-sanctioned forms of violence). Guilt reactions may take the form of (a) high anticipation of being punished by the target group; (b) preoccupation with the question of whether or not the action was correct; and (c) affective disturbances, which may range from completely conscious feelings of guilt to vague feelings of uneasiness with no awareness of the source of the disturbance. One of the typical ways in which people attempt to reduce or counteract such guilt reactions is to attribute evil and immoral intentions to the target toward which their violence had been directed. Such attributions may enable a guilt-ridden person to justify the violent action to himself and to others; it may also involve a projection mechanism which operates as an unconscious technique for warding off guilt (3). The perception of the target as being extremely threatening and evil would have the double effect of (a) increasing the tendency to attack violently again in an effort to weaken the target, (b) decreasing one's willingness to work out compromises with the target group, and (c) altering the conception of humanitarian objectives in such a way as to exclude members of the target group.

2. Participating in any violent action may have the effect of weakening the internal superego controls which are the product of normal socialization. Superego controls are often based on exaggerated conceptions and partially unconscious fantasies about the possible consequences of performing the forbidden act. In psychotherapy a characteristic sequence of changes occurs when patients overcome anxiety or guilt reactions in the sexual sphere or in connection with socially aggressive behavior. After they have once "tested out" the new (noninhibited) mode of action, they are left with less exaggerated conceptions and fantasies about the consequences of such behavior. Thus the inhibition tends to be gradually extinguished. The same sort of process seems to go on among combat soldiers whose inhibitions about killing the enemy begin to lessen after the first time they are induced to perform the disturbing act of shooting at enemy soldiers. A similar learning process may go on in connection with each instance of group-

sanctioned violent action such that the person's automatic superego controls are lessened and he becomes capable of indulging in more and more extreme forms of violence.

3. Social contagion effects may occur within a group or organization such that when a highly respected leader or member of the group uses a violent means under highly "justified" circumstances, other members of the group become less inhibited about engaging in similar acts of violence. This contagion may be partly the product of learning that the violent means is not disapproved, if it is used without criticism by the standard-bearers of morality within the group. Unconscious processes of identification may also facilitate the contagion effect. While, in the first instances, violence is applied by group leaders only after careful judgment, in subsequent instances the followers will be much more ready to indulge in violence without such a careful appraisal of whether or not it is justified. Thus the attitude may gradually develop that violent means are acceptable and even desirable, provided only that they are used in the service of the group's cause.

Constructive Effects of Abstaining from Violence

The next question is the converse of the one just discussed: Why and under what conditions would *abstaining* from the use of sanctioned violent action be expected to have positive effects—e.g., decreasing the probability that violent means will be used in the future, increasing the probability that the group will work toward achieving its original humanitarian goals, and increasing members' adherence to the positive social objectives and moral standards which the group sponsors?

Some of the answers to this question may involve the same psychological mechanisms and social contagion effects specified in the preceding section. However, there may also be some processes that are of a different character, and for this reason we feel that the question of the constructive effects of non-violent action should be considered separately from the question of the "corrupting" effects of violent action. In the discussion which follows, we shall indicate additional mechanisms that may come into play when members of a group adhere to a group decision to abstain from using violent means under conditions where such means are considered to be an acceptable or expected form of behavior.

In many persons, participation in sanctioned violence may serve as a means of reducing conscious and unconscious fears of being passively manipulated by others or of being exposed to damaging attacks and deprivations at the hands of one's rivals. To the extent that such fears are based on misconceptions or exaggerated fantasies about the magnitude of the danger, a given act of abstaining from sanctioned violence may involve a process of *emotional relearning* (similar to that referred to in the preceding section in connection with the lowering of superego control). In this instance, however, the process would be equivalent to that which goes on when a hyperaggressive patient undergoes psychotherapy. Sooner or later he tries out a passive, non-aggressive way of responding to the therapist and discovers that the dangers of passivity which he had so greatly feared do not actually materialize. Similarly, when the members of a group adhere to a group decision to behave in a conciliatory rather than a hostile way, their anticipations about the dangerous consequences of non-violence may be brought more into line with reality. If their fear of being passive is thereby reduced to some extent, they will no longer be so strongly motivated to engage in violence on future occasions when confronted with a choice between violent and non-violent means of struggle against their opponents.

Guilt mechanisms may also play an important role in the internalization of non-violent norms. Insofar as any act of violence (whether sanctioned or not) generates some degree of guilt, at least a slight degree of emotional tension would be experienced by the average group member whenever he *anticipates* engaging in a future act of violence. A reduction in emotional tension might occur if, at the time when the group member is experiencing anticipatory guilt, a communication from a group leader or an expression of group consensus conveys the idea that the group's goals can be better achieved by abstaining from violence and by using an effective form of non-violent action instead. The decision to accept the recommendation would be reinforced by the *reduction of anticipatory guilt*. The reward value of the decision might be enhanced if the ideology of the group included the norm that violence is a morally inferior form of action which should be avoided as much as possible. Even if only lip service is given to this norm, the group member may experience a heightening of self-esteem in addition to guilt reduction if he anticipates that others in his group will approve of his decision to abstain from violence. If each act of abstention is rewarded in this way, a new attitude will gradually tend to develop such that the person becomes increasingly more predisposed to decide or vote in favor of non-violent means. Perhaps under these conditions, good moral "practice makes perfect."

Attitude Changes Produced by Treating Opponents as Potential Allies

Many of the positive ethical means to which we have referred involve more than merely abstaining from violence. Among the examples which we have cited are such means as displaying an attitude of trust toward the members of opposing groups, maintaining friendly personal interactions with them, and seeking to understand their motives and attitudes by deliberately empathizing with them. Although somewhat different rationales for the various positive means have been put forth by their proponents, all of them seem to point in the general direction of replacing a hostile, competitive, antagonistic approach by a policy of treating opponents as potential friends or allies. The hypotheses which follow pertain to the use of any positive ethical means or combination of such means, provided that they are employed on the basis of adhering to this general policy.

Just as in the case of using violent means, *social contagion* effects may occur when positive ethical means are used. But the factors which facilitate the contagion may be somewhat different. Because hostility and violent aggressive action is a very elementary impulsive form of reaction to people who interfere with the attainment of important group objectives, many persons may remain unaware of alternative ways of dealing with opponents and of overcoming the frustrations engendered by their opposition. Thus, whenever violent group action is regarded as the socially accepted mode of response to this type of frustration, many members of the group may gain sudden enlightenment if a respected leader or subgroup calls attention to the possibility of using an alternative approach. If the group decides to try out the proposed alternatives, even if its success remains ambiguous, those members of the group who have a relatively low need for aggression may also learn that the new means is less energy-consuming and less disagreeable than the traditional means. In this way, a process of acculturation may take place whereby a social technique evolved through the intelligence and ingenuity of others comes to be adopted by people who had formerly accepted, more or less unthinkingly, a general policy of dealing with opponents in a hostile manner.

Other psychological changes may also mediate the effects of adopting, on a tentative basis, the use of positive ethical means. Whenever a member of a group accepts a group decision to use an *unconventional* friendly approach to rivals, he is likely to feel it necessary to justify the fact that he is deviating from the expected course of action (e.g., "Why am I willing to allow these people to provoke us so much without our hating and punishing them?"). The need for such justification may sometimes arise from exposure to cross-pressures resulting from conflicting (pro-hostility) norms held by other groups with which one is affiliated. Or the need for justifying may come from internalized standards—e.g., awareness that one is deviating from the ego ideal associated with sex role ("Am I a sissy?"). In any case, the need to justify the policy of treating opponents as potential allies would motivate the person (a) to take account of the positive attitudes and human qualities of the rivals; (b) to minimize the hostile intentions of the rivals; and (c) to predict that the friendly positive approach will be more successful than an antagonistic approach would be. Thus the effort to justify an act of friendly treatment may lead to cognitive restructurings and a shift in motivational pressures, which could contribute to two types of attitude change: (1) reduced hostility toward the rival group and (2) more favorable evaluations of the desirability of using positive means in general.

Nor are the beneficial effects of non-violence confined to the members of the group pursuing this policy. As group members take into consideration the positive attitudes of members of the out-group and stop reacting toward them as if they were deadly enemies, the out-group itself is under less pressure to be defensively aggressive. Thus the opponents may be influenced to engage in fewer acts of provocative hostility, and, in the long-run, some of their leaders and part of the membership may even become motivated to live up to the other group's view of them as potential allies.

Some Further Consequences and Causes of Institutionalized Aggression

In the preceding discussion we have considered in some detail the psychological mechanisms which may account for the corrupting effects of violence and the constructive effects of abstaining from violence. In this section attention will be given to further consequences of the operation of these basic mechanisms and to supplementary social-psychological processes which make institutionalized aggression the persistent problem of organized society. Though our major concern is with socially sanctioned aggression, it is important for theoretical reasons to differentiate between personal hostility and institutionalized forms of aggression with respect to both causes and effects. The recipient and the initiator of social aggression may be affected differentially if the violence is a sanctioned institutional practice or if it is the release of personal aggression. The two violent actions may be alike in physical character, but they are not necessarily perceived, experienced, or reacted to as the same. Personal aggression may be felt by the recipient as more of an attack upon his ego than the institutional action; it may lead to personalized resentment, more immediate resort to counteraggression, and perhaps less long-term effect. Institutionalization may leave the individual no easily identifiable target of a personal nature for counteraggression; it may confront the individual with sufficient force that he has no way of striking back. It may lead to displaced aggression against a convenient scapegoat, to intropunitiveness, to apathetic acceptance, or to repressed hostility. These consequences can occur in response to personal acts of aggression, but they are less likely to occur where the personal target is easily identifiable and where countermeasures are within the grasp of the individual.

Our major problem today is not protection against the hostile elements among us as individuals capable of violence. Our major problem is with institutionalized forms of violence, as in conflicts between organized groups and nations. Such institutionalized aggression is accentuated by the presence of hostile people in certain situations, but the correlation between the amount and intensity of group conflict and the amount of latent hostility would not be high save under very special conditions. There are situations, however, in which the interaction of the two—personal hostility and institutional aggression—is of far-reaching significance, as in the opportunities which institutional channels may offer for the expression of latent hostility.

There are three psychological dangers in the institutionalization of violence which are worthy of special investigation: (1) the release of latent hostilities under conditions of social sanction of violence, (2) the apathetic condoning of any institutionally approved practice, and (3) the perpetuation and intensification of institutional violence.

1. The Release of Latent Hostilities under Conditions of the Sanctioning of Violence

In Western society the antisocial nature of acts of aggression is communicated to children very early in the socialization process. Aggressive acts toward others are repeatedly censured and punished. If there is lack of understanding by parent and child in this process and continued frustration of the child, there may be repression, but retention, of the hostility. As a result, the adult will be burdened with strong latent hostility which comes into continual conflict with his superego standards. These standards are reinforced by perception of the social norms of the group which proscribes personal acts of violence. The presence of others and the presence of authority represent the stimulus situation which inhibits the aggressions of the individual. But, then, a curious reversal occurs. In certain contexts acts of violence are legitimized and sanctioned by groups and institutions. In times of war almost all sources of authority within each nation assert that it is noble and proper to kill for one's country. The social support for the antisocial action generally has three elements: the justification of a moral purpose, the justification of legitimacy, and the justification that others approve. Since the traditional inhibitor of violence has been the social environment, violence can assume intense and bizarre forms when the inhibitor is transformed into the facilitator. This is the classic theory of crowd behavior (6). But whereas in the crowd the social support is limited and temporary, in organized groups the support is more extensive and continuing. Thus within the areas where aggression is socially sanctioned, individuals can resolve their conflicts by indulging their worst impulses and by attaining social recognition and reward for so doing.

One danger in the social sanctioning of violence is that the release of hostility will go far beyond the bounds of what is sanctioned. Supposedly appropriate force is invoked on an objective and impersonal basis to accomplish the group's purpose. In practice, however, the way is open for abuses of various sorts. To the extent that latent hostility does exist in the members of the group and their leaders, there will be a tendency to push beyond the necessary force to accomplish the group goal because leaders directly and followers vicariously enjoy the opportunity to release repressed impulses. An extreme illustration would be the use of terror by the German Nazis to maintain power for the Nazi party, which was then pushed to the point of attempts to exterminate entire groups and alleged races. The classic argument against the use of corporal punishment in the schools is the possibility of sadism when the punisher can use aggression disguised as socially approved and necessary discipline. In a preceding section it was indicated that the use of violence leads to further violence through weakening the internal superego controls. The inhibition against the expression of aggression becomes extinguished. This is especially true in the area of institutional aggression, where social support makes it easy to violate the basic social prohibitions. Such social support makes it possible to rationalize away guilt feelings and makes similar violence easier in the future.

We are really dealing in these examples with an interaction of institutional and personal aggression. Our contention is that people may perceive and react differently to personal, as against institutional, violence. The former is more identifiable and leads to more personalized resentment, since it is felt to be a direct attack upon the self. The latter induces in its victims more displacement and more generalized hostility. Frequently, however, in the case of sanctioned violence which permits the expression of latent hostility we have a pattern combining both types of aggression. The chances are that this combination will be perceived by its victims as the most unjust of all aggression. There is a tendency to personalize actions which are in any way injurious to the self. In this instance, however, the afflicted individual is right, since there is personal animus in his punisher. But, unlike purely personal aggression, there is no recourse to any form of counteraggression, since the punishment is legal and proper. Moreover, the victim has limited opportunity for even verbally blowing off steam against his opponent. The result is often intense generalized hatred. Where the situation becomes completely intolerable, it may result in identification with the aggressor.

This combination of institutional violence and personal aggression is one reason why group conflicts become intensified over time and become difficult of solution, even when there is a good objective basis for solution. The scars left by a strike in which both company and union have used force are of this character. The company guards given free rein to their destructive impulses may have abused their power in a manner which the strikers never forget. And the strong-arm squad of the union may have acted similarly toward strike breakers. Both sides feel that the other side has taken advantage of a group struggle to perpetuate a personal outrage. Some of the bitterest memories of World War II are not of massive destruction by heavy artillery and bombers but the use of the cloak of military necessity for the expression of personal sadism.

2. The Apathetic Condoning of Institutional Practices

Another danger lies in the passive acceptance of any violence perpetrated by one's own group or even by a rival group if it has some legal sanction. This is a different response from vicarious indulgence in one's own impulses toward violence and has not received adequate attention. Since the act of force is institutionally sanctioned, it is perceived by many as an objective event. There is no sense of personal outrage, even

if the action is directed at deviant group members. This passive acceptance of violence sanctioned by the group, which would otherwise be
regarded as basically wrong, is often the result of a compartmentalization in thinking and attitude. It is related to psychological factors mentioned in our previous discussion of the means-end problem, which also
involves a compartmentalization such that the individual is not compelled to face up to consequences of his behavior. When this compartmentalization is carried to an extreme, it means that there is one
morality for the individual and a completely different morality for the
group. Since the group standard can be justified by very remote goals,
any action which the group leaders suggest must be accepted. When the
German people passively accepted the violence perpetrated in Nazi concentration camps, it was probably not because of their higher level of
latent hostility or sadism but because of their compartmentalization of
morality. What was legal and sanctioned by the authorities was right,
whether or not it was consistent with their own personal standards of
morality.

In general, people as group members will condone actions by
group representatives which they will not approve of for themselves as
individuals. In time the punishments used against individual members
by the group will tend to be brought into line with the punishments
approved of by members in their personal lives. Brutal forms of physical punishment tend to be dropped from public institutions after they
are no longer approved of in interpersonal relations. But in the area of
group actions in relation to other groups we permit types of behavior
that we do not countenance among individuals. Though such a dual set
of standards can be defended, the danger is that the justification comes
to rest not on practice and its consequences but on a social myth which
asserts the unquestioned prerogative of the institution qua institution.
The corrupting effect of such condoning of institutional aggression can
be seen in war and postwar periods when encroachments are made on
individual and civil liberties. Self-seeking politicians under some cloak
of governmental authority can carry such threats to an extreme and still
secure the acquiescence of many people, since such institutional attacks
against our enemies, external and internal, are assumed to be their
legitimate function.

3. The Perpetuation and Intensification of Institutional Violence

Personal aggression, lacking institutional supports, is sporadic and variable. When violence becomes an accepted part of the practice of an

organization, it not only is perpetuated but tends to grow much like other parts of the organization. This perpetuation and intensification of institutional aggression comes about in three ways: (*a*) the setting-up of specialized roles, (*b*) role adaptation, or the effects of taking roles upon personalities, and (*c*) role selection, or the tendency toward a fit between unusual roles and personality types.

(*a*) By creating special roles, organizational structures do not rely upon chance factors for the performance of various functions but make such performance the systematic work of trained experts. In addition to the motivation intrinsic to the role, the institution enlists a variety of organization motivations such as monetary rewards, promotion or upgrading, group acceptance, etc. Moreover, in any sector of an organization people occupying given roles tend to make their role functions as important as possible, partly because of self-interest in their careers, partly for the encouragement of morale, and partly because of the psychological prominence of their own tasks compared to others they know less about. The armed forces or the FBI is like any other part of a bureaucratic structure in seeking bigger appropriations and more personnel.

(*b*) In the earlier discussion of the effects of instrumental actions upon group goals it was pointed out that such actions affect the value system of the individual. Role behavior, like any other form of behavior, leads to its rationalization. Personal values are brought into line with the individual's action. What he does, he may do as his job, but after a time he sees this as necessary, important, and desirable. Even in those cases where the role is not originally congenial to the personality pattern of its occupant, remaining in the role results in modification of the personality. To be a member of a combat force and to hold pacifist values produces intense internal conflict. If the individual cannot readily escape from the behavioral demands of a role, he will tend to accept the rationalization provided by the organization in order to dull the sharp edges of the conflict. In time, this acceptance undermines old values and builds up a new value system. Thus role adaptation means not only carrying out the required behavior but justifying it as a desirable course of action. Every occupational and professional group develops an ideology which is supportive of its practices to the extent of occasional idealization of its functions. In the same fashion the military, police, and custodial vocations develop values consonant with their behavior.

(*c*) There is a tendency toward a fit between unusual institutional roles and basic personality patterns. The general notion of the fit between bureaucratic roles and personality has probably been overdone, but there is a good deal of truth in the thesis when we are dealing with unusual roles which call for atypical patterns of motivation and behavior. The censor of pornographic literature may sometimes be suspected of enjoying his duty. When an institution permits violence as part of its function, people will be attracted to this role who derive satisfactions from the nature of the work. Thus there is a self-selection process for brutal roles. In the police forces of some American cities, among prison guards, and in the strong-arm squads of some labor unions there will be individuals who gravitate to and remain in these roles (when there are equally well-paid positions open to them elsewhere) who are of a special personality type. Before the professionalization of American police forces there were many cities in which it was not always easy to distinguish between the member of the third-degree squad and the criminal he was bringing to justice.

Institutional support for roles of violence can be a corrupting factor within an organization far more than is generally realized. Even though not all roles of violence are filled by persons with strong needs to discharge sadistic or hostile impulses, such personalities can readily dominate their part of the organization. Less congenial personalities for these roles will tend to drop out over time. The more brutal individuals will remain and, through their continuity in the organization and their greater motivation, will set the pattern of accepted practice. Moreover, their mutual reinforcement of one another may intensify brutal practices and perpetuate them. The history of some concentration camps illustrates this trend. Brutal practices in prisons and among police forces have been difficult to uproot because it would mean the wholesale dismissal of large groups of people—those guilty of flagrant violations and those who are virtual accomplices in such violations.

V. SUMMARY

The purpose of this paper was to show the applicability of the research methods of behavioral science to problems of group conflict and interpersonal hostility. The particular frame of reference employed is that of social psychology. Applications from the concepts and techniques of this field are made to certain aspects of the use of violence and of constructive methods in achieving group goals. A section on methods, moreover, outlines both a general strategy for research investigation and the more specific techniques called for at the tactical level. Some of the normative propositions from Arne Naess's analysis of the Gandhian ethical system are examined as the basis for empirical studies. Particular attention is given to the effects of the use of violence and of abstaining from violence in terms of the psychological processes involved. The concluding part of the paper discusses factors making for the perpetuation and intensification of institutionalized aggression.

REFERENCES

1. Allport, G. W. *Personality: A Psychological Interpretation.* New York: Henry Holt & Co., 1937.

2. Dewey, J. *Human Nature and Conduct.* New York: Henry Holt & Co., 1922.

3. Flugel, J. C. *Man, Morals, and Society.* London: Duckworth, 1945.

4. Gandhi, M. K. *Non-Violence in Peace and War.* Vols. I, VIII, XI. Ahmedabad, 1942-49.

5. Kelman, H. C. "Societal, Attitudinal, and Structural Factors in International Relations," *Journal of Social Issues*, IX (1955), 42-56.

6. Martin, E. D. *The Behavior of Crowds.* New York: Harper & Bros., 1920.

7. Naess, Arne. "A Systematization of Gandhian Ethics of Conflict Resolution," *Journal of Conflict Resolution*, I (1958), 140-55.

14 Modification of Severe Disruptive and Aggressive Behavior Using Brief Timeout and Reinforcement Procedures

—*DARREL E. BOSTOW and J. B. BAILEY*

Several techniques for the modification of deviant behaviors have been developed in recent years (Ullmann and Krasner, 1965, Ulrich, Stachnik, and Mabry, 1966). There is some evidence to suggest that one technique, timeout, which involves the temporary suspension of the subject's usual activities, often is effective in eliminating severe problem behaviors in applied settings. Wolf, Risley, and Mees (1964) demonstrated that tantrums and self-destructive behavior in an autistic child could be effectively reduced by placing him alone in a room each time the behavior occurred and removing him only when the tantrums subsided. Hamilton, Stephens, and Allen (1967) used a similar procedure by confining severely retarded patients to a timeout area for from 30 min to 2 hr after each incidence of aggressive or destructive behavior. The timeout procedure greatly reduced the aggressive and destructive behaviors of five patients. Tyler and Brown (1967) put delinquents who resided in a training cottage in timeout for 15 min for each act of misbehavior around a pool table, and demonstrated that this technique effectively reduced the undesirable behavior. Risley (1968) found, however, that confining a severely deviant girl to her room for 10 min for

Reprinted by permission from the *Journal of Applied Behavior Analysis*, 2, 1969, 31-37. Copyright 1969 by the Society for the Experimental Analysis of Behavior, Inc. This research, carried out at Caro State Home and Training School, Caro, Michigan, was supported by a Michigan Department of Mental Health grant to Roger E. Ulrich, Psychology Department, Western Michigan University. The authors are indebted to the Superintendent, Dr. Bettye McFarland, and Program Director, Dr. Marjorie Clos, for their full cooperation. We are especially grateful to the Nursing Service Supervisors Mrs. Partlo, Mrs, O'Connor, Mrs. Terbush, and Mrs. Bailey, without whose full cooperation this project would not have been possible. We also extend our thanks to Evelyn Barber, Leona Bailey, and Diane Bostow for their assistance in the conduct of this study. Preparation of this manuscript was partially supported by NICHHD grant HD-00183-03 to the Bureau of Child Research, University of Kansas.

climbing on top of furniture and in dangerous places had no effect on her rate of climbing.

In these studies the duration of timeout ranged from 10 min to 2 hrs. The present research describes the use of brief timeout (2 min) in conjunction with reinforcement for acceptable behavior to eliminate extreme disruptive and aggressive behavior in two institutionalized retarded patients. Research by Holz, Azrin, and Ayllon (1963) suggested that the simultaneous application of reinforcement for acceptable behavior may enhance the effectiveness of a timeout procedure.

The present experiments were carried out in the normal ward situation. The procedures involved each subject serving as his own control using a reversal design (Baer, Wolf, and Risley, 1968).

EXPERIMENT I

Subject and Setting

Ruth B., a 58-yr-old wheel-chair patient in a large state hospital resided in an infirmary ward with approximately 50 other nonambulatory patients. She was brought to the experimenters' attention as a result of her loud and abusive verbal behavior which, according to the institution's records, had been a source of irritation to the staff and to the other patients for several years. The staff reported that they were forced to spend an inordinate amount of time complying with her demands and attending to nearby patients who became disruptive as a result of Ruth's excessive verbal outbursts.

The experimenters observed the patient for several hours and observed that: like most other patients, Ruth awoke at an early hour, was moved from her bed into a wheel chair, and taken to her usual place in the day room. Her cursing and verbal tirades centered around demands for various articles of clothing, favorite objects which she frequently dropped, and complaints of rough handling by the attendants. She seemed to be a particular problem at mealtimes, screaming violently until her tray was brought, until she had her second cup of coffee, or her tray was removed, and so on. Ruth posed another daily problem by frequently refusing to allow the attendant who brought her medicine to come near. If she refused medicines she had to be forcibly restrained while being given an intra-muscular injection of tranquilizer to calm her. Throughout this period, the attendants were frequently observed to reason with Ruth and to reassure her of their good intentions.

Procedure

The loud vocal responses were measured by a Concord Model 330 voice-operated portable tape recorder. The volume control was set at 5.5 on a scale of 10.[1] Each above-threshold noise started the recorder and produced a distinct "blip" sound on the tape. The frequency of "blips" served as the measure of the dependent variable, i.e., a vocalization was an utterance sufficiently loud to activate the tape recorder. Once activated, the recorder ran for 2 sec after the sound had dropped below threshold. Stopwatches were used to measure the length of the session, length of timeouts, and time between reinforcers. After each session the tape was reviewed and the number of "blips" counted. Reliability checks were made by two observers listening to the tape. The number of "blips" each minute was recorded independently by each observer. The per cent agreement for each of the 60 min of the session was calculated by dividing the smaller by the larger number. The 60 percentages were averaged to determine the per cent agreement for an entire session. The subject's vocalizations were sampled in the above manner during daily 1-hr sessions which began in the morning when the subject was wheeled to her favorite table in the day room, where she typically remained for the rest of the day. The only part of the recorder that remained in the subject's view was the microphone. Other patients, whose behavior might also activate the recorder, were moved to another part of the day room. The experimental design consisted of four conditions.

Baseline I

The subject received tranquilizing medications twice during the first 18 sessions of Baseline I. These were prescribed by the ward physician in an attempt to control the violent outbursts. After this period, five days were used to establish the baseline level of vocalizations without any drugs. Throughout this phase, attendants and staff were requested and observed not to alter their normal routine with respect to the subject.

Timeout + DRO I

In this phase, each time Ruth's vocalizations activated the recorder during the day, the experimenters wheeled her to a nearby corner of the day room (about 10 ft), took her out of her wheel chair, and placed her on the floor. This entire operation took approximately 10 sec to complete. She remained on the floor for a minimum of 2 min, after which a 15-sec interval of silence was required before she was placed back in her chair. If she screamed continuously through the 2-min period, she was left on the floor until she was quiet for 15 sec. If she screamed while being lifted from the floor she was placed down on the floor again and the 15-sec interval started once more. During timeout periods the timer which timed the 1-hr session was stopped. Vocalizations during the timeout periods were not counted in the total responses for that session. The session timer was not restarted until Ruth had been returned to her table.

In addition to timeout for loud vocalizations, the subject was provided with the things for which she usually screamed, only after periods of remaining quiet (i.e., not activating the recorder). These reinforcers for appropriate behavior were provided on an increasing time-interval schedule; i.e., if she remained quiet for the first 5 min of the session the first reinforcer was delivered. She then had to remain quiet for 10 min before the next reinforcer (e.g., breakfast, second cup of coffee, or juice) was delivered. Five minutes were thus added each time a reinforcer was given, up to a maximum of 30 min. For the remainder of the day, if Ruth remained quiet she would receive some treat, favored object, or attention at least every 30 min. If she screamed or shouted during the interval before a reinforcer was delievered she was placed on the floor and that interval was reset. It was not begun again until she had been placed back in her chair and wheeled to her table. This schedule is one which differentially reinforces other behavior (DRO) than that being punished. (Because the interval between reinforcers grew by an increasing interval each time a reinforcer was delivered, it might technically be called an Escalating DRO schedule.)

This procedure was continued throughout each day (although data were taken only for the first hour) from the time the patient arose until she was put back into bed at night.

Baseline II

The attendants were instructed to interact with the subject as they had during the initial Baseline condition and were observed to do so. The subject's vocalizations were recorded as before, during the 1-hr period in the morning immediately after Ruth arose, but she received no scheduled consequences for screaming or remaining quiet. The condition was in effect for four days.

Timeout + DRO II

This was a replication of the Timeout + DRO I condition.

Results

During the Baseline condition, a tranquilizing drug (Prolixin Enanthate) was prescribed by the ward physician in an attempt to reduce Ruth's vocalizations. As shown in Fig. 1, response frequency was sharply reduced when 12.5 mg of the drug was administered just before Session 6. The patient remained stuporous for a number of days after the injection; contrary to the anticipated effects of the drug, the reduction of vocalizations was only temporary and the rate rose steadily to the highest point (Session 17). One-half the amount of the same tranquilizer was given just before Session 18 but the effect seen with the first amount was not replicated. For the last six sessions of Baseline, vocalizations averaged 86 per session.

The timeout and reinforcement procedures were employed starting with Session 24 and continued through Session 28. Figure 1 shows

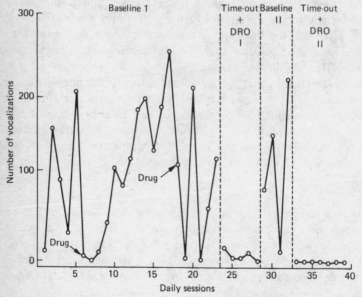

Figure 1. Number of vocalizations per 1-hr session under Baseline and Timeout + DRO conditions.

the gross and immediate effect of the procedures which reduced vocalizations to one in Session 28.

A return to Baseline conditions was begun with Session 29, during which 79 vocalizations occurred. The number of responses rose to 229 on the last day of Baseline II (well within the range observed during Baseline I). It thus appeared that not only could the high rate be eliminated but that reinstating previous conditions produced a corresponding increase in the undesirable yelling.

The timeout and reinforcement condition was instituted again, starting with Session 33. This time the subject came in contact with the timeout contingency only once and thus received almost all the scheduled reinforcements each day for six of the seven days. During Session 37, Ruth made no vocalizations loud enough to trip the recorder and received all the scheduled reinforcers.

Ruth learned that if she simply sat quietly, most of her needs would be met in a short while. In addition, she would occasionally raise her hand and whisper a request to an attendant. She became less of a problem at medicine time after she had been put on the floor several times for screaming at the aides who brought her medicines.

The screaming, shouting, and loud cursing, which the staff had tolerated for years thus came under control of this combination of timeout and reinforcement for appropriate behavior. The reliability of observer agreement of the number of "blips" per minute ranged from 50 to 100% and averaged 84% per session.

EXPERIMENT 2

Subject

Dennis M. was a 7-yr old and had been admitted to the institution 18 months before the beginning of this study. Before Dennis was admitted, he lived with his parents and two siblings. The parents reported that while at home Dennis exhibited severe disruptive behavior, such as attacking other persons and breaking furniture. The supervisor for the ward reported that in the hospital setting, Dennis was so aggressive that he could not be kept with other children in the day room or on the playground. Instead, he was tied to a door in a hallway, where he was able to strike only occasional passersby. To control Dennis' aggressive behavior, several tranquilizing drugs had been prescribed; however, at the beginning of this study none had been discovered to be very effective.

Procedure

A timeout booth, measuring 4 by 2 by 5.5 ft high and constructed of 0.5-in. plywood, was placed in a corner of the day room. A latch on the door of the timeout booth prevented the subject from escaping, but the booth was open at the top so that the subject could be observed if necessary. Milk, cookies, and carbonated drinks, which the subject had been observed to take readily, were used as reinforcers. Stopwatches, counters, and data sheets were used for recording data during each session.

The observers used record sheets marked off at 1-min intervals. Aggressive behavior was defined as any bite, hit, kick, scratch, or head butt directed against another patient or an attendant. Reliability checks were made by having two observers record the aggressive behavior independently. Agreement of the two records was measured by comparing the total number of aggressive behaviors per session (the smaller number was divided by the larger to give a per cent figure).

Each daily 30-min session began when the subject was brought from where he was customarily tied, to the day room. The day room, approximately 20 by 25 ft, was devoid of any objects except for a few chairs, a small diaper changing table, and a television mounted 7 ft above the floor. There were usually 12 to 15 other patients of about the same age and physical size as the subject and at least one attendant in the room while sessions were being conducted. The experimental design consisted of four conditions.

Baseline I

The Baseline condition was in effect during seventeen 30-min sessions while Dennis received one or two tranquilizing drugs each day. He received neither drug in the last five days of Baseline I. The experimenter counted the number of times the subject hit, bit, kicked, scratched, or butted any other person, while he was in the day room. The attendants were requested not to alter their normal routine, and to ignore the subject if he came up to them. The experimenters observed this request to be met. At the end of the session, Dennis was re-tied in his usual place in the hall outside the supervisor's office.

Timeout + DRO I

During the timeout phase the subject was brought to the day room as usual; however, each time an aggressive response occurred he was quickly picked up and placed into the timeout booth by one of the experimenters. The total time required to put him in timeout was less than 5 sec from the moment a response occurred. Nothing was said to the subject at these times. Dennis remained in the booth for 2 min after each response. The session timer was stopped until he was released so that he was allowed the same total amount of time (30 min) in the room as during the Baseline phase. In addition to the timeout for aggressive responses, Dennis received a small amount of milk or carbonated drink, or a bite of cookie each time 2 min elapsed with no aggressive behavior regardless of what else he was doing. The attendants were again requested not to alter their normal routine in the day room and were observed to carry out the request.

Baseline II

The subject was taken to the day room and the sessions were run exactly as in Baseline I. The number of aggressive responses was recorded for 10 sessions.

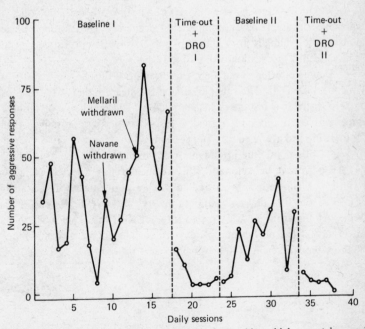

Figure 2. Number of aggressive responses (i.e., hits, bites, kicks, scratches, and head butts) per 30-min session under Baseline and Timeout + DRO conditions.

Timeout + DRO II

This was a replication of the timeout and DRO condition and lasted for five sessions.

Results

The number of aggressive responses varied from 4 to 57 during the first eight sessions of baseline. Just before session 9, one of Dennis' tranquilizers (Navane) was withdrawn. As shown in Fig. 2, the number of aggressive responses did not appear to increase appreciably. Before Session 13, the other tranquilizer, Mellaril, was withdrawn and the number of aggressive responses was somewhat higher thereafter. There is presently, however, not enough data to indicate precisely what role each of the drugs played in suppressing the aggressive behavior.

As shown in Fig. 2, on the first day of the timeout and reinforcement procedure, the number of aggressive responses was reduced to 16. The frequency then declined regularly until aggressive responses were reduced to five per session on the last day of this condition.

When Baseline II was reestablished, the original high rate of aggressive behavior never completely recovered. The number of responses rose gradually to a high of 41 in Session 31 and totaled 29 for Session 33, the last session of Baseline II.

Dennis emitted seven aggressive responses during the first day of Timeout + DRO II. The rate per session fell steadily thereafter to a low of zero aggressive responses in Session 38. The reliability of observer agreement on number of aggressive responses per session ranged from 75 to 100% and averaged 92% per session.

At this point, the attendants were trained in the use of the timeout for aggressive behavior and reinforcement for appropriate behavior. Dennis was allowed to be untied for longer periods each day, starting with 3 hr the first day and building up to a full day at the end of the week. At that time he was free to interact with the other boys in the ward and on the playground all day. According to the informal report of the attendants he had to be put into timeout only two or three times per day.

DISCUSSION

The present results showed that severe and chronic behavior problems can be significantly reduced by providing consistent and immediate consequences for them. According to the institutional records, Ruth's disruptive screaming and shouting had been a problem for at least 5 yr, and Dennis' aggressive behavior (hitting, biting, etc.) for at least 18 months. For both patients, the frequency of problem behaviors was reduced to near-zero level in less than a week when brief timeout and reinforcement procedures were applied. Furthermore, the treatment procedures were shown to have a continuing role in the management of the behaviors, since both disruptive behaviors increased when the contingencies were removed. Although no objective data were gathered, it did appear that administration of the treatment procedures could be transferred to the ward staff, thus allowing them to handle both patients without the use of drugs or restraints.

Casual observation of the social interactions of both subjects indicated that many more acceptable behaviors occurred during treatment phases than during the baseline phases. During treatment conditions of Exp. 1, one patient who had frequently been the focus of Ruth's violent verbal outbursts would move close to Ruth and show her toys and point to pages in books. Often, this patient would handle articles belonging to Ruth. Ordinarily, such interactions would have provoked outbursts form the subject; however, during treatment sessions Ruth responded by nodding and making comments such as "That's nice", or "Yes, I see it".

Dennis seemed to exhibit a similar increase in acceptable social interactions. During periods when the timeout and DRO contingency were in effect he would occasionally approach other children to hug and embrace them. The staff reported never having seen this behavior before (possibly because the opportunity was restricted by his being restrained 24 hr per day). He was not as active as during Baseline conditions and would spend a great deal of time sitting and observing other children in the day room. Further research is needed to measure more adequately such effects and to determine those procedures which maximize these desirable "side effects".

This demonstration that brief, "non-painful," and easily administered consequences can prove to be extremely effective in reducing severe and even violent behaviors, may provide an attractive alternative to the use of electric shock for such purposes. Most hospital personnel will drug or restrain a violent or aggressive patient rather than use electric shock, even though shock has proved repeatedly to be highly effective (Lovaas, Schaeffer, and Simmons, 1965; Tate and Baroff, 1966; Whaley and Tough, 1968) and apparently to be devoid of undesirable side effects (Risley, 1968).

Timeout appears to be more acceptable because it can be of short duration (both present studies used 2-min timeouts), is not injurious to the patient, and closely resembles the rather common use of seclusion rooms as punishers in institutions. Reinforcement for desirable behavior is also readily accepted, since nursing service personnel are highly skilled at providing aid and comfort to patients and need do little more than make such personal attention contingent upon desired behavior.

The present results, showing abrupt reductions in the frequency of undesirable behaviors, are similar to those of Hamilton et al. (1967) and Tyler and Brown (1967). These studies used longer timeouts (varying from 10 min to 2 hr) than used here (2 min). It is difficult to compare directly the effects produced by timeout alone (Hamilton et al. and Tyler and Brown) and the timeout in conjunction with an alternative reinforcement schedule used in the present study. Holz, Azrin, and Ayllon (1963) observed that when an alternative response was made available for obtaining reinforcement, mild punishment was completely effective. Further research is needed to determine whether the effectiveness of the present procedure is related to the effects observed in the Holz et al. (1963) study. Additional research is also needed to determine the parameters of timeout that are most effective and the functional properties, if any, that distinguish an escalating DRO from the standard DRO procedure.

REFERENCES

Ayllon, T. Intensive treatment of psychotic behavior by stimulus satiation and food reinforcement. *Behavior Research and Therapy*, 1963, 1, 53-61.

Ayllon, T. Some behavioral problems associated with eating in chronic schizophrenic patients. In *Case Studies in behavior modification*, L. P. Ullmann, and L. Krasner (Eds.), New York: Holt, Rinehart and Winston, Inc., 1966. Pp. 73-77.

Ayllon, T. and Michael J. The psychiatric nurse as a behavioral engineer. *Journal of the Experimental Analysis of Behavior*, 1959, 2, 323-334.

Baer, D. M., Wolf, M. M., and Risley, R. T. Some current dimensions of applied behavior analysis. *Journal of Applied Behavior Analysis*, 1968, 1, 91-97.

Hamilton, J., Stephens, L., and Allen, P. Controlling aggressive and destructive behavior in severely retarded institutionalized residents. *American Journal of Mental Deficiency*, 1967, 7, 852-856.

Holz, W. C., Azrin, N. H., and Ayllon, T. Elimination of behavior of mental patients by response-produced extinction. *Journal of the Experimental Analysis of Behavior*, 1963, 6, 407-412.

Lovaas, O. I., Schaeffer, B., and Simmons, J. Q. Building social behavior in autistic children by use of electric shock. *Journal of Experimental Research in Personality*, 1965, 1, 99-109.

Risley T. The effects and side effects of punishing the autistic behaviors of a deviant child. *Journal of Applied Behavior Analysis*, 1968, 1, 21-34.

Tate, B. G. and Baroff, G. S. Aversive control of self-injurious behavior in a psychotic boy. *Behavior Research and Therapy*, 1966, 4, 281-287.

Tyler, V. O. and Brown, G. D. The use of swift, brief isolation as a group control device for institutionalized delinquents. *Behavior Research and Therapy*, 1967, 5, 1-9.

Ullmann, L. P. and Krasner, L. *Case studies in behavior modification*, New York: Holt, Rinehart and Winston, Inc., 1966.

Ulrich, R., Stachnik, T., and Mabry, J. (Eds.), *Control of human behavior*. Glenview, Ill.: Scott, Foresman, 1966.

Whaley, D. and Tough, J. Treatment of a self-injuring mongoloid with shock-induced suppression and avoidance. *Michigan Mental Health Research Bulletin*, 1968, 2, 33-35.

Wolf, M. M., Risley, T., and Mees, H. Application of operant conditioning procedures to the behavior problems of an autistic child. *Behavior Research and Therapy*, 1964, 1, 305-312.

FOOTNOTE

[1] A General Radio Co. sound level meter type 1551-C was used to ascertain noise-level readings. Ambient noise level in the experimental area was measured at approximately 55 db. With the subject located 6 ft. from the recorder microphone, a noise level of 72 db was the lowest value sufficient to operate the recorder.

15   Treatment of Aggression: Aggression in Childhood

Aggression in childhood is primarily constructive and only secondarily destructive, hostile, and guilt-arousing.

Freud offered the theory that human instincts may be classified as love (libido) and aggression, the first as the erotic instinct being directed toward the furtherance of life, both in object or love relationship and reproduction; and the second or death instinct expressing itself as an ego instinct primarily directed against the ego itself, but also as "an instinct of destruction directed against the external world and other living organisms" (1). Paul Schilder, in the recently published book, *The Goals and Desires of Man*, refuted Freud's theory of the death instinct and primary aggression, showing that activity and aggression are primarily constructive, are an expression of an interest in the construction of the object and in object relationship. Where his conclusions have related to children they were based on work done with Sylvan Keiser, Frank J. Curran, David Wechsler and myself (4) (5) (6), and included studies on aggressiveness in children, which led to a repeated emphasis on the basically constructive tendencies in human activities, specifically in children.

However, the belief in instinctual destructive aggression in children has been advanced by many in the field of child psychology, psychopathology and pedagogy. The child is seen as instinctively destructive, and needing to acquire inhibitions, sublimation and discipline for inborn infantile hostility and death wishes.

Melanie Klein (7) postulates that the infant before the end of the first year has aggressive impulses against the mother, especially the inside of the mother, and that this is the origin of the sense of guilt although primarily directed against the organism itself and giving rise to anxiety as an awareness of a sense of danger.

From the *American Journal of Orthopsychiatry, 13,* 1943, 392-399. Reprinted by permission of the American Orthopsychiatric Association.

Anna Freud expresses her belief in the instinctual aggression of children in the Anna Freud-Dorothy Burlingham report of the Hampstead Nurseries of April 1942 (8). "It is one of the recognized aims of education to deal with the aggressiveness of the child's nature, i.e., in the course of the first four or five years, to change the child's own attitude towards these impulses in himself. . . . They are usually first restricted and then suppressed by commands and prohibitions, a little later repressed, which means it disappears from the child's consciousness. . . . The danger (i.e., from present war) lies in the fact that the destruction raging in the outer world may meet the very real aggressiveness which rages in the inside of the child. Children have to be safe-guarded against the primitive forces of the war, not because horrors and atrocities are so strange to them but because we want them at this decisive stage of their development to overcome and estrange themselves from the primitive and atrocious wishes of their infantile nature." Even Anna Freud's own reports of London children in these war times do not seem to confirm these theories. My own experiences in regard to New York children's reaction to the war will be discussed later.

As a result of our experience with problem children at Bellevue, we have understood "the behavior of children as a continuous process of trial and error, which leads to constructions and configurations as a basis for action. Behavior difficulties and neuroses (whether or not in terms of hostile or destructive aggression) are interruptions in this constructive, psychological process" (9).

The studies on the physical aggressiveness in children by Paul Schilder and myself in 1936 (4) led to these conclusions: aggressiveness has close relation to motor drives and undoubtedly has its foundation in the organic structure. The aggressiveness on an organic basis (with hyperkinesis) is mostly more diffuse, whereas psychic traumas lead to aggression in relation to a specific situation. Deprivation of love and food increases the aggressive tendencies in children. Aggression results from experimentation with objects in the external world. Destruction may become a direct aim but it is never a final one; construction is the final aim. The psychological situation of the child leads to the final crystallization of his aggressive attitudes which are gradually organized into the socially accepted concepts and ideologies then available. Studies by Paul Schilder and David Wechsler (6) have also shown that the child's concept of death includes the belief that the dead will live again. In children's play, killed soldiers always get up and fight again.

My contribution today will include (1) an analysis of my clinical experience with hostile and destructive aggression in children as a protest against frustrating experiences which interfere with the normal process of personality development, and (2) a discussion of children's reactions to the war as illustrative of their aggressive and social tendencies and attitudes.

It is natural that aggression should be the most common complaint against the children who are sent to the Bellevue Children's Observation Ward. There have been about 4000 in the eight years I have been there. Several children, ages eleven to eighteen months, biologically and intellectually normal, have been referred to us because of dangerous head banging, body rocking, screaming, chewing up their cribs, biting themselves and others, and a spitting and gagging resistance to feeding. Their aggressiveness against people and other objects in the environment was limited only by their inability to reach them. The picture they present closely resembles in overt behavior the infantile fantasy life which is postulated by Anna Freud. The problems were usually similar: a deserted child, who has been given nursery crib care in a temporary shelter pending placement, following an earlier period of gross neglect. The child had reached the end of the first year at a time when he should be getting on his feet, striving toward independent locomotion, independent feeding, habit training and beginning language development, but on the background of a secure relationship and identification with a loved mother. These children are driven by their impulses to action and to experience the patterned behavior suitable for their maturation level, and are frustrated by the severe physical restrictions, their own lack of emotional maturity, and the opportunity for newly organized habits and play. Their activity is characterized by a hostile aggression directed against themselves and against the narrow environment in which they are confined. A change in behavior occurs rather rapidly when the child is given a foundation of reliable, affectionate care *from one individual* and is pushed forward into active patterned behavior in locomotion, play with toys, habit training and contacts with other people.

Somewhat similar but less severe problems in children of this age occur because mothers are either oversolicitous in giving their child the right routine physical care and try not to "spoil the child," or the mothers are ambivalent and do not recognize their actual rejection of their child. In either case there is produced a barrier between mother and child, with too much physical confinement. The objective concept of John Watson (10) that the infant reacts with rage to his futile efforts to overcome a frustrating restraint to his motor activity is descriptive of the situation.

Melanie Klein's theory that anxiety springs directly from aggression is conclusively disproved in our clinical experience with children raised in institutions through the first three or four years without any object relationship with any one person. Such children are characteristically aggressive psychopathic personalities, totally lacking in anxiety. Extensive experience with all types of personality deviations in children, due to various degrees of deprivations in love object relationships in the infantile period, convincingly shows that anxiety is possible only in children who *have experienced* a relationship and identification with a personal love object; a threat to this relationship is the source of anxiety. The aggression of psychopathic children is typically diffuse. It aims at getting some satisfaction from the outside physical or social world. It is also destructive; the object has no value for them, since they have no capacity for object relation or for identification.

My own conception (11) of the hyperkinesis and aggressiveness of the postencephalitic children whom I have seen is that this increased activity is an effort to overcome the incapacity of the brain to integrate the perceptual patterns. There is an intense need to perceive, contact, and experience the social and physical world, and sometimes literally to incorporate it in order to get meanings and values which can be utilized in an effort to integrate. It should also be emphasized that these children with pathology in the midbrain, resulting in increased aggressive behavior and aggressive fantasies, do not as a rule experience anxiety, although their aggressiveness is of the most impulsive uninhibited type.

One of the most illuminating cases in the study of the aggressive reaction to frustration was a four year old Negro girl brought to us because of baffling temper tantrums and exhausting fits of rage. This was the chief concern of the foster mother, although the child could not talk and she was aware of some motor disability because of a head shaking. The child sought the protection of walls or corners or squatted on the floor if she could not find a basic security, preferably in some motherly lap. Any effort at activity directed at normal locomotion, self-feeding, play or gestures intimating personal relationships would precipitate her aggressive outbursts. When we realized that the child was suffering from showers of myoclonic epileptic spasms, precipitated by an initiation of activity and simultaneously interfering with the activity, the significance of her frustration was appreciated. Her aggressiveness, which as often as not was turned against herself, was her protest against a total inability to experience new patterns of behavior which her developing brain demanded. Here the frustration was localized at the origin of the motor impulse while the perceptual experiences of the child, indicating the level of maturation, remained normal. Her aggressiveness could be reduced as soon as we understood the problem and permitted her a base for her own motor activities, preferably a maternal lap with the reassuring emotional personal relationship. Thus we could make available to her patterns of experience which she could utilize with minimum stimulus and minimal demands on her impaired motor functions.

This case serves to remind us that in every child there are discrepancies in the development of the various fields of action which vary from period to period, and that frustration arising from inadequate functioning of the body itself, from the child's limited capacity to investigate the physical world, and limited experiences in social relationships, are an inevitable part of reality. These are undoubtedly the source of temper tantrums which belong to the age of most rapid growth and the source of some amount of hostile and destructive aggression with which every child must deal.

From the age of three to six, and sometimes eight, the sources of hostile and destructive aggression begin to draw on the more intricate social relationships of members of the family group, relating to the Oedipus situation. A six year old boy of sexually mismated parents, with an older sister disabled by encephalitis, was brought to us with a description of behavior which led to the question of organic brain pathology in the boy also. He clung to his mother, refusing to go to school, but demonstrating hyperkinetic and aggressive behavior which was destructive of everything about him and all of his relationships. It was found that he had been the object of satisfaction for his mother's erotic needs to a very unusual degree. He responded to increased sexual stimulation by secural activity. There were secual fantasies mostly in relation to his mother, and aggressive fantasies against the father, who was a marine. During occasional visits home he had sexual relations with the mother, the boy being in the bed. The boy responded, as he told us, "When I saw it I felt mad; I got an ache in my head; I banged the walls and rang the outside bells so that all the people in the house had to get up, and went out and started fights." There was exaggeration in all his aggressive fantasies and dreams, his destructiveness, his refusal to go to school and clinging to his mother, insisting that he could only go to sleep with his hand on her breast. The combination of infantile dependence on the mother and aggressiveness against the father, with more or less overt sexual drives and refusal to grow up into the role of a school boy, presents an exaggerated picture of the Oedipus situation. Similar problems of a lesser degree can occur in any family group due to lesser difficulties in the various relationships between the members of the family, creating frustrations in the child which must find some solution.

In years of trying to understand numerous apparently impulsive and unmotivated fights between children, I have found that many of them, at least under our observation, are explained by one child with the statement that another child cursed his mother. As often as not the only source of this perceptual experience was the mere proximity of two tense, anxious, frustrated children, separated for the time being from their very insecure love object (the mother), igniting the unresolved Oedipus conflict. Each identified himself with the other protesting, frustrated child, and seemed to project both the cursing of the mother and the aggression against themselves out on the other child, and find some mutual relief in a physical combat which is also a physical contact.

Frequently Negro boys of unmarried mothers, living under the conditions of totally inadequate cultural, social, recreational and educational opportunities which our present society offers them, are sent to us because of menacing outbursts of violence and aggression on a background of blocked, inhibited, mute behavior. These boys respond wholeheartedly to offered opportunities for patterned activity in all the fields which are necessary for the normal development of a boy, especially in personal relationships.

In the latent stage other factors which lead to aggressiveness are the reading disabilities in children in our large educational systems. I see this problem as a developmental lag in patterned behavior mostly in the symbolic language field, but overflowing into other fields of behavior. It is easy to understand that a disorganized over-activity will lead to more errors than successes in the trial and error method of constructive aggression, especially in the highly competitive social milieu of the school child.

Some of the most all-out destructively aggressive behavior we have seen was in boys and girls of late prepuberty or beginning puberty who have spent their life in trying to run back to aggressively rejecting mothers. Sent from one institution to another, ultimately finding themselves on our wards as a last resort, has brought out a "to the death" defiance. Whenever we attempt any effort directed at their own welfare, such as attending a dental clinic, or a staff conference, going out of the hospital with a social worker to get new clothes, or especially a return to court for disposition or to some more confining institution, there results precipitous chair throwing, head butting, kicking, pounding and screaming, refusal to eat, threats to jump from the roof, and all kinds of open homicidal and suicidal threats. Some of these emotional outbursts threaten the life of the child. One girl passed into a state of benign stupor and others have shown states of exhausting excitement, although it was certain that they were not psychotic. Their one unmodifiable demand was to go home to their mothers. These children protest against entering into adolescence when even their infantile period was totally frustrating and all of their life experience had given them no substance for development. Their latent stage is usually characterized by the type of asocial and aggressive behavior readily expressed during that period. No satisfactory solution has ever been found. Sometimes by some compromise the child may be placed with the mother. This solves the blind emotional conflict for the time, but there soon occurs a more highly organized form of asocial behavior which makes more confining supervised care necessary.

The treatment of aggression in childhood should be constructive and preventive, i.e., the child should be treated constructively in order to prevent or ameliorate the developmental discrepancies in whatever field they may lie.

I have not attempted to demonstrate every type of childhood destructive and hostile aggression, but rather to give illustrative examples at various age levels in order to show certain characteristic problems of the developmental crises. I wanted also to indicate that the sources of frustration may arise from every field of the child's totally developing personality—such as brain pathology disturbing motor functions or impulse control or perceptual integration; developmental lags in language; deprivation in love objects in the infantile period, and relative deprivations in love in the child-parent relationships; and deprivation in social, educational and cultural opportunities. Many other conditions might have been emphasized.

Accordingly, I conclude that aggression in childhood is a symptom complex resulting from deprivations which cause developmental discrepancies in the total personality structures, such that the constructive patterned drives for action of the child find inadequate means of satisfaction, and result in amplification or disorganization of the drives into hostile or destructive aggression.

The war's effect on aggressiveness in children is a timely subject. John Frosch, William Q. Wolfson and myself have already made reports on the effect of the war on the psychology and behavior of children brought to Bellevue Children's Observation Ward. The results were not those which would be anticipated from Anna Freud's theories, that "the danger lies in the fact that the destruction raging in the outer world may meet the real aggressiveness which rages in the inside of the child."

Our studies showed no marked anxieties or increased aggression directly as the result of the war. Clinical pictures solely determined by the war were absent. Children with previous neurotic complexes showed some tendency to weave them into the war situation. There was some evidence that the war had some influence on the play and fantasy life of children in that they seemed to be experimenting with new structural patterns and the family group was threatened. There was a tendency to reactivate children who have previously been threatened. The children's art work was studied by analysis of drawings of boats before and after the onset of war. It was found that the boat was a universal symbol for the mother, the father was in the background, usually as the sun, and the child was inside the boat. After the declaration of war there were more highly integrated compositions with emphasis on the father symbols in the family scene. There was also more emphasis on identifying the boat, with a flag, symbol or name, and, symbolically, the family unit and therefore the things it stood for, or its ideology. In other words, there was evidence of an increase in constructive psychological processes.

Consideration of a group of children similar to those reported on by John Frosch and myself in May 1941, for these first months of the second year of the war (December 1942–January 1943), indicates some changes in the children's psychological reactions. For the majority of children in New York City, the threats against the family setting have not materialized, and the child has given up any concern the war may have caused at first. A child's composite war ideology is: yes, there is a war against Japan and Germany. There is a good deal of doubt as to the cause of the war. It is a bad thing because too many people are killed; it should be stopped but the best way to stop it is to fight with our superior strength; although some suggest that we just quit. The child does not want to go to war, but if he were of age he would go. The Japs and Germans should be killed—this, however, refers to the soldiers; the civilians should be saved. There is no evidence that the children get any satisfaction in fantasying the aggressiveness in the war; no expression of hatred, and little of fear.

In their dreams several children have taken part in the war, accompanying soldiers, although they themselves remain children, and kill the enemy in hand to hand combat with gun or knife. Sometimes the enemy has been German, sometimes Jap, and once Indian. After each death the child sat down and cried because he felt that the enemy soldier wanted his life, too. Then the child said to himself—no, he is not sorry, because the enemy is bad, he started it. In each case the child was killed at the end of his dreams. A seven year old aggressive girl dreamt that her brother joined the army—so she joined the marines and shouted, "Here I am ready to kill the Japs and Germans but I'd rather not." However, this was an enuretic dream and when she awoke and found she had not wet the bed she said, "It is all right to join the marines if you don't go to sea." Thus one sees how the children weave the war into their own personal problems.

In children, socially maladjusted due to family inadequacies, the child's chief aim is for identification with the war effort. A psychopathic foundling boy fantasied or confabulated a soldier father in uniform, a Red Cross nurse for a mother in Pearl Harbor, and himself a Polish refugee in all sorts of heroic experiences in winning this war. Children excuse their parent's inadequacies on the basis that they are so busy as minute men selling war bonds, or as students in first aid courses, that they cannot take care of them. A boy seeking to find some way to accept his mother's lovers found security in this relationship when the lovers went to war, by referring to them as "uncles," and accepting them as objects of identification. A boy of mixed parentage, with a Catholic Italian father, and a German Jewish mother, found himself in a whirlpool of conflicts from which he emerged stating that each parent denied his religion and that both hated Jews and Catholics alike; that he had uncles who were Filipinos and heroes in the war, and that he himself was a Filipino. The problems of these children can be summarized as a need to identify themselves with whatever constructive patterns the world at war has to offer them, to compensate for their own disturbed life's experiences. When the identification is difficult the child must deal with excessive unorganized impulses which lead to aggressiveness, hostility and anxiety. The increased activity in intersocial relationships in the world reflects itself in the increased activity in psychological processes in the child which, if they find a pattern for expression, are more likely to be constructive than destructive.

BIBLIOGRAPHY

1. Freud, Sigmund. *The ego and the id*. Hogarth Press, London, 1935.

2. ————. *Beyond the pleasure principle*. Hogarth Press, London, 1935.

3. Schilder, Paul. *Goals and desires of man*. Columbia University Press, New York, 1942.

4. Bender, Lauretta, Schilder, Paul, and Keiser, Sylvan. Studies in aggressiveness. *Genetic Psychology Monograph*, XVIII: No. 5 and 6, pp. 361-564, 1936.

5. Curran, Frank J., and Schilder, Paul. A constructive approach to the problems of childhood and adolescence. *J. Criminal Psychopathology*, II: pp. 125-142, 305-320, 1940-41.

6. Schilder, Paul, and Wechsler, David. Attitude of children toward death. *J. Genetic Psychology*, XLV: pp. 406-451, 1934.

7. Klein, Melanie. *The psychoanalysis of children*. Hogarth Press, London, 1937.

8. Anna Freud-Dorothy Burlingham. *Reports from Hampstead nurseries*. April, 1942. Foster Parents Plan for Children, 55 West 42nd St., N.Y.C.

9. Schilder, Paul. Vita and bibliography by Paul Schilder. *J. Criminal Psychopathology*, II: pp. 221-234, 1940.

10. Watson, John. *Behaviorism*. W. W. Norton and Co., New York, 1925.

11. Bender, Lauretta. Post-encephalitic behavior disorders in childhood. Josephine G. Neal, *Encephalitis*, Grune and Stratton, New York, 1942.

12. Bender, Lauretta, and Frosch, John. Children's reactions to the war. *Am. J. Orthopsychiatry*, XII: 3, pp. 571-586, 1942.

13. Bender, Lauretta, and Wolfson, William Q. The significance of the nautical theme in the art of children. *Am. J. Orthopsychiatry*, XIII, 1943.

16　The Moral Equivalent of War

—WILLIAM JAMES

The war against war is going to be no holiday excursion or camping party. The military feelings are too deeply grounded to abdicate their place among our ideals until better substitutes are offered than the glory and shame that come to nations as well as to individuals from the ups and downs of politics and the vicissitudes of trade. There is something highly paradoxical in the modern man's relation to war. Ask all our millions, north and south, whether they would vote now (were such a thing possible) to have our war for the Union expunged from history and the record of a peaceful transition to the present time substituted for that of its marches and battles, and probably hardly a handful of eccentrics would say yes. Those ancestors, those efforts, those memories and legends, are the most ideal part of what we now own together, a sacred spiritual possession worth more than all the blood poured out. Yet ask those same people whether they would be willing in cold blood to start another civil war now to gain another similar possession, and not one man or woman would vote for the proposition. In modern eyes, precious though wars may be, they must not be waged solely for the sake of the ideal harvest. Only when forced upon one, only when an enemy's injustice leaves us no alternative, is a war now thought permissible.

　　It was not thus in ancient times. The earlier men were hunting men, and to hunt a neighboring tribe, kill the males, loot the village and possess the females, was the most profitable, as well as the most exciting, way of living. Thus were the more martial tribes selected, and in chiefs and peoples a pure pugnacity and love of glory came to mingle with the more fundamental appetite for plunder.

　　Modern war is so expensive that we feel trade to be a better avenue to plunder; but modern man inherits all the innate pugnacity and all the love of glory of his ancestors. Showing war's irrationality and horror is of no effect upon him. The horrors make the fascination.

Reprinted by permission of Alexander R. James, Literary Executor. This article first appeared in a 1910 publication of the Association for International Conciliation.

War is the *strong* life; it is life *in extremis*; war-taxes are the only ones men never hesitate to pay, as the budgets of all nations show us.

History is a bath of blood. The Iliad is one long recital of how Diomedes and Ajax, Sarpedon and Hector *killed*. No detail of the wounds they made is spared us, and the Greek mind fed upon the story. Greek history is a panorama of jingoism and imperialism—war for war's sake, all the citizens being warriors. It is horrible reading, because of the irrationality of it all—save for the purpose of making 'history'—and the history is that of the utter ruin of civilization in intellectual respects perhaps the highest the earth has ever seen.

Those wars were purely piratical. Pride, gold, women, slaves, excitement, were their only motives. In the Peloponnesian war, for example, the Athenians ask the inhabitants of Melos (the island where the 'Venus of Milo' was found), hitherto neutral, to own their lordship. The envoys meet, and hold a debate which Thucydides gives in full, and which, for sweet reasonableness of form, would have satisfied Matthew Arnold. 'The powerful exact what they can,' said the Athenians, 'and the weak grant what they must.' When the Meleans say that sooner than be slaves they will appeal to the gods, the Athenians reply: 'Of the gods we believe and of men we know that, by a law of their nature, wherever they can rule they will. This law was not made by us, and we are not the first to have acted upon it; we did but inherit it, and we know that you and all mankind, if you were as strong as we are, would do as we do. So much for the gods; we have told you why we expect to stand as high in their good opinion as you.' Well, the Meleans still refused, and their town was taken. 'The Athenians,' Thucydides quietly says, 'thereupon put to death all who were of military age and made slaves of the women and children. They then colonized the island, sending thither five hundred settlers of their own.'

Alexander's career was piracy pure and simple, nothing but an orgy of power and plunder, made romantic by the character of the hero. There was no rational principle in it, and the moment he died his generals and governors attacked one another. The cruelty of those times is incredible. When Rome finally conquered Greece, Paulus Aemilius was told by the Roman Senate to reward his soldiers for their toil by 'giving' them the old kingdom of Epirus. They sacked seventy cities and carried off a hundred and fifty thousand inhabitants as slaves. How many they killed I know not; but in Etolia they killed all the senators, five hundred and fifty in number. Brutus was 'the noblest Roman of them all', but to reanimate his soldiers on the eve of Philippi he similarly promises to give them the cities of Sparta and Thessalonica to ravage, if they win the fight.

Such was the gory nurse that trained societies to cohesiveness. We inherit the warlike type; and for most of the capacities of heroism that the human race is full of we have to thank this cruel history. Dead men tell no tales, and if there were any tribes of other type than this they have left no survivors. Our ancestors have bred pugnacity into our bone and marrow, and thousands of years of peace won't breed it out of us. The popular imagination fairly fattens on the thought of wars. Let public opinion once reach a certain fighting pitch, and no ruler can withstand it. In the Boer War both governments began with bluff but couldn't stay there; the military tension was too much for them. In 1898 our people had read the word 'war' in letters three inches high for three months in every newspaper. The pliant politician McKinley was swept away by their eagerness, and our squalid war with Spain became a necessity.

At the present day, civilized opinion is a curious mental mixture. The military instincts and ideals are as strong as ever, but are confronted by reflective criticisms which sorely curb their ancient freedom. Innumerable writers are showing up the bestial side of military service. Pure loot and mastery seem no longer morally avowable motives, and pretexts must be found for attributing them solely to the enemy. England and we, our army and navy authorities repeat without ceasing, arm solely for 'peace', Germany and Japan it is who are bent on loot and glory. 'Peace' in military mouths today is a synonym for 'war expected'. The word has become a pure provocative, and no government wishing peace sincerely should allow it ever to be printed in a newspaper. Every up-to-date dictionary should say that 'peace' and 'war' mean the same thing, now *in posse*, now *in actu*. It may even reasonably be said that the intensely sharp competitive *preparation* for war by the nations *is the real war*, permanent, unceasing; and that the battles are only a sort of public verification of the mastery gained during the 'peace'-interval.

It is plain that on this subject civilized man has developed a sort of double personality. If we take European nations, no legitimate interest of any one of them would seem to justify the tremendous destructions which a war to compass it would necessarily entail. It would seem as though common sense and reason ought to find a way to reach agreement in every conflict of honest interests. I myself think it our bounden duty to believe in such international rationality as possible.

But, as things stand, I see how desperately hard it is to bring the peace-party and the war-party together, and I believe that the difficulty is due to certain deficiencies in the program of pacificism which set the militarist imagination strongly, and to a certain extent justifiably, against it. In the whole discussion both sides are on imaginative and sentimental ground. It is but one utopia against another, and everything one says must be abstract and hypothetical. Subject to this criticsm and caution, I will try to characterize in abstract strokes the opposite imaginative forces, and point out what to my own very fallible mind seems the best utopian hypothesis, the most promising line of conciliation.

In my remarks, pacifist though I am, I will refuse to speak of the bestial side of the war-*regime* (already done justice to by many writers) and consider only the higher aspects of militaristic sentiment. Patriotism no one thinks discreditable; nor does anyone deny that war is the romance of history. But inordinate ambitions are the soul of every patriotism, and the possibility of violent death the soul of all romance. The militarily patriotic and romantic-minded everywhere, and especially the professional military class, refuse to admit for a moment that war may be a transitory phenomenon in social evolution. The notion of a sheep's paradise like that revolts, they say, our higher imagination. Where then would be the steeps of life? If war had ever stopped, we should have to re-invent it, on this view, to redeem life from flat degeneration.

Reflective apologists for war at the present day all take it religiously. It is a sort of sacrament. Its profits are to the vanquished as well as to the victor; and quite apart from any question of profit, it is an absolute good, we are told, for it is human nature at its highest dynamic. Its 'horrors' are a cheap price to pay for rescue from the only alternative supposed, of a world of clerks and teachers, of co-education and zoophily, of 'consumers' leagues' and 'associated charities', of industrialism unlimited, and feminism unabashed. No scorn, no hardness, no valor any more! Fie upon such a cattleyard of a planet!

So far as the central essence of this feeling goes, no healthy-minded person, it seems to me, can help to some degree partaking of it. Militarism is the great preserver of our ideals of hardihood, and human life with no use for hardihood would be contemptible. Without risks or prizes for the darer, history would be insipid indeed; and there is a type of military character which everyone feels that the race should never cease to breed, for everyone is sensitive to its superiority. The duty is incumbent on mankind, of keeping military characters in stock—of keeping them, if not for use, then as ends in themselves and as pure pieces of perfection—so that Roosevelt's weaklings and mollycoddles may not end by making everything else disappear from the face of nature.

This natural sort of feeling forms, I think, the innermost soul of army-writings. Without any exception known to me, militarist authors take a highly mystical view of their subject, and regard war as a biological or sociological necessity, uncontrolled by ordinary psychological checks and motives. When the time of development is ripe the war must come, reason or no reason, for the justifications pleaded are invariably fictitious. War is, in short, a permanent human *obligation*. General Homer Lea, in his recent book *The Valor of Ignorance*,* plants himself squarely on this ground. Readiness for war is for him the essence of nationality, and ability in it the supreme measure of the health of nations.

Nations, General Lea says, are never stationary—they must necessarily expand or shrink, according to their vitality or decreptitude. Japan now is culminating; and by the fatal law in question it is impossible that her statesmen should not long since have entered, with extraordinary foresight, upon a vast policy of conquest—the game in which the first moves were her wars with China and Russia and her treaty with England, and of which the final objective is the capture of the Philippines, the Hawaiian Islands, Alaska, and the whole of our coast west of the Sierra Passes. This will give Japan what her ineluctable vocation as a state absolutely forces her to claim, the possession of the entire Pacific Ocean; and to oppose these deep designs we Americans have, according to our author, nothing but our conceit, our ignorance, our commercialism, our corruption, and our feminism. General Lea makes a minute technical comparison of the military strength which we at present could oppose to the strength of Japan, and concludes that the islands, Alaska, Oregon, and Southern California, would fall almost without resistance, that San Francisco must surrender in a fortnight to a Japanese investment, that in three or four months the war would be over, and our republic, unable to regain what it had heedlessly neglected to protect sufficiently, would then 'disintegrate', until perhaps some Caesar should arise to weld us again into a nation.

A dismal forecast indeed! Yet not unplausible, if the mentality of Japan's statesmen be of the Caesarian type of which history shows so many examples, and which is all that General Lea seems able to imagine. But there is no reason to think that women can no longer be the mothers of Napoleonic or Alexandrian characters; and if these come in Japan and find their opportunity, just such surprises as *The Valor of Ignorance* paints may lurk in ambush for us. Ignorant as we still are of the inner-most recesses of Japanese mentality, we may be foolhardy to disregard such possiblities.

Other militarists are more complex and more moral in their considerations. The *Philosophie des Krieges*, by S. R. Steinmetz, is a good example. War, according to this author, is an ordeal instituted by God, who weighs the nations in its balance. It is the essential form of the State, and the only function in which peoples can employ all their powers at once and convergently. No victory is possible save as the resultant of a totality of virtues, no defeat for which some vice or weakness is not responsible. Fidelity, cohesiveness, tenacity, heroism, conscience, education, inventiveness, economy, wealth, physical health and vigor—there isn't a moral or intellectual point of superiority that doesn't tell, when God holds his assizes and hurls the peoples upon one another. *Die Weltgeschichte ist das Weltgericht*;[1] and Dr. Steinmetz does not believe that in the long run chance and luck play any part in apportioning the issues.

The virtues that prevail, it must be noted, are virtues anyhow, superiorities that count in peaceful as well as in military competition; but the strain on them, being infinitely intenser in the latter case, makes war infinitely more searching as a trial. No ordeal is comparable to its winnowings. Its dread hammer is the welder of men into cohesive states, and nowhere but in such states can human nature adequately develop its capacity. The only alternative is 'degeneration'.

Dr. Steinmetz is a conscientious thinker, and his book, short as it is, takes much into account. Its upshot can, it seems to me, be summed up in Simon Patten's word, that mankind was nursed in pain and fear, and that the transition to a 'pleasure-economy' may be fatal to a being wielding no powers of defence against its disintegrative influences. If we speak of the *fear of emancipation from the fear-régime*, we put the whole situation into a single phrase; fear regarding ourselves now taking the place of the ancient fear of the enemy.

Turn the fear over as I will in my mind, it all seems to lead back to two unwillingnesses of the imagination, one aesthetic, and the other moral; unwillingness, first to envisage a future in which army-life, with its many elements of charm, shall be forever impossible, and in which the destinies of peoples shall nevermore be decided as quickly, thrillingly, and tragically, by force, but only gradually and insipidly by 'evolution'; and, secondly, unwillingness to see the supreme theater of human strenuousness closed, and the splendid military aptitudes of men doomed to keep always in a state of latency and never show themselves in action. These insistent unwillingnesses, no less than other aesthetic and ethical insistencies, have, it seems to me, to be listened to and respected. One cannot meet them effectively by mere counter-insistency on war's expensiveness and horror. The horror makes the thrill; and when the question is of getting the extremest and supremest out of human nature, talk of expense sounds ignominious. The weakness of so much merely negative criticism is evident—pacificism makes no converts from the military party. The military party denies neither the bestiality nor the horror, nor the expense; it only says that these things tell but half the story. It only says that war is *worth* them; that, taking human nature as a whole, its wars are its best protection against its weaker and more cowardly self, and that mankind cannot *afford* to adopt a peace-economy.

Pacificists ought to enter more deeply into the aesthetical and ethical point of view of their opponents. Do that first in any controversy, says J. J. Chapman, *then move the point*, and your opponent will follow. So long as antimilitarists propose no substitute for war's disciplinary function, no *moral equivalent* of war, analogous, as one might say, to the mechanical equivalent of heat, so long they fail to realize the full inwardness of the situation. And as a rule they do fail. The duties, penalties, and sanctions pictured in the utopias they paint are all too weak and tame to touch the military-minded. Tolstoy's pacificism is the only exception to this rule, for it is profoundly pessimistic as regards all this world's values, and makes the fear of the Lord furnish the moral spur provided elsewhere by the fear of the enemy. But our socialistic peace-advocates all believe absolutely in this world's values; and instead of the fear of the Lord and the fear of the enemy, the only fear they reckon with is the fear of poverty if one be lazy. This weakness pervades all the socialistic literature with which I am acquainted. Even in Lowes Dickinson's exquisite dialogue, high wages and short hours are the only forces invoked for overcoming man's distaste for repulsive kinds of labor. Meanwhile men at large still live as they always have lived, under a pain-and-fear economy—for those of us who live in an ease-economy are but an island in the stormy ocean—and the whole atmosphere of present-day utopian literature tastes mawkish and dishwatery to people who still keep a sense for life's more bitter flavors. It suggests, in truth, ubiquitous inferiority.

Inferiority is always with us, and merciless scorn of it is the keynote of the military temper. 'Dogs, would you live forever?' shouted Frederick the Great. 'Yes,' say our utopians, 'let us live forever, and raise our level gradually.' The best thing about our 'inferiors' today is that they are as tough as nails, and physically and morally almost as insensitive. Utopianism would see them soft and squeamish, while militarism would keep their callousness, but transfigure it into a meritorious characteristic, needed by 'the service', and redeemed by that from the suspicion of inferiority. All the qualities of a man acquire dignity when he knows that the service of the collectivity that owns him needs them. If proud of the collectivity, his own pride rises in proportion. No collectivity is like an army for nourishing such pride; but it has to be confessed that the only sentiment which the image of pacific cosmopolitan industrialism is capable of arousing in countless worthy breasts is shame at the idea of belonging to *such* a collectivity. It is obvious that the United States of America as they exist today impress a mind like General Lea's as so much human blubber. Where is the sharpness and precipitousness, the contempt for life, whether one's own, or another's? Where is the savage 'yes' and 'no', the unconditional duty? Where is the conscription? Where is the blood-tax? Where is anything that one feels honored by belonging to?

Having said thus much in preparation, I will now confess my own utopia. I devoutly believe in the reign of peace and in the gradual advent of some sort of a socialistic equilibrium. The fatalistic view of the war-function is to me nonsense, for I know that war-making is due to definite motives and subject to prudential checks and reasonable criticisms, just like any other form of enterprise. And when whole nations are the armies, and the science of destruction vies in intellectual refinement with the sciences of production, I see that war becomes absurd and impossible from its own monstrosity. Extravagant ambitions will have to be replaced by reasonable claims, and nations must make common cause against them. I see no reason why all this should not apply to yellow as well as to white countries, and I look forward to a future when acts of war shall be formally outlawed as between civilized peoples.

All these beliefs of mine put me squarely into the antimilitarist party. But I do not believe that peace either ought to be or will be permanent on this globe, unless the states pacifically organized preserve some of the old elements of army-discipline. A permanently successful peace-economy cannot be a simple pleasure-economy. In the more or less socialistic future toward which mankind seems drifting we must still subject ourselves collectively to those severities which answer to our real position upon this only partly hospitable globe. We must make new energies and hardihood continue the manliness to which the military mind so faithfully clings. Martial virtues must be the enduring cement; intrepidity, contempt of softness, surrender of private interest, obedience to command, must still remain the rock upon which states are built—unless, indeed, we wish for dangerous reactions against commonwealths fit only for contempt, and liable to invite attack whenever a center of crystallization for military-minded enterprise gets formed anywhere in their neighborhood.

The war-party is assuredly right in affirming and reaffirming that the martial virtues, although originally gained by the race through war, are absolute and permanent human goods. Patriotic pride and ambition in their military form are, after all, only specifications of a more general competitive passion. They are its first form, but that is no reason for supposing them to be its last form. Men now are proud of belonging to a conquering nation, and without a murmur they lay down their persons and their wealth, if by so doing they may fend off subjection. But who can be sure that *other aspects of one's country* may not, with time and education and suggestion enough, come to be regarded with similarly effective feelings of pride and shame? Why should men not some day feel that it is worth a blood-tax to belong to a collectivity superior in *any* ideal respect? Why should they not blush with indignant shame if the community that owns them is vile in any way whatsoever? Individuals, daily more numerous, now feel this civic passion. It is only a question of blowing on the spark till the whole population gets incandescent, and on the ruins of the old morals of military honor, a stable system of morals of civic honor builds itself up. What the whole community comes to believe in grasps the individual as in a vise. The war-function has grasped us so far; but constructive interests may some day seem no less imperative, and impose on the individual a hardly lighter burden.

Let me illustrate my idea more concretely. There is nothing to make one indignant in the mere fact that life is hard, that men should toil and suffer pain. The planetary conditions once for all are such, and we can stand it. But that so many men, by mere accidents of birth and opportunity, should have a life of *nothing else* but toil and pain and hardness and inferiority imposed upon them, should have *no* vacation, while others natively no more deserving never get any taste of this campaigning life at all—*this* is capable of arousing indignation in reflective minds. It may end by seeming shameful to all of us that some of us have nothing but campaigning, and others nothing but unmanly ease. If now—and this is my idea—there were, instead of military conscription a conscription of the whole youthful population to form for a certain number of years a part of the army enlisted against *Nature*, the injustice would tend to be evened out, and numerous other goods to the commonwealth would follow. The military ideals of hardihood and discipline would be wrought into the growing fiber of the people; no one would remain blind as the luxurious classes now are blind, to man's relations to the globe he lives on, and to the permanently sour and hard foundations of his higher life. To coal and iron mines, to freight-trains, to fishing fleets in December, to dish-washing, clothes-washing, and window-washing, to road-building and tunnel-making, to foundries and stoke-holes, and to the frames of skyscrapers, would our gilded youths be drafted off, according to their choice, to get the childishness knocked out of them, and to come back into society with healthier sympathies and soberer ideas. They would have paid their blood-tax, done their own part in the immemorial human warfare against nature; they would tread the earth more proudly, the women would value them more highly, they would be better fathers and teachers of the following generation.

Such a conscription, with the state of public opinion that would have required it, and the many moral fruits it would bear, would preserve in the midst of a pacific civilization the manly virtues which the military party is so afraid of seeing disappear in peace. We should get toughness without callousness, authority with as little criminal cruelty as possible, and painful work done cheerily because the tudy is temporary, and threatens not, as now, to degrade the whole remainder of one's life. I spoke of the 'moral equivalent' of war. So far, war has been the only force that can discipline a whole community, and until an equivalent discipline is organized, I believe that war must have its way. But I have no serious doubt that the ordinary prides and shames of social man, once developed to a certain intensity, are capable of organizing such a moral equivalent as I have sketched, or some other just as effective for preserving manliness of type. It is but a question of time, of skillful propagandism, and of opinion-making men seizing historic opportunities.

The martial type of character can be bred without war. Strenuous honor and disinterestedness abound elsewhere. Priests and medical men are in a fashion educated to it, and we should all feel some degree of it imperative if we were conscious of our work as an obligatory service to the state. We should be *owned*, as soldiers are by the army, and our pride would rise accordingly. We could be poor, then, without humiliation, as army officers now are. The only thing needed henceforward is to inflame the civic temper as past history has inflamed the military temper. H. G. Wells, as usual, sees the center of the situation. 'In many ways,' he says, 'military organization is the most peaceful of activities. When the contemporary man steps from the street of clamorous, insincere advertisement, push, adulteration, underselling, and intermittent employment into the barrackyard, he steps on to a higher social plane, into an atmosphere of service and cooperation and of infintely more honorable emulations. Here at least men are not flung out of employment to degenerate because there is no immediate work for them to do. They are fed and drilled and trained for better services. Here at least a man is supposed to win promotion by self-forgetfulness and not by self-seeking. And beside the feeble and irregular endowment of research by commercialism, its little short-sighted snatches at profit by innovation and scientific economy, see how remarkable is the steady and rapid development of method and appliances in naval and military affairs! Nothing is more striking than to compare the progress of civil conveniences which has been left almost entirely to the trader, to the progress in military apparatus during the last few decades. The house-appliances of today, for example, are little better than they were fifty years ago. A house of today is still almost as ill-ventilated, badly heated by wasteful fires, clumsily arranged and furnished as the house of 1858. Houses a couple of hundred years old are still satisfactory places of residence, so little have our standards risen. But the rifle or battleship of fifty years ago was beyond all comparison inferior to those we possess; in power, in speed, in convenience alike. No one has a use now for such superannuated things.'

Wells adds that he thinks that the conception of order and discipline, the tradition of service and devotion, of physical fitness, unstinted exertion and universal responsibility, which universal military duty is now teaching European nations, will remain a permenent acquisition, when the last ammunition has been used in the fireworks that celebrate the final peace. I believe as he does. It would be simply preposterous if the only force that could work ideals of honor and standards of efficiency into English or American natures should be the fear of being killed by the Germans or the Japanese. Great indeed is

Fear; but it is not, as our military enthusiasts believe and try to make us believe, the only stimulus known for awakening the higher ranges of men's spiritual energy. The amount of alteration in public opinion which my utopia postulates is vastly less than the difference between the mentality of those black warriors who pursued Stanley's party on the Congo with thier cannibal warcry of 'Meat! Meat!' and that of the 'general-staff' of any civilized nation. History has seen the latter interval bridged over: the former one can be bridged over much more easily.

FOOTNOTES

* Lea, *The Valor of Ignorance*. New York: Harper & Bros., 1909. THE EDITORS

[1] *Universal history is the tribunal of humanity.*

17 The Psychiatric Aspects of War and Peace

—*FRANZ ALEXANDER*

As the manifestation of man's destructive impulses, war appears as a legitimate subject of psychology. It would seem somewhat more dubious whether or not it also belongs within the more restricted scope of psychiatry—a field devoted to abnormal behavior. Certainly historical experience would not justify considering war an abnormal phenomenon.

In a letter to Einstein, Freud wrote: "Conflict of interest among mankind is in the main usually decided by the use of force. This is true of the whole animal kingdom from which mankind should not be excluded."[1] This is not an opinion but a terse statement of fact. From the point of view of psychiatry, it would seem that the pacifist, who thinks of the elimination of war as an actual possibility, might be considered a neurotic. He might easily be called a dreamer, subject to wishful thinking, who does not dare to face reality and who escapes into fantasy.

Freud in his letter to Einstein is fully aware of this. He sees the

From the *American Journal of Sociology*, 1941, *46*, 504-520. Reprinted by permission of The University of Chicago Press.

essence of cultural development in the "strengthening of the intellect, which begins to master the instincts, and the turning inward of the tendency of aggression with all its advantageous and dangerous results."[2] He even questions whether or not culture "is leading to the extinction of mankind since it encroaches upon the sexual function in more than one way, and even today uncultured races and the backward elements of the population increase more rapidly than the highly cultivated."[3] But no matter whether we consider the cultural domestication of the human animal as progress or degeneration, the simple statement remains unassailable that, as long as powerful nations or groups of people continue to pursue their interests by force, by destruction, and by subjugation of their opponents, pacifism must be considered as a morbid phenomenon—in fact, the rationalization of self-destructive wishes. In the present state of affairs, as long as pacifism is not being shared by all inhabitants of the earth, it is a renunciation of self-defense. If self-preservation is a fundamental and normal attribute of life, pacifism must be considered a morbid phenomenon—the manifestation of self-destructiveness. It cannot be excused even as the result of unclear thinking. Anyone who is blind to the ubiquitous manifestations of human aggressiveness in the past and present can be rightly considered a man who does not face reality. If he is not of subnormal intelligence—unable to grasp events around him—his inability to face facts must be of emotional origin, and he may be considered a neurotic.

All this preamble is to refute those who seek the causes of war in a specific emotional disturbance of the masses and refer to it as a mass psychosis. We might even go one step farther and question the existence of a real peace. Since we know that war has always been the usual way of settling conflicts between groups, we might ask how peace is possible at all. It cannot be assumed that during the brief periods of peace there were no conflicting interests between different national, economic, and political groups. The answer seems to be that, at least as far as the history of ancient and western civilization is concerned, periods of peace were nothing but preparations for a coming war. Those who may doubt the validity of this pessimistic interpretation of history should be reminded of the experiences of their own lifetime. Anyone who grew up before the World War* in any of the central European states should have come to the conviction that that period of peace was nothing but a preparation for the World War. A casual glance at different national budgets, compulsory military service, and the different moves of the diplomatic chess game must have convinced everyone of the validity of this statement. Nobody who lived in central Europe after the World War could have overlooked the fact that all these years were devoted to a preparation for revenge, a preparation which increased in tempo with the speed of a geometrical progression.

A student of European and ancient history can differentiate rightly only between periods of actual and latent war, the latter being misleadingly called periods of peace. This statement can hardly be refuted, no matter how disinclined we may be to accept it in this simple formulation. We struggle against this depressing view because it is so difficult to look upon human events objectively without wishful distortion. We easily confuse the state of affairs we should like to have with existing conditions. In the light of this formulation the history of civilized mankind is a history of wars interrupted by preparations for more wars.

In contrast to this pessimistic and realistic evaluation of the past, which is common among European historians, stands a more hopeful attitude, frequently expressed by statesmen, publicists, and social scientists in this country. These authors like to consider Europe a congested area populated by people who are divided into small economic, racial, and political units and, on account of this unfortunate situation, live in permanent strife with each other. According to them war is characteristic for that part of the world; that is to say, it is the disease of the old world but is not necessarily a fundamental, inevitable phenomenon of social life. The experience of this country, where 122,000,000 people live in a large economic unit in peace with each other and with neighboring nations,† would indicate that under the right conditions real peace is a possible state of affairs. The more cynical observer may answer: "Of course peace is possible while there are no adjacent groups with conflicting interests or the power relations between them are extremely unequal. But wait until the oceans no longer serve as imposing barriers. Asiatic, European, and American nations will become, from the sociological point of view, adjacent nations competing in the world-markets for world-trade." In fact, the American public is divided into two groups: those who with a keen sense already anticipate that the protective value of geographical distance will shrink and finally disappear, and those who complacently continue to put their faith in the vastness of the oceans. The introduction of compulsory military service and our expenditures for national defense are the best proofs of the prevalence of the more pessimistic view that peace is nothing but preparation for war.

No matter how convincing these arguments may be, American history could be considered as an experiment of nature demonstrating that peace which is not a preparation for war is a conceivable condition of human affairs. In this continent, since the Revolutionary War and its aftermath, the War of 1812, not considering the Civil War, there was no preparation for foreign wars and no standing armies worth mentioning. The Mexican and Spanish wars were mere incidental military operations. There was no emotional anticipation of a coming war—quite different from the state of mind in which people of Europe have been raised for centuries.

Since war is the more common phenomenon and real peace the extremely rare exception, it seems more promising to approach the problem of peace and war not by asking what are the causes of war but by studying the causes or, more precisely, the conditions of peace. If war is a permanent phenomenon of human history one might more easily expect an understanding of its deeper roots if one tries to establish those unusual conditions under which peace can exist. Since the advent of dynamic psychology it is customary to approach the problem of peace and war primarily from the point of view of the individual. It is, however, quite obvious that one cannot expect from the study of the individual's psychological makeup to establish those specific conditions under which nations can live beside each other peacefully, contrary to their habit of conducting wars against each other. One cannot well assume that people deriving from the same European stock here in America have, in a brief period, changed their psychological structure to such a degree as to explain the considerably long period of peace in this country. It is obvious that those social conditions, both national and international, under which the same types of individuals live must be responsible for war and peace. Since war is the common state of affairs, the question arises as to what those sociopolitical conditions are under which a peaceful attitude among population and leaders will prevail or, in other words, under which peaceful leaders and attitudes become popular. In this approach we are considering for the moment basic human nature as the constant factor and the sociological factors as the variables.

Contrary to a current view that war is the unavoidable manifestation of man's innate destructiveness, we have good reason to believe that both war and peace are compatible with human nature. In order to demonstrate the validity of this view the example of the Western Hemisphere appears promising, particularly the history of the United

States. It would seem that whenever organized groups (in this case the individual states) live beside each other without conflicting interests peace is possible. If in addition their interests are complementary to each other (in this case the economic conquest of a vast continent) the chances of a peaceful co-existence even increase. Under such conditions it is possible for the adjacent groups to fuse into one greater economic and political unit.

In the case of the Western Hemisphere the lack of conflicting interests is the explanation of peaceful development. This is borne out by the fact that as soon as a conflict of interests had arisen between two more or less distinguishable parts of the country, as between the South and the North, war became unavoidable. Of course the Civil War was but a single interruption of a long and peaceful development. The fact that the conflicting interests could not be adjusted by peaceful negotiation and compromise but had to be decided by the victory of the more powerful North led as an immediate result to the unavoidable impoverishment of the vanquished South. Since then a slow but gradual process of readjustment is in progress—quite different from the usual post-war situations in European history, where a peace settlement was nothing but the prelude to a war of revenge.

Obviously there must be powerful irrational forces counteracting the rational course of development which would lead to co-operation and organization of neighboring groups. The example of post-war Europe is most revealing as to the nature of these irrational forces. Small economic units which formerly were united in a larger political and economic framework became separated by the Treaty of Trianon, although they lacked all necessary prerequisites for separate existence. The treaties of Versailles and Trianon were the most disastrous steps toward creating such a morbid state of affairs. Small national units which had existed in the framework of the Austro-Hungarian empire were split up into independent nations. Instead of the rational needs of economic life, national pride and unrealistic racial constructions became the basis of the new order. This procedure was essentially the same as if the United States were suddenly to be divided according to its national groups into a little Czechoslovakia, Germany, Poland, Italy, Ireland, England, and Palestine. Obviously, militant nationalism was the irrational factor which hindered a peaceful reorganization of the European nations into an integrated politico-economic system.

All this would indicate that the process of social disintegration is the direct effect of certain dynamically powerful but irrational emotional factors hindering the creation of larger economic and social entities which would allow people of different language and race to live with each other in a peaceful co-operation.

If it were only a question of reason, it would not be difficult at all to work out a plan of organization, in any given continent, which would give different national units the security of economic and national existence. It is not for objective rational reasons that the Pan-Europe dream of a Briand, a Stresemann, or a Coudenhove-Kallergi could not be realized. It is obvious that the recognition of these irrational, emotional factors are of utmost importance. As Freud has stated, an enemy which one does not see cannot be defeated. The knowledge of those emotional factors which oppose a friendly and rational organization of international relationships is the first step toward understanding the problem of war and peace.

If real peace has ever existed in the history of mankind, which seems somewhat questionable, these emotional obstacles must have been absent during these periods. In order to find an answer we should turn our attention to the rare periods of peace and compare them with the usual state of war. History, however, does not give us any opportunity for such a study. American history appears more hopeful in this respect, but here again the objection might be raised that the relatively long periods of peace in international affairs was due to the geographical isolation of the United States and to the fact that the power relationships between the states and the adjacent nations have been so unequal. Furthermore, the conflict of interests between these different national units are smaller than anywhere else. The rarity of the population of the American continent may be quoted as an added factor. It also must be considered that great portions of this continent are still unexploited, which makes economic expansion possible without creating conflicts of interest among competing groups. As these favorable conditions are changing, as the uninhabited parts become populated, and the geographical isolation is diminishing, the possibility of war is increasing.

All this would indicate that it is not the lack of the disintegrating irrational emotional forces which is responsible for a relatively peaceful state of affairs in the Western Hemisphere but the lack of sufficiently strong conflicts. As far as such conflicts exist, the geographical obstacles against solving them by war are insuperable. If we accept Freud's statement that the conflict of interests among mankind is usually decided by the use of force, we should not be surprised to see absence of war where the conflict of interests is not excessive or where, owing to external circumstances, they cannot be settled by war. Therefore the study of the Western Hemisphere as compared with Europe does not offer a solution to our problem. As far as the psychological factors are concerned, we must find another approach. We have to look for an example of peaceful co-existence of human beings in which the conflicts are not decided by the use of force, in spite of the fact that there are conflicting interests, and the use of force is not physically impossible.

Every well-organized national unit offers such an example. Freud's above-quoted statement, like most generalizations, is ambiguous. He stated that a conflict of interest among mankind is usually decided by the use of force; but he obviously referred only to a conflict of interests in international relationships and not to conflicts between members of the same nation. Such conflicts usually are not settled by the use of force—certainly not by warlike violence—but by law. This discrepancy between the settling of intergroup conflicts and personal conflicts within a group is indeed an encouraging fact. If a man is able to live in a social system in which conflicting interests are decided by mediation and compromise, as it is done all over the civilized world and also in the so-called primitive societies, it seems to be possible that such a peaceful organization could be extended to regulate group relationships also.

Of course within a nation conflicting interests occasionally are settled by violence—as in civil wars. But while in the history of international relationship peace is the exception and war the rule, in the internal life of a nation peace is the rule and civil war the exception. The solution to our question as to the conditions of peace obviously lies in this contrast. Why do people accept a peaceful order in their national life but revert to violence to settle their international conflicts? Einstein in his correspondence with Freud on this problem seems to look primarily for an external reason. Internal order is based on the enforcement of law. Since international law cannot be enforced at present, international courts like the League of Nations are doomed to failure. Law without power of enforcement is ineffective. No international police force is conceivable at present which could match the force of any powerful single nation.

Freud, on the other hand, is inclined to see the difficulty in the biological nature of man—in his innate destructiveness. According to him, destruction for its own sake is one of the strongest human motive forces. The only condition under which human beings can live peacefully is by deflecting this destructiveness to an outward object. In a sense, therefore, national peace is dependent upon international wars. Unquestionably, both Einstein and Freud have much in favor of their arguments, yet it seems to me neither of them gave a fully satisfactory analysis of the problem.

It is easier to see the flaw in Einstein's argument. The majority of people do not need intimidation by police force in order not to commit homicide or not to steal. Within a nation police force is not the sole supporter of peace and order. Were it not for the internal psychological control of the human conscience (superego), the existing police force would not be able to maintain order. Police force is needed only for checking the least adjusted members of society or those who on account of their social status have the least advantages from social order. This internal inhibitory system, the conscience, is not inherited but is the result of an educational process. It is powerfully supported by religion. The force of this internal inhibition became so strong that in our modern wars the majority of people have a genuine distaste for killing their fellowmen who belong to another nation. It is partially external coercion and partly the realization of the unwelcome necessity for violence which makes men fight. It is obvious, therefore, that the lack of enforcement of international law is not an ultimate obstacle against peace. If men belonging to different national units would be further domesticated so that their consideration for the lives of people of other nations would be equally as strong as that for their own compatriots, peace would be possible. There is no reason to assume that the domestication can go only thus far and no farther. One could imagine a kind of education which would lead to the development of an international or world-conscience. Education, of course, is not a matter of individual action, and therefore such an education could be introduced only after an international organization of nations is established. At present the domestication of the human individual—his social conscience—officially stops at the national boundaries. Therefore contemporary man can only be complimented for the fact that his conscience does not meticulously accept national boundaries, and only under the influence of propaganda and often even of alcohol can he be persuaded to kill.

From this perspective it would seem that only after historical events have led to the extinction of small groups and their inclusion into larger ones could a further step in the progress of human domestication take place. In the history of nations this gradual extension of group boundaries can be clearly observed. Until as late as the middle of the nineteenth century, Italy was not yet a national entity but was divided into independent portions. For a long time almost every city was a unit in itself, fighting the neighboring city. The citizen was, in the original sense of the word, a member of a city and not of a nation. Instead of nationalism there was the local patriotism of members belonging to the same city. A transformation of this particularistic state into a nation was accomplished as late as in the second half of the last century. Fascism is obviously a last step in the development of a national consciousness. Not much different is the history of Germany, where only Hitler abolished the rudimentary remnants of independence of the constituent parts of the Reich. It scarcely can be considered as an accident that totalitarianism developed in just those two countries where national unity and consciousness have developed later than in the rest of the great European nations.

This process of integration of more or less independent units within a nation seems to follow the pattern that the powerful part becomes dominant and subjugates the others. Gradually the difference between a dominant part, which is the crystallization point of the union, and the rest of the new unit disappears. As soon as a single German unit was formed out of Prussia, Saxony, Baden, Bavaria, and Württemberg, the crystallization point of this union, Prussia, ceased to remain the leading portion. The country became one homogeneous state. In this sense the present European war appears as the violent birth of a United States of Europe in which one dominant part will subjugate the rest and thus serve as the crystallization point of a new larger entity. The psychological expression of this new state of affairs will necessarily be the development of a European nationalism in the place of a German, French, English, or Spanish. The future historian of a not too remote age, whose nationality will be "European," will evaluate the present European war in the same way as we today evaluate the centuries-long war between the cities of Venice, Genoa, Milan, Florence, Pisa, Rome, and Naples. It is, of course, still an open question whether England or Germany will be the crystallization point of this future Pan-Europe.

Thus, if we turn to historical events, we come to the conclusion that peace can exist only between members of organized groups and that the formation of such groups leads through wars and subjugation of the weaker by the stronger. A more just organization within the new larger group is, then, a matter of later development. These historical mass events are decisive and determine the later internal changes in the individual members of the group. Human beings do not become at first more social and then create larger social organizations, but they become forced into larger national units by wars and subjugation, and only then do they become more social in adjusting themselves to the life within this larger organization. After such larger organizations have been established, the question of enforcing the law does not exist any longer. Within a well-organized group there is no difficulty in enforcing the law which protects—if not with theoretical equality, yet to some degree at least—all members of the group. Police and its internal ally, the conscience, together secure social order.

One should not, of course, one-sidedly overlook the other aspect of this process. The human material which is to become organized into larger social units is of great significance. A creation of such a larger entity is possible only if the process of domestication of the members has advanced to a certain stage while still living in smaller groups. This process of domestication is the basis of every group life; it starts in early childhood. The overcoming of the oedipus complex is obviously the first and most significant step in this development. It means the adjustment of the individual to the first society—the family. Also in the prehistoric past, the brothers' renunciation of the use of violence against the father and against one another was obviously the first step toward the building of the first social cell—the family—under the authority of the father. With the help of a similar principle of marital tabus and renunciation of violence between male competitors the next social unit, the clan, became possible. Later clan spirit is gradually replaced by national solidarity. This is the phase of social evolution in which the human race is at present. In becoming a social being man at first must give up for the sake of the other members of the family some of his personal narcissism. Then he must abandon clan spirit (clan narcissism) and replace it with patriotism (national narcissism). Obviously the next step toward world-peace is overcoming nationalism and exchanging it for a kind of highly diluted narcissism: an all-encompassing humanism. In this progressive process of extension of self-love to object love, every step consists in the desperate struggle to overcome the previous stage. Every psychoanalyst knows that in early life individual narcissism is the greatest obstacle against accepting and recognizing the rights of the other siblings. History shows that particularistic clan spirit was the greatest obstacle against the creation of a unified nation. The fight of the peers against the centralistic tendencies of the kings in English and Hungarian history, the struggle of the Italian cities under the leadership of one or two prominent families for their independence, are the most convincing examples. Against the rational necessity for the fusion of smaller units into greater ones, there is the irrational tendency of the smaller units to maintain their independent existence. This particularistic tendency is not fully abandoned even after the smaller entities have been fused into a high organization. In times of stress, when the larger unit is exposed to external attacks and is defeated, the dormant individualistic tendencies of the formerly independent constituents become mobilized and a process of disintegration sets in. The decline of the Austro-Hungarian empire is an exquisite example. It is only natural that any organization will have little stability when the constituent parts are dissatisfied because the mutual sacrifices which every member has to make for the sake of social order are unequally distributed. The cohesion of such an organization is secure only by the superior suppressing force of the powerful exploiting group and not by the co-operation of all the constituents. Only if the constituent groups voluntarily give up their independent existence for the sake of greater security and enjoy the advantages of the coalition, as in an ideal democracy, will the larger social organism become a stable institution. It is obvious that such a democratic state requires a more socialized kind of human material than any other social order based on suppression and coercion. Only human material which has progressed far beyond the more primitive stages of personal, clan, and class narcissism can live in a social order that is based on free consent and enlightened self-interest. According to the testimony of history such an advanced state of affairs can develop only gradually. At first the constituent group must be forced into a larger social unit by superior power of a dominant group.[4] Only then follows a further socialization of the personality in a long process of adjustment of the individual to his new social climate.

The human material of the group which by conquest of the weaker ones becomes the crystallization point of a larger entity is by no means irrelevant. The future history of Europe will be entirely different if the coming Pan-Europa is organized under British or German predominance. From a distant historical point of view the end effect might be the same. The road to it, however, would certainly be smoother and less tortuous in the first than in the second case.

This seems to be the answer to Einstein's statement concerning the difficulties of enforcing international law. This difficulty is merely a relative one. So long as independent Italian cities existed, there was no way of enforcing the law which protected all citizens of the Appenine Peninsula. As soon as this part of the world became organized into one national state, there was no longer any difficulty in enforcing laws which insured peace and order within the new nation.

Now we may turn to Freud's views concerning the innate destructiveness of man as the insuperable obstacle against peace. His departure seems unassailable. The internal peace of a nation is possible only at the price of war with the neighbor. Men must hate and destroy something. Only if they have a common external object of hate can they be saved from each other's destructiveness. Leaders know this instinctively, and, in order to avert revolution, they instigate war. The question is whether this is a universal principle or is true only under certain conditions.

We deal here obviously with a dynamic principle of collective life which, like every dynamic principle, expresses a quantitative relationship. "Man is innately destructive—war is the manifestation of destructiveness—*ergo*, man will always have to conduct wars" is the kind of scholastic deductive syllogism which may well serve as harmless intellectual gymnastics for those amateur philosophers who try to revive prescientific thinking of the Middle Ages, but it is useless for learning something about the complex phenomena and the world surrounding us. Freud did not follow this philosophical procedure which he so thoroughly disavowed in his writings. Although he assumed that man has an innate destructiveness, he studied the manifestation of this destructiveness under different conditions and found that this destructiveness might be diverted toward external objects, thus allowing people within the group to maintain internal peace. Where he did not go far enough into the differential analysis of the problem was in discussing the relation of destructiveness to war. Killing is certainly not the only manifestation of human aggression; litigation by law in the defense of one's own interests can also be considered one of its expressions. Neither can a political campaign be equated to war. Even intellectual endeavors to understand and master the environment can be considered as sublimated manifestations of aggressive impulses. The mere fact that man has aggressive impulses does not permit us to postulate that war is unavoidable.

This is not the place to discuss Freud's problematic death-instinct theory. The relationship of aggressiveness to self-preservation is still, however, an open question. The existence of an innate destructiveness for its own sake, which goes beyond the limits of self-preservation, has not been convincingly demonstrated. Not that such a destructiveness for its own sake, in the form of sadism, would not exist as a secondary phenomenon. There is more and more evidence that sadism must be considered a secondary erotization of an originally only self-preservative aggressiveness. It seems that in morbid development, when large quantities of inhibited hostile impulses accumulate, these may be drained by sadistic behavior which serves merely as a gratification and not for self-preservation. Should this view prove correct, there is hope that when individual interests are well preserved by reasonable social order, international peace is a possible state of affairs.

Obviously, democracy comes nearest to a peaceful solution of the problem of maintaining social order within one nation; an international league of nations organized along the principles of democracy is at least a theoretically conceivable solution of international peace. Of course such an international democratic organization of all the inhabitants of the earth would mean the gradual disappearance of political and economic boundaries between national or continental entities. We do not possess sufficient knowledge of sociodynamics to predict that historical development will necessarily lead to a continuous enlargement of politico-economic units until all inhabitants of the earth are organized under one system. History shows that, apart from the tendency toward the emergence of large social entities, there is also in operation the opposite principle of disintegration and separation leading to the breaking-up of empires. One after another—the Roman Empire, the Holy Roman Empire, that of Genghis Khan, of Islam, of the Hapsburgs, and of Napoleon—have been dissolved. Possibly at present we are witnessing the dissolution of the British Empire. The estimation of the quantitative relationship between these upbuilding and disintegrating forces of history is beyond our present knowledge.

Peace is obviously the function of many factors—two of them outstanding. Peace among large numbers of human beings is possible if they are united in one large and well-integrated social organization, usually under the leadership of one powerful dominant element. This is the *Pax Romana* or *Pax Britannica* type of peace. The other factor is represented by the technique by which conflicting interests are settled among the members of one social organization and between different organizations. This technique, whether it is based on violence or negotiation, does not depend on the innate qualities of man but on his mentality, dependent upon the prevailing ideology which influences the prevailing education. From this point of view the democratic type of society and ideology seems to offer the greatest chances toward peace. *Pax Romana* was based on the suppressing force of that dominant power which was the crystallization point of the larger organization. Because such peace is based primarily on power, it is of temporary nature. At the beginning *Pax Britannica* also was unquestionably based on superior force and only gradually assumed a more democratic structure. When conflicting interests are not settled by their free expression, negotiation and compromise, the particularistic tendencies of the constituent parts, which originally were independent, is great. The astonishing cohesion of the British Empire, which we are witnessing at present, is due to the fact that its constituent parts are members of the Empire by their own choice and also because the whole Empire is threatened by a common enemy. Only a democratic, international organization of the world would warrant lasting world-peace.

The experience of history does not entitle us to hope that such a state of affairs, even if conceivable, could be achieved without at first one powerful group subduing the weaker ones and coercing them to participate in a larger organization. Later a gradual co-ordination of conflicting interests may ensue. The further domestication of man will be an adjustment of the individual to a new and more encompassing social organization which at first must be created by force. The cardinal question is whether or not a democratic or an autocratic nation will be the crystallization point of such a world-organization. The League of Nations failed because it did not recognize that at the beginning it must rely on armed force. At the beginning a powerful nation or a group of nations must enforce international law by their superior force. At the beginning the league cannot be based on theoretical justice or the equal right of all participants. At first it must function as an enlightened benevolent tyranny only to progress to the status of constitutional monarchy and finally to democracy. The League of Nations was an anemic institution theoretically conceived in utter disregard for the dynamic principles of history and mass psychology. A future league of nations will have to accept the historical principle that at first only with power can masses and groups be welded into a social organization. Its armed force can be reduced to something like a police power only after this new order has found an internal, psychological ally in the form of an international conscience.

It is of great importance that the superior power enforcing the new supernational organization should be a democracy, the most progressed form of social organization. The tragic paradox of history is, however, that democracies, which should serve as the model of the future supernational organization because of their nature, are averse to revert to force. Totalitarian states, on the other hand, are full of internal unresolved tensions and in order to survive are driven toward dynamic expansive imperialism. Therefore they are destined to create new supernational organizations. They are more likely to make the first necessary step toward coalition by coercion. Their ideology of force, suppression, and racial superiority, however, makes them unsuited to create a stable organization of the subjected nations. The only way out of this dilemma is that democracies finally recognize their historical vocation to assume leadership toward a new league of nations. This must be based at first both on justice and on armed force, the latter to be discarded only gradually at the same pace as the indispensable internal psychological ally in man's personality gains strength. This internal ally, a slowly growing product of education, is an advanced form of humanism which does not stop at economic, linguistic, or racial borders.

FOOTNOTES

* World War I. THE EDITORS

† This is a dated statement. The population of the United States has increased considerably (to 200,000,000) and the state of internal peace referred to by the author is hardly the case today. THE EDITORS

[1] *Why War?* (Paris: International Institute of Intellectual Co-operation, League of Nations, 1933), p. 3. [See also this book, Reading 1.]

[2] Ibid., p. 9.

[3] Ibid.

[4] Hitler with his astute political instinct is keenly aware of this historical principle. He writes in *Mein Kampf* (New York: Reynal & Hitchcock, 1940): "Thus also the founding of the German Reich was by itself not the result of some common intention along common ways but rather the result of a conscious—sometimes also unconscious—wrestling for hegemony out of which struggle Prussia finally emerged victorious. . . . For who in German lands would seriously have believed two hundred years ago that the Prussia of the Hohenzollerns and not Hapsburg would some day become the germ cell, founder and teacher of the new Reich?" (op. cit., pp. 754-55). And in another place in discussing how federated states develop into a united state he says: "But the very construction of the Reich did not take place on the basis of the free will of the identical actions of the individual states, but as the effect of the hegemony of one State among them, Prussia" (ibid., p. 831).

18   On War and Peace in Animals and Man:
An Ethologist's Approach to the Biology of Aggression

——N. TINBERGEN

In 1935 Alexis Carrel published a best seller, *Man—The Unknown* (*1*). Today, more than 30 years later, we biologists have once more the duty to remind our fellowmen that in many respects we are still, to ourselves, unknown. It is true that we now understand a great deal of the way our bodies function. With this understanding came control: medicine.

The ignorance of ourselves which needs to be stressed today is ignorance about our behavior—lack of understanding of the causes and effects of the function of our brains. A scientific understanding of our behavior, leading to its control, may well be the most urgent task that faces mankind today. It is the effects of our behavior that begin to endanger the very survival of our species and, worse, of all life on earth. By our technological achievements we have attained a mastery of our environment that is without precedent in the history of life. But these achievements are rapidly getting out of hand. The consequences of our "rape of the earth" are now assuming critical proportions. With short-sighted recklessness we deplete the limited natural resources, including even the oxygen and nitrogen of our atmosphere (*2*). And Rachel Carson's warning (*3*) is now being followed by those of scientists, who give us an even gloomier picture of the general pollution of air, soil, and water. This pollution is seriously threatening our health and our food supply. Refusal to curb our reproductive behavior has led to the population explosion. And, as if all this were not enough, we are waging war on each other—men are fighting and killing men on a massive scale. It is because the effects of these behavior patterns, and of attitudes that determine our behavior, have now acquired such truly lethal potentialities that I have chosen man's ignorance about his own behavior as the subject of this paper.

Reprinted from *Science*, 1968, *160*, 1411-1418. Copyright 1968 by the American Association for the Advancement of Science. Dr. Tinbergen is professor of animal behavior, Department of Zoology, University of Oxford, Oxford, England.

I am an ethologist, a zoologist studying animal behavior. What gives a student of animal behavior the temerity to speak about problems of human behavior? Of course the history of medicine provides the answer. We all know that medical research uses animals on a large scale. This makes sense because animals, particularly vertebrates, are, in spite of all differences, so similar to us; they are our blood relations, however distant.

But this use of zoological research for a better understanding of ourselves is, to most people, acceptable only when we have to do with those bodily functions that we look upon as parts of our physiological machinery—the functions, for instance, of our kidneys, our liver, our hormone-producing glands. The majority of people bridle as soon as it is even suggested that studies of animal behavior could be useful for an understanding, let alone for the control, of our own behavior. They do not want to have their own behavior subjected to scientific scrutiny; they certainly resent being compared with animals, and these rejecting attitudes are both deep-rooted and of complex origin.

But now we are witnessing a turn in this tide of human thought. On the one hand the resistances are weakening, and on the other, a positive awareness is growing of the potentialities of a biology of behavior. This has become quite clear from the great interest aroused by several recent books that are trying, by comparative studies of animals and man, to trace what we could call "the animal roots of human behavior." As examples I select Konrad Lorenz's book *On Aggression* (4) and *The Naked Ape* by Desmond Morris (5). Both books were best sellers from the start. We ethologists are naturally delighted by this sign of rapid growth of interest in our science (even though the growing pains are at times a little hard to endure). But at the same time we are apprehensive, or at least I am.

We are delighted because, from the enormous sales of these and other such books, it is evident that the mental block against self-scrutiny is weakening—that there are masses of people who, so to speak, want to be shaken up.

But I am apprehensive because these books, each admirable in its own way, are being misread. Very few readers give the authors the benefit of the doubt. Far too many either accept uncritically all that the authors say, or (equally uncritically) reject it all. I believe that this is because both Lorenz and Morris emphasize our knowledge rather than our ignorance (and, in addition, present as knowledge a set of statements which are after all no more than likely guesses). In themselves brilliant, these books could stiffen, at a new level, the attitude of certainty, while what we need is a sense of doubt and wonder, and an urge to investigate, to inquire.

POTENTIAL USEFULNESS OF ETHOLOGICAL STUDIES

Now, in a way, I am going to be just as assertive as Lorenz and Morris, but what I am going to stress is how much we do not know. I shall argue that we shall have to make a major research effort. I am of course fully aware of the fact that much research is already being devoted to problems of human, and even of animal, behavior. I know, for instance, that anthropologists, psychologists, psychiatrists, and others are approaching these problems from many angles. But I shall try to show that the research effort has so far made insufficient use of the potential of ethology. Anthropologists, for instance, are beginning to look at animals, but they restrict their work almost entirely to our nearest relatives, the apes and monkeys. Psychologists do study a larger variety of animals, but even they select mainly higher species. They also ignore certain major problems that we biologists think have to be studied. Psychiatrists, at least many of them, show a disturbing tendency to apply the *results* rather than the *methods* of ethology to man.

None of these sciences, not even their combined efforts, are as yet parts of one coherent science of behavior. Since behavior is a life process, its study ought to be part of the mainstream of biological research. That is why we zoologists ought to "join the fray." As an ethologist, I am going to try to sketch how my science could assist its sister sciences in their attempts, already well on their way, to make a united, broad-fronted, truly biological attack on the problems of behavior.

I feel that I can cooperate best by discussing what it is in ethology that could be of use to the other behavioral sciences. What we ethologists do not want, what we consider definitely wrong, is uncritical application of our results to man. Instead, I myself at least feel that it is our method of approach, our rationale, that we can offer (6), and also a little simple common sense, and discipline.

The potential usefulness of ethology lies in the fact that, unlike other sciences of behavior, it applies the method or "approach" of biology to the phenomenon behavior. It has developed a set of concepts and terms that allow us to ask:

(1) In what ways does this phenomenon (behavior) influence the survival, the success of the animal?

(2) What makes behavior happen at any given moment? How does its "machinery" work?

(3) How does the behavior machinery develop as the individual grows up?

(4) How have the behavior systems of each species evolved until they became what they are now?

The first question, that of survival value, has to do with the effects of behavior; the other three are, each on a different time scale, concerned with its causes.

These four questions are, as many of my fellow biologists will recognize, the major questions that biology has been pursuing for a long time. What ethology is doing could be simply described by saying that, just as biology investigates the functioning of the organs responsible for digestion, respiration, circulation, and so forth, so ethology begins now to do the same with respect to behavior; it investigates the functioning of organs responsible for movement.

I have to make clear that in my opinion it is the comprehensive, integrated attack on all four problems that characterizes ethology. I shall try to show that to ignore the questions of survival value and evolution—as, for instance, most psychologists do—is not only short-sighted but makes it impossible to arrive at an understanding of behavioral problems. Here ethology can make, in fact is already making, positive contributions.

Having stated my case for animal ethology as an essential part of the science of behavior, I will now have to sketch how this could be done. For this I shall have to consider one concrete example, and I select aggression, the most directly lethal of our behaviors. And, for reasons that will become clear, I shall also make a short excursion into problems of education.

Let me first try to define what I mean by aggression. We all understand the term in a vague, general way, but it is, after all, no more than a catchword. In terms of actual behavior, aggression involves approaching an opponent, and, when within reach, pushing him away, inflicting damage of some kind, or at least forcing stimuli upon him that subdue him. In this description the effect is already implicit: such behavior tends to remove the opponent, or at least to make him change his behavior in such a way that he no longer interferes with the attacker. The methods of attack differ from one species to another, and so do the weapons that are used, the structures that contribute to the effect.

Since I am concentrating on men fighting men, I shall confine myself to intraspecific fighting, and ignore, for instance, fighting between predators and prey. Intraspecific fighting is very common among animals. Many of them fight in two different contexts, which we can call "offensive" and "defensive." Defensive fighting is often shown as a last resort by an animal that, instead of attacking, has been fleeing from an attacker. If it is cornered, it may suddenly turn round upon its enemy and "fight with the courage of despair."

Of the four questions I mentioned before, I shall consider that of the survival value first. Here comparison faces us right at the start with a striking paradox. On the one hand, man is akin to many species of animals in that he fights his own species. But on the other hand he is, among the thousands of species that fight, the only one in which fighting is disruptive.

In animals, intraspecific fighting is usually of distinctive advantage. In addition, all species manage as a rule to settle their disputes without killing one another; in fact, even bloodshed is rare. Man is the only species that is a mass murderer, the only misfit in his own society.

Why should this be so? For an answer, we shall have to turn to the question of causation: What makes animals and man fight their own species? And why is our species "the odd man out"?

CAUSATION OF AGGRESSION

For a fruitful discussion of this question of causation I shall first have to discuss what exactly we mean when we ask it.

I have already indicated that when thinking of causation we have to distinguish between three subquestions, and that these three differ from one another in the stretch of time that is considered. We ask, first: Given an adult animal that fights now and then, what makes each outburst of fighting happen? The time scale in which we consider these recurrent events is usually one of seconds, or minutes. To use an analogy, this subquestion compares with asking what makes a car start or stop each time we use it.

But in asking this same general question of causation ("What makes an animal fight?") we may also be referring to a longer period of time; we may mean "How has the animal, as it grew up, developed this behavior?" This compares roughly with asking how a car has been constructed in the factory. The distinction between these two subquestions remains useful even though we know that many animals continue their development (much slowed down) even after they have attained adulthood. For instance, they may still continue to learn.

Finally, in biology, as in technology, we can extend this time scale even more, and ask: How have the animal species which we observe today—and which we know have evolved from ancestors that were different—how have they acquired their particular behavior systems during this evolution? Unfortunately, while we know the evolution of cars because they evolved so quickly and have been so fully recorded, the behavior of extinct animals cannot be observed, and has to be reconstructed by indirect methods.

I shall try to justify the claim I made earlier, and show how all these four questions—that of behavior's survival value and the three subquestions of causation—have to enter into the argument if we are to understand the biology of aggression.

Let us first consider the short-term causation; the mechanism of fighting. What makes us fight at any one moment? Lorenz argues in his book that, in animals and in man, there is an internal urge to attack. An individual does not simply wait to be provoked, but, if actual attack has not been possible for some time, this urge to fight builds up until the individual actively seeks the opportunity to indulge in fighting. Aggression, Lorenz claims, can be spontaneous.

But this view has not gone unchallenged. For instance, R. A. Hinde has written a thorough criticism (7), based on recent work on aggression in animals, in which he writes that Lorenz's "arguments for the spontaneity of aggression do not bear examination" and that "the contrary view, expressed in nearly every textbook of comparative psychology . . . " is that fighting "derives principally from the situation"; and even more explicitly: "There is no need to postulate causes that are purely internal to the aggressor" (7, p. 303). At first glance it would seem as if Lorenz and Hinde disagree profoundly. I have read and reread both authors, and it is to me perfectly clear that loose statements and misunderstandings on both sides have made it appear that there is disagreement where in fact there is something very near to a common opinion. It seems to me that the differences between the two authors lie mainly in the different ways they look at internal and external variables. This in turn seems due to differences of a semantic nature. Lorenz uses the unfortunate term "the spontaneity of aggression." Hinde takes this to mean that external stimuli are in Lorenz's view not necessary at all to make an animal fight. But here he misrepresents Lorenz, for nowhere does Lorenz claim that the internal urge ever makes an animal fight "in vacuo"; somebody or something is attacked. This misunderstanding makes Hinde feel that he has refuted Lorenz's views by saying that "fighting derives principally from the situation." But both authors are fully aware of the fact that fighting is started by a number of variables, of which some are internal and some external. What both authors know, and what cannot be doubted, is that fighting behavior is not like the simple slot machine that produces one platform ticket every time one threepenny bit is inserted. To mention one animal example: a male stickleback does not always show the full fighting behavior in response to an approaching standard opponent; its response varies from none at all to the optimal stimulus on some occasions, to full attack on even a crude dummy at other times. This means that its internal state varies, and in this particular case we know from the work of Hoar (8) that the level of the male sex hormone is an important variable.

Another source of misunderstanding seems to have to do with the stretch of time that the two authors are taking into account. Lorenz undoubtedly thinks of the causes of an outburst of fighting in terms of seconds, or hours—perhaps days. Hinde seems to think of events which may have happened further back in time; an event which is at any particular moment "internal" may well in its turn have been influenced previously by external agents. In our stickleback example, the level of male sex hormone is influenced by external agents such as the length of the daily exposure to light over a period of a month or so (9). Or, less far back in time, its readiness to attack may have been influenced by some experience gained, say, half an hour before the fight.

I admit that I have now been spending a great deal of time on what would seem to be a perfectly simple issue: the very first step in the analysis of the short-term causation, which is to distinguish at any given moment between variables within the animal and variables in the environment. It is of course important for our further understanding to unravel the complex interactions between these two worlds, and in particular the physiology of aggressive behavior. A great deal is being discovered about this, but for my present issue there is no use discussing it as long as even the first step in the analysis has not led to a clearly expressed and generally accepted conclusion. We must remember that we are at the moment concerned with the human problem: "What makes men attack each other?" And for this problem the answer to the first stage of our question is of prime importance: Is our readiness to start an attack constant or not? If it were—if our aggressive behavior were the outcome of an apparatus with the properties of the slot machine—all we would have to do would be to control the external situation: to stop providing threepenny bits. But since our readiness to start an attack is variable, further studies of both the external and the internal variables are vital to such issues as: Can we reduce fighting by lowering the population density, or by withholding provocative stimuli? Can we do so by changing the hormone balance or other physiological variables? Can we perhaps in addition control our development in such a way as to change the dependence on internal and external factors in adult man? However, before discussing development, I must first return to the fact that I have mentioned before, namely, that man is, among the thousands of other species that fight, the only mass murderer. How do animals in their intraspecific disputes avoid bloodshed?

THE IMPORTANCE OF "FEAR"

The clue to this problem is to recognize the simple fact that aggression in animals rarely occurs in pure form; it is only one of two components of an adaptive system. This is most clearly seen in territorial behavior, although it is also true of most other types of hostile behavior. Members of territorial species divide, among themselves, the available living space and opportunities by each individual defending its home range against competitors. Now in this system of parceling out living space, avoidance plays as important a part as attack. Put very briefly, animals of territorial species, once they have settled on a territory, attack intruders, but an animal that is still searching for a suitable territory or finds itself outside its home range withdraws when it meets with an already established owner. In terms of function, once you have taken possession of a territory, it pays to drive off competitors; but when you are still looking for a territory (or meet your neighbor at your common boundary), your chances of success are improved by avoiding such established owners. The ruthless fighter who "knows no fear" does not get very far. For an understanding of what follows, this fact, that hostile clashes are controlled by what we could call the "attack-avoidance system," is essential.

When neighboring territory owners meet near their common boundary, both attack behavior and withdrawal behavior are elicited in both animals; each of the two is in a state of motivational conflict. We know a great deal about the variety of movements that appear when these two conflicting, incompatible behaviors are elicited. Many of these expressions of a motivational conflict have, in the course of evolution, acquired signal function; in colloquial language, they signal "Keep out!" We deduce this from the fact that opponents respond to them in an appropriate way: instead of proceeding to intrude, which would require the use of force, trespassers withdraw, and neighbors are contained by each other. This is how such animals have managed to have all the advantages of their hostile behavior without the disadvantages: they divide their living space in a bloodless way by using as distance-keeping devices these conflict movements ("threat") rather than actual fighting.

GROUP TERRITORIES

In order to see our wars in their correct biological perspective one more comparison with animals is useful. So far I have discussed animal species that defend individual or at best pair territories. But there are also animals which possess and defend territories belonging to a group, or a clan (*10*).

Now it is an essential aspect of group territorialism that the members of a group unite when in hostile confrontation with another group that approaches, or crosses into their feeding territory. The uniting and the aggression are equally important. It is essential to realize that group territorialism does not exclude hostile relations on lower levels when the group is on its own. For instance, within a group there is often a peck order. And within the group there may be individual or pair territories. But frictions due to these relationships fade away during a clash between groups. This temporary elimination is done by means of so-called appeasement and reassurance signals. They indicate "I am a friend," and so diminish the risk that, in the general flare-up of anger, any animal "takes it out" on a fellow member of the same group (*11*). Clans meet clans as units, and each individual in an intergroup clash, while united with its fellow-members, is (as in interindividual clashes) torn between attack and withdrawal, and postures and shouts rather than attacks.

We must now examine the hypothesis (which I consider the most likely one) that man still carries with him the animal heritage of group territoriality. This is a question concerning man's evolutionary origin, and here we are, by the very nature of the subject, forced to speculate. Because I am going to say something about the behavior of our ancestors of, say, 100,000 years ago, I have to discuss briefly a matter of methodology. It is known to all biologists (but unfortunately unknown to most psychologists) that comparison of present-day species can give us a deep insight, with a probability closely approaching certainty, into the evolutionary history of animal species. Even where fossil evidence is lacking, this comparative method alone can do this. It has to be stressed that this comparison is a highly sophisticated method, and not merely a matter of saying that species A is different from species B (*12*). The basic procedure is this. We interpret differences between really allied species as the result of adaptive divergent evolution from common stock, and we interpret similarities between nonallied species as adaptive convergencies to similar ways of life. By studying the adaptive functions of species characteristics we understand how natural selection can have produced both these divergencies and convergencies. To mention one striking example: even if we had no fossil evidence, we could, by this method alone, recognize whales for what they are—mammals that have returned to the water, and, in doing so, have developed some similarities to fish. This special type of comparison, which has been applied so successfully by students of the structure of animals, has now also been used, and with equal success, in several studies of animal behavior. Two approaches have been applied. One is to see in what respects species of very different origin have convergently adapted to a similar way of life. Von Haartman (*13*) has applied this to a study of birds of many types that nest in holes—an anti-predator safety device. All such hole-nesters center their territorial fighting on a suitable nest hole. Their courtship consists of luring a female to this hole often with the use of bright color patterns). Their young gape when a general darkening signals the arrival of the parent. All but the most recently adapted species lay uniformly colored, white or light blue eggs that can easily be seen by the parent.

An example of adaptive divergence has been studied by Cullen (*14*). Among all the gulls, the kittiwake is unique in that it nests on very narrow ledges on sheer cliffs. Over 20 peculiarities of this species have been recognized by Mrs. Cullen as vital adaptations to this particular habitat.

These and several similar studies (*15*) demonstrate how comparison reveals, in each species, systems of interrelated, and very intricate adaptive features. In this work, speculation is now being followed by careful experimental checking. It would be tempting to elaborate on this, but I must return to our own unfortunate species.

Now, when we include the "Naked Ape" in our comparative studies, it becomes likely (as has been recently worked out in great detail by Morris) that man is a "social Ape who has turned carnivore" (*16*). On the one hand he is a social primate; on the other, he has developed similarities to wolves, lions and hyenas. In our present context one thing seems to stand out clearly, a conclusion that seems to me of paramount importance to all of us, and yet has not yet been fully accepted as such. As a social, hunting primate, man must originally have been organized on the principle of group territories.

Ethologists tend to believe that we still carry with us a number of behavioral characteristics of our animal ancestors, which cannot be eliminated by different ways of upbringing, and that our group territorialism is one of those ancestral characters. I shall discuss the problem of the modifiability of our behavior later, but it is useful to point out here that even if our behavior were much more modifiable than Lorenz maintains, our cultural evolution, which resulted in the parceling-out of our living space on lines of tribal, national, and now even "bloc" areas, would, if anything, have tended to enhance group territorialism.

GROUP TERRITORIALISM IN MAN?

I put so much emphasis on this issue of group territorialism because most writers who have tried to apply ethology to man have done this in the wrong way. They have made the mistake, to which I objected before, of uncritically extrapolating the results of animal studies to man. They try to explain man's behavior by using facts that are valid only of some of the animals we studied. And, as ethologists keep stressing, no two species behave alike. Therefore, instead of taking this easy way out, we ought to study man in his own right. And I repeat that the message of the ethologists is that the methods, rather than the results, of ethology should be used for such a study.

Now, the notion of territory was developed by zoologists (to be precise, by ornithologists, *17*), and because individual and pair territories are found in so many more species than group territories (which are particularly rare among birds), most animal studies were concerned with such individual and pair territories. Now such low-level territories do occur in man, as does another form of hostile behavior, the peck order. But the problems created by such low-level frictions are not serious; they can, within a community, be kept in check by the apparatus of law and order; peace within national boundaries can be enforced. In order to understand what makes us go to war, we have to recognize that man behaves very much like a group-territorial species. We too unite in the face of an outside danger to the group; we "forget our differences." We too have threat gestures, for instance, angry facial expressions. And all of us use reassurance and appeasement signals, such as a friendly smile. And (unlike speech) these are universally understood; they are cross-cultural; they are species-specific. And, incidentally, even within a group sharing a common language, they are often more reliable guides to a man's intentions than speech, for speech (as we know now) rarely reflects our true motives, but our facial expressions often "give us away."

If I may digress for a moment: it is humiliating to us ethologists that many nonscientists, particularly novelists and actors, intuitively understand our sign language much better then we scientists ourselves do. Worse, there is a category of human beings who understand intuitively more about the causation of our aggressive behavior: the great demagogues. They have applied this knowledge in order to control our behavior in the most clever ways, and often for the most evil purposes. For instance, Hitler (who had modern mass communication at his disposal, which allowed him to inflame a whole nation) played on both fighting tendencies. The "defensive" fighting was whipped up by his passionate statements about "living space," "encirclement," Jewry, and Freemasonry as threatening powers which made the Germans feel "cornered." The "attack fighting" was similarly set ablaze by playing the myth of the Herrenvolk. We must make sure that mankind has learned its lesson and will never forget how disastrous the joint effects have been—if only one of the major nations were led now by a man like Hitler, life on earth would be wiped out.

I have argued my case for concentrating on studies of group territoriality rather than on other types of aggression. I must now return, in this context, to the problem of man the mass murderer. Why don't we settle even our international disputes by the relatively harmless, animal method of threat? Why have we become unhinged so that so often our attack erupts without being kept in check by fear? It is not that we have no fear, nor that we have no other inhibitions against killing. This problem has to be considered first of all in the general context of the consequences of man having embarked on a new type of evolution.

CULTURAL EVOLUTION

Man has the ability, unparalleled in scale in the animal kingdom, of passing on his experiences from one generation to the next. By this accumulative and exponentially growing process, which we call cultural evolution, he has been able to change his environment progressively out of all recognition. And this includes the social environment. This new type of evolution proceeds at an incomparably faster pace than genetic evolution. Genetically we have not evolved very strikingly since Cro-Magnon man, but culturally we have changed beyond recognition, and are changing at an ever-increasing rate. It is of course true that we are highly adjustable individually, and so could hope to keep pace with these changes. But I am not alone in believing that this behavioral adjustability, like all types of modifiability, has its limits. These limits are imposed upon us by our hereditary constitution, a constitution which can only change with the far slower speed of genetic evolution. There are good grounds for the conclusion that man's limited behavioral adjustability has been outpaced by the culturally determined changes in his social environment, and that this is why man is now a misfit in his own society.

We can now, at last, return to the problem of war, of uninhibited mass killing. It seems quite clear that our cultural evolution is at the root of the trouble. It is our cultural evolution that has caused the population explosion. In a nutshell, medical science, aiming at the reduction of suffering, has, in doing so, prolonged life for many individuals as well—prolonged it to well beyond the point at which they produce offspring. Unlike the situation in any wild species, recruitment to the human population consistently surpasses losses through mortality. Agricultural and technical know-how have enabled us to grow food and to exploit other natural resources to such an extent that we can still feed (though only just) the enormous numbers of human beings on our crowded planet. The result is that we now live at a far higher density than that in which genetic evolution has molded our species. This, together with long-distance communication, leads to far more frequent, in fact to continuous, intergroup contacts, and so to continuous external provocation of aggression. Yet this alone would not explain our increased tendency to kill each other; it would merely lead to continuous threat behavior.

The upsetting of the balance between aggression and fear (and this is what causes war) is due to at least three other consequences of cultural evolution. It is an old cultural phenomenon that warriors are both brainwashed and bullied into all-out fighting. They are brainwashed into believing that fleeing—originally, as we have seen, an adaptive type of behavior—is despicable, "cowardly." This seems to me due to the fact that man, accepting that in moral issues death might be preferable to fleeing, has falsely applied the moral concept of "cowardice" to matters of mere practical importance—to the dividing of living space. The fact that our soldiers are also bullied into all-out fighting (by penalizing fleeing in battle) is too well known to deserve elaboration.

Another cultural excess is our ability to make and use killing tools, especially long-range weapons. These make killing easy, not only because a spear or a club inflicts, with the same effort, so much more damage than a fist, but also, and mainly, because the use of long-range weapons prevents the victim from reaching his attacker with his appeasement, reassurance, and distress signals. Very few aircrews who are willing, indeed eager, to drop their bombs "on target" would be willing to strangle, stab, or burn children (or, for that matter, adults) with their own hands; they would stop short of killing, in response to the appeasement and distress signals of their opponents.

These three factors alone would be sufficient to explain how we have become such unhinged killers. But I have to stress once more that all this, however convincing it may seem, must still be studied more thoroughly.

There is a frightening, and ironical paradox in this conclusion: that the human brain, the finest life-preserving device created by evolution, has made our species so successful in mastering the outside world that it suddenly finds itself taken off guard. One could say that our cortex and our brainstem (our "reason" and our "instincts") are at loggerheads. Together they have created a new social environment in which, rather than ensuring our survival, they are about to do the opposite. The brain finds itself seriously threatened by an enemy of its own making. It is its own enemy. We simply have to understand this enemy.

THE DEVELOPMENT OF BEHAVIOR

I must now leave the question of the moment-to-moment control of fighting, and, looking further back in time, turn to the development of aggressive behavior in the growing individual. Again we will start from the human problem. This, in the present context, is whether it is within our power to control development in such a way that we reduce or eliminate fighting among adults. Can or cannot education in the widest sense produce nonaggressive men?

The first step in the consideration of this problem is again to distinguish between external and internal influences, but now we must apply this to the growth, the changing, of the behavioral machinery during the individual's development. Here again the way in which we phrase our questions and our conclusions is of the utmost importance.

In order to discuss this issue fruitfully, I have to start once more by considering it in a wider context, which is now that of the "nature-nurture" problem with respect to behavior in general. This has been discussed more fully by Lorenz in his book *Evolution and Modification of Behaviour* (*18*); for a discussion of the environmentalist point of view I refer to the various works of Schneirla (see *19*).

Lorenz tends to classify behavior types into innate and acquired or learned behavior. Schneirla rejects this dichotomy into two classes of behavior. He stresses that the developmental process, of behavior as well as of other functions, should be considered, and also that this development forms a highly complicated series of interactions between the growing organism and its environment. I have gradually become convinced that the clue to this difference in approach is to be found in a difference in aims between the two authors. Lorenz claims that "we are justified in leaving, at least for the time being, to the care of the experimental embryologists all those questions which are concerned with the chains of physiological causation leading from the genome to the development of . . . neurosensory structures" (*18*, p. 43). In other words, he deliberately refrains from starting his analysis of development prior to the stage at which a fully coordinated behavior is performed for the first time. If one in this way restricts one's studies to the later stages of development, then a classification in "innate" and "learned" behavior, or behavior components, can be considered quite justified. And there was a time, some 30 years ago, when the almost grotesquely environmentalist bias of psychology made it imperative for ethologists to stress the extent to which behavior patterns could appear in perfect or near-perfect form without the aid of anything that could be properly called learning. But I now agree (however belatedly) with Schneirla that we must extend our interest to earlier stages of development and embark on a full program of experimental embryology of behavior. When we do this, we discover that interactions with the environment can indeed occur at early stages. These interactions may concern small components of the total machinery of a fully functional behavior pattern, and many of them cannot possibly be called learning. But they are interactions with the environment, and must be taken into account if we follow in the footsteps of the experimental embryologists, and extend our field of interest to the entire sequence of events which lead from the blueprints contained in the zygote to the fully functioning, behaving animal. We simply have to do this if we want an answer to the question to what extent the development of behavior can be influenced from the outside.

When we follow this procedure the rigid distinction between "innate" or unmodifiable and "acquired" or modifiable behavior patterns becomes far less sharp. This is owing to the discovery, on the one hand, that "innate" patterns may contain elements that at an early stage developed in interaction with the environment, and, on the other hand, that learning is, from step to step, limited by internally imposed restrictions.

To illustrate the first point, I take the development of the sensory cells in the retina of the eye. Knoll has shown (20) that the rods in the eyes of tadpoles cannot function properly unless they have first been exposed to light. This means that, although any visually guided response of a tadpole may well, in its integrated form, be "innate" in Lorenz's sense, it is so only in the sense of "nonlearned," not in that of "having grown without interaction with the environment." Now it has been shown by Cullen (21) that male sticklebacks reared from the egg in complete isolation from other animals will, when adult, show full fighting behavior to other males and courtship behavior to females when faced with them for the first time in their lives. This is admittedly an important fact, demonstrating that the various recognized forms of learning do not enter into the programing of these integrated patterns. This is a demonstration of what Lorenz calls an "innate response." But it does not exclude the possibility that parts of the machinery so employed may, at an earlier stage, have been influenced by the environment, as in the case of the tadpoles.

Second, there are also behavior patterns which do appear in the inexperienced animal, but in an incomplete form, and which require additional development through learning. Thorpe has analyzed a clear example of this: when young male chaffinches reared alone begin to produce their song for the first time, they utter a very imperfect warble; this develops into the full song only if, at a certain sensitive stage, the young birds have heard the full song of an adult male (22).

By far the most interesting aspect of such intermediates between innate and acquired behavior is the fact that learning is not indiscriminate, but is guided by a certain selectiveness on the part of the animal. This fact has been dimly recognized long ago; the early ethologists have often pointed out that different, even closely related, species learn different things even when developing the same behavior patterns. This has been emphasized by Lorenz's use of the term "innate teaching mechanism." Other authors use the word "template" in the same context. The best example I know is once more taken from the development of song in certain birds. As I have mentioned, the males of some birds acquire their full song by changing their basic repertoire to resemble the song of adults, which they have to hear during a special sensitive period some months before they sing themselves. It is in this sensitive period that they acquire, without as yet producing the song, the knowledge of "what the song ought to be like." In technical terms, the bird formed a *Sollwert* (*23*) (literally, "should-value," an ideal) for the feedback they receive when they hear their own first attempts. Experiments have shown (*24*) that such birds, when they start to sing, do three things: they listen to what they produce; they notice the difference between this feedback and the ideal song; and they correct their next performance.

This example, while demonstrating an internal teaching mechanism, shows, at the same time, that Lorenz made his concept too narrow when he coined the term "innate teaching mechanism." The birds have developed a teaching mechanism, but while it is true that it is internal, it is not innate; the birds have acquired it by listening to their father's song.

These examples show that if behavior studies are to catch up with experimental embryology our aims, our concepts, and our terms must be continually revised.

Before returning to aggression, I should like to elaborate a little further on general aspects of behavior development, because this will enable me to show the value of animal studies in another context, that of education.

Comparative studies, of different animal species, of different behavior patterns, and of different stages of development, begin to suggest that wherever learning takes a hand in development, it is guided by such *Sollwerte* or templates for the proper feedback, the feedback that reinforces. And it becomes clear that these various *Sollwerte* are of a bewildering variety. In human education one aspect of this has been emphasized in particular, and even applied in the use of teaching machines: the requirement that the reward, in order to have maximum effect, must be immediate. Skinner has stressed this so much because in our own teaching we have imposed an unnatural delay between, say, taking in homework, and giving the pupil his reward in the form of a mark. But we can learn more from animal studies than the need for immediacy of reward. The type of reward is also of great importance, and this may vary from task to task, from stage to stage, from occasion to occasion; the awards may be of almost infinite variety.

Here I have to discuss briefly a behavior of which I have so far been unable to find the equivalent in the development of structure. This is exploratory behavior. By this we mean a kind of behavior in which the animal sets out to acquire as much information about an object or a situation as it can possibly get. The behavior is intricately adapted to this end, and it terminates when the information has been stored, when the animal has incorporated it in its learned knowledge. This exploration (subjectively we speak of "curiosity") is not confined to the acquisition of information about the external world alone; at least mammals explore their own movements a great deal, and in this way "master new skills." Again, in this exploratory behavior, *Sollwerte* of expected, "hoped-for" feedbacks play their part.

Without going into more detail, we can characterize the picture we begin to get of the development of behavior as a series, or rather a web, of events, starting with innate programing instructions contained in the zygote, which straightaway begin to interact with the environment; this interaction may be discontinuous, in that periods of predominantly internal development alternate with periods of interaction, or sensitive periods. The interaction is enhanced by active exploration; it is steered by selective *Sollwerte* of great variety; and stage by stage this process ramifies; level upon level of ever-increasing complexity is being incorporated into the programing.

Apply what we have heard for a moment to playing children (I do not, of course, distinguish sharply between "play" and "learning"). At a certain age a child begins to use, say, building blocks. It will at first manipulate them in various ways, one at a time. Each way of manipulating acts as exploratory behavior: the child learns what a block looks, feels, tastes like, and so forth, and also how to put it down so that it stands stably.

Each of these stages "peters out" when the child knows what it wanted to find out. But as the development proceeds, a new level of exploration is added: the child discovers that it can put one block on top of the other; it constructs. The new discovery leads to repetition and variation, for each child develops, at some stage, a desire and a set of *Sollwerte* for such effects of construction, and acts out to the full this new level of exploratory behavior. In addition, already at this stage the *Sollwert* or ideal does not merely contain what the blocks do, but also what, for instance, the mother does; her approval, her shared enjoyment, is also of great importance. Just as an exploring animal, the child builds a kind of inverted pyramid of experience, built of layers, each set off by a new wave of exploration and each directed by new sets of *Sollwerte*, and so its development "snowballs." All these phases may well have more or less limited sensitive periods, which determine when the fullest effect can be obtained, and when the child is ready for the next step. More important still, if the opportunity for the next stage is offered either too early or too late, development may be damaged, including the development of motivational and emotional attitudes.

Of course gifted teachers of many generations have known all these things (25) or some of them, but the glimpses of insight have not been fully and scientifically systematized. In human education, this would of course involve experimentation. This need not worry us too much, because in our search for better educational procedures we are in effect experimenting on our children all the time. Also, children are fortunately incredibly resilient, and most grow up into pretty viable adults in spite of our fumbling educational efforts. Yet there is, of course, a limit to what we will allow ourselves, and this, I should like to emphasize, is where animal studies may well become even more important than they are already.

CAN EDUCATION END AGGRESSION?

Returning now to the development of animal and human aggression, I hope to have made at least several things clear: that behavior development is a very complex phenomenon indeed; that we have only begun to analyze it in animals; that with respect to man we are, if anything, behind in comparison with animal studies; and that I cannot do otherwise than repeat what I said in the beginning: we must make a major research effort. In this effort animal studies can help, but we are still very far from drawing very definite conclusions with regard to our question: To what extent shall we be able to render man less aggressive through manipulation of the environment, that is, by educational measures?

In such a situation personal opinions naturally vary a great deal. I do not hesitate to give as my personal opinion that Lorenz's book *On*

Aggression, in spite of its assertativeness, in spite of factual mistakes, and in spite of the many possibilities of misunderstandings that are due to the lack of a common language among students of behavior—that this work must be taken more seriously as a positive contribution to our problem than many critics have done. Lorenz is, in my opinion, right in claiming that elimination, through education, of the internal urge to fight will turn out to be very difficult, if not impossible.

Everything I have said so far seems to me to allow for only one conclusion. Apart from doing our utmost to return to a reasonable population density, apart from stopping the progressive depletion and pollution of our habitat, we must pursue the biological study of animal behavior for clarifying problems of human behavior of such magnitude as that of our aggression, and of education.

But research takes a long time, and we must remember that there are experts who forecast worldwide famine 10 to 20 years from now; and that we have enough weapons to wipe out all human life on earth. Whatever the causation of our aggression, the simple fact is that for the time being we are saddled with it. This means that there is a crying need for a crash program, for finding ways and means for keeping our inter-group aggression in check. This is of course in practice infinitely more difficult than controlling our intranational frictions; we have as yet not got a truly international police force. But there is hope for avoiding all-out war because, for the first time in history, we are afraid of killing ourselves by the lethal radiation effects even of bombs that we could drop in the enemy's territory. Our politicians know this. And as long as there is hope, there is every reason to try and learn what we can from animal studies. Here again they can be of help. We have already seen that animal opponents meeting in a hostile clash avoid bloodshed by using the expressions of their motivational conflicts as intimidating signals. Ethologists have studied such conflict movements in some detail (*26*), and have found that they are of a variety of types. The most instructive of these is the redirected attack; instead of attacking the provoking, yet dreaded, opponent, animals often attack something else, often even an inanimate object. We ourselves bang the table with our fists. Redirection includes something like sublimation, a term attaching a value judgment to the redirection. As a species with group territories, humans, like hyenas, unite when meeting a common enemy. We do already sublimate our group aggression. The Dutch feel united in their fight against the sea. Scientists do attack their problems together. The space program—surely a mainly military effort—is an up-to-date example. I would not like to claim, as Lorenz does, that redirected attack exhausts the aggressive urge. We know from soccer matches and from animal work how aggressive behavior has two simultaneous, but opposite effects: a waning effect, and one of self-inflammation, of mass hysteria, such as recently seen in Cairo. Of these two the inflammatory effect often wins. But if aggression were used successfully as the motive force behind nonkilling and even useful activities, self-stimulation need not be a danger; in our short-term cure we are not aiming at the elimination of aggressiveness, but at "taking the sting out of it."

Of all sublimated activities, scientific research would seem to offer the best opportunities for deflecting and sublimating our aggression. And, once we recognize that it is the disrupted relation between our own behavior and our environment that forms our most deadly enemy, what could be better than uniting, at the front or behind the lines, in the scientific attack on our own behavioral problems?

I stress "behind the lines." The whole population should be made to feel that it participates in the struggle. This is why scientists will always have the duty to inform their fellowmen of what they are doing, of the relevance and the importance of their work. And this is not only a duty, it can give intense satisfaction.

I have come full circle. For both the long-term and the short-term remedies at least we scientists will have to sublimate our aggression into an all-out attack on the enemy within. For this the enemy must be recognized for what it is: our unknown selves, or, deeper down, our refusal to admit that man is, to himself, unknown.

I should like to conclude by saying a few words to my colleagues of the younger generation. Of course we all hope that, by muddling along until we have acquired better understanding, self-annihilation either by the "whimper of famine," or by the "bang of war" can be avoided. For this, we must on the one hand trust, on the other hand help (and urge) our politicians. But it is no use denying that the chances of designing the necessary preventive measures are small, let alone the chances of carrying them out. Even birth control still offers a major problem.

It is difficult for my generation to know how seriously you take the danger of mankind destroying his own species. But those who share the apprehension of my generation might perhaps, with us, derive strength from keeping alive the thought that has helped so many of us in the past when faced with the possibility of imminent death. Scientific research is one of the finest occupations of our mind. It is, with art and religion, one of the uniquely human ways of meeting nature, in fact, the most active way. If we are to succumb, and even if this were to be ultimately due to our own stupidity, we could still, so to speak, redeem our species. We could at least go down with some dignity, by using our brain for one of its supreme tasks, by exploring to the end.

REFERENCES

1. A. Carrel, *L'Homme, cet Inconnu* (Librarie Plon, Paris, 1935).

2. AAAS Annual Meeting, 1967 [see *New Scientist 37*, 5 (1968)].

3. R. Carson, *Silent Spring* (Houghton Mifflin, Boston, 1962).

4. K. Lorenz, *On Aggression* (Methuen, London, 1966).

5. D. Morris, *The Naked Ape* (Jonathan Cape, London, 1967).

6. N. Tinbergen, *Z. Tierpsychol. 20*, 410 (1964).

7. R. A. Hinde, *New Society 9*, 302 (1967).

8. W. S. Hoar, *Animal Behaviour 10*, 247 (1962).

9. B. Baggerman, in *Symp. Soc. Exp. Biol. 20*, 427 (1965).

10. H. Kruuk, *New Scientist 30*, 849 (1966).

11. N. Tinbergen, *Z. Tierpsychol. 16*, 651 (1959); *Zool. Mededelingen 39*, 209 (1964).

12. ——, *Behaviour 15*, 1-70 (1959).

13. L. von Haartman, *Evolution 11*, 339 (1957).

14. E. Cullin, *Ibis 99*, 275 (1957).

15. J. H. Crook, *Symp. Zool. Soc. London 14*, 181 (1965).

16. D. Freeman, *Inst. Biol. Symp. 13*, 109 (1964); D. Morris, Ed., *Primate Ethology* (Weidenfeld and Nicolson, London, 1967).

17. H. E. Howard, *Territory in Bird Life* (Murray, London, 1920); R. A. Hinde et al., *Ibis 98*, 340-530 (1956).

18. K. Lorenz, *Evolution and Modification of Behaviour* (Methuen, London, 1966).

19. T. C. Schneirla, *Quart. Rev. Biol. 41*, 283 (1966).

20. M. D. Knoll, *Z. Vergleich. Physiol. 38*, 219 (1956).

21. E. Cullen, *Final Rept. Contr. AF 61 (052)-29*, USAFRDC, 1-23 (1961).

22. W. H. Thorpe, *Bird-Song* (Cambridge Univ. Press, New York, 1961).

23. E. von Holst and H. Mittelstaedt, *Naturwissenschaften 37*, 464 (1950).

24. M. Konishi, *Z. Tierpsychol. 22*, 770 (1965); F. Nottebohm, *Proc. 14th Intern. Ornithol. Congr.* 265-280 (1967).

25. E. M. Standing, *Maria Montessori* (New American Library, New York, 1962).

26. N. Tinbergen, in *The Pathology and Treatment of Sexual Deviation*, I. Rosen, Ed. (Oxford Univ. Press, London, 1964), pp. 3-23; N. B. Jones, *Wildfowl Trust 11th Ann. Rept.*, 46-52 (1960); P. Sevenster, *Behaviour, Suppl. 9*, 1-170 (1961); F. Rowell, *Animal Behaviour 9*, 38 (1961).

19   Possible Substitutes for War

—ANTHONY STORR

The psychiatrist who, like myself, specializes in psychotherapy, must necessarily feel ill at ease when speaking to such an audience as this, expecially when his place on the programme is such that he has followed so many scientists whose opinions are backed by experiment and verification.*

The doctor who seeks to help the individual human being is always at a disadvantage compared with the biologist, the anthropologist, or the historian. For his conclusions are bound to be limited by the human material with which he deals. His conception of human nature is closely circumscribed by the consulting room, and, although the view from a Harley Street window may take in the activities of a Dr. Ward or a Miss Keeler in the neighbouring mews, it cannot comprehend the wider vistas which unfold themselves before those whose concern is with the destinies of nations, of races, or of species. Remember then that what I have to say is limited because my experience is limited, and that in discussing the natural history of aggression, my observations are based upon my experience of treating a relatively small number of human beings who may themselves be unrepresentative of even Western society in general.

My subject is entitled "Possible Substitutes for War", and this title itself implies that war has not always been regarded as wholly evil. For if a thing is totally bad, we should surely seek simply to abolish it, rather than to look for substitutes for it. The very fact that we might need substitutes for war implies that war satisfies a need in our human

From *The Natural History of Aggression* (J. D. Carthy and F. J. Ebling, Eds.), London: Academic Press, 1964. Reprinted by permission of A. D. Peters and Co.

nature, and thus has something valuable about it. We don't look for substitutes for cholera or the plague, although I suppose certain groups of persons might regard contraception in the same light, or even as a worse evil than these epidemics. Until the invention of nuclear weapons, however, there can be no doubt that war satisfied deeply felt needs, and this is why men have been so reluctant to abandon it. If we are to discuss possible substitutes for war, it is important first to examine the satisfactions which war used to provide.

One of these satisfactions was certainly an increased sense of identity with others. American research into human reactions to disaster has shown that, under extreme threat, human beings cling more closely together. Distinctions of class, of age, of status, all tend to disappear. Faced with a common enemy, whether this be flood or fire or human opponent, we become brothers in a way which never obtains in ordinary life. It is a great thing to have an enemy, for it is only then that we discover our neighbour; only then that we can transcend the barriers of class, of education or of creed which generally divide us, and which descend once more when external danger ceases to threaten us. The comradeship of war, the fact that, under conditions of stress, our capacity for identification with our fellows is increased, has been one reason for the continued popularity of war. There can be few people who do not recall something of the increased warmth they both showed towards and received from their fellow men after exposure to some such common danger as a night's bombing of London; and there are many who look back to the days of the blitz with nostalgia, as is evidenced by the eagerness with which, even twenty years after, they are prepared to recall those sleepless nights and smoking dawns. Hand in hand with this increased feeling of companionship, of identity with the group, goes a diminished sense of individual responsibility. Every psychiatrist is familiar with cases of men who broke down under stress of peace, because they were unable to resume personal responsibility for their lives or to take decisions which, during war, had generally been taken for them. In a democratic society we tend to assume that every adult individual is capable of independence and rational decision. But, in wartime, innumerable individuals are thankful to abandon the burden of conscious choice, to submerge their individuality in a crowd, and to take their orders from above. To be fed, clothed, and delivered from immediate anxiety is such a relief to many characters that these advantages amply compensate them for the loss of liberty which is their inevitable accompaniment.

Nor is this relief confined to the weak or to the neurotic:

decision-making is a burden to all of us, which is why we tend to give high financial rewards to the decision-makers in our society. In wartime, life is simplified for everyone, for the collective decision has been taken that the enemy must be defeated; and all other decisions and values are subsidiary to this one. Those people, therefore, who have some difficulty in finding any overruling purpose to which to devote their lives and who are unsatisfied by the mundane incentives which motivate the average person, find an almost religious satisfaction in devoting themselves to one main objective, and in orientating their lives in submission to the single wartime aim of victory.

The psychological advantages of a feeling of group solidarity, of relief from personal responsibility, and of the incentive given by a sense of purpose merit further discussion, but this conference is principally concerned with aggression, and there can be little doubt that, in the past, an important function of war was to provide an opportunity for the apparently justifiable discharge of those aggressive impulses which seem so inescapable a part of human nature.

Elsewhere I have tried to define what I mean by aggression, and find it very difficult. At the risk of repetition, I must briefly summarize my argument. Like most psychotherapists nowadays, to whatever school they may belong, I cannot accept Freud's idea of a "death instinct". Freud's conception of aggression was that it was primarily self-destructive; an instinct which, to quote his own words, was "trying to bring living matter back into an inorganic condition". The aggressiveness which men show towards the external world was, in Freud's view, a secondary phenomenon. The death instinct was the primary urge, a kind of personification of the second law of thermodynamics. Although decay and death are certainly our lot, and although, as Eddington puts it, time's arrow cannot be reversed, I cannot believe in an instinct which is self-destructive. Entropy constantly increases, but whatever the forces are within us which lead to our final dissolution, they are not to be subsumed under the same heading as the instincts which serve to preserve us or to encourage us to reproduce. Surely the very concept of instinct is of a pattern of behaviour which is of some value to the organism in question.

Nor is it possible to accept the idea that man's aggression is merely a response to frustration. No one doubts that frustration increases aggressiveness: you have only to try and drive a car in London at 5:30 p.m. on any weekday evening to discover this. But to imagine that, if only we had all had ideally loving parents and the serenest of possible childhoods we would not be aggressive creatures, is to be totally unrealistic about human nature.

Throughout history men have had a vision of the millennium; a condition in which perfect peace would prevail, men would agree with each other, and we should all be splendidly co-operative, creative and free. This vision can be found in Greek mythology, in Ovid, in Isaiah, and even in Bertrand Russell. Men think that a world without war would be a world without aggression; and envisage a future in which we should all reach undreamed-of heights of prosperity and achievement, under the beneficent care of a single world government.

Such visions are based on the idea that we can somehow get rid of our aggression. If only, these prophets allege, inequality between nations was abolished, or capitalism overthrown, or Esperanto universally spoken, or birth control everywhere adopted—then, at last, we should be able to live at peace with one another, and our true nature, pacific, gentle and loving, become universally manifest.

The idea that we can get rid of aggression seems to me to be nonsense. Surely everything we know of the behaviour of men in groups contradicts this conception. Innumerable attempts have been made to found communities in which there would be no cause for strife—but strife always breaks in; and where it is least expected, it is usually most destructive. My favourite example is the psychoanalytic movement. No sooner had Freud established himself when Adler broke away and then Jung. Later, the Freudian school was split in two by the heterodoxy of Melanie Klein, and the Jungian group has been similarly riven. Those of you in academic life will know of many comparable examples. Man's aggression is more than a response to frustration—it is an attempt to assert himself as an individual, to separate himself from the herd, to find his own identity.

As you will understand, the conception of aggression which I am advancing is one which lays stress upon its positive aspects. My observation of individuals leads me to suppose that aggression only becomes really dangerous when it is suppressed or disowned. The man who is able to assert himself is seldom vicious; it is the weak who are most likely to stab one in the back.

This positive aspect of aggression is seldom emphasized. Owing to an accident of history, Freud, who has had more influence than any other individual upon man's view of his own nature, did not admit the existence of a separate aggressive drive until fairly late in his life, and then found it so difficult to fit into his scheme that he was led into the blind alley of the death instinct conception. Other psychologists, however, notably Alfred Adler, have always recognized that a striving for superiority or urge for power, or desire for self-assertion, exists, and common sense demands that some such drive be recognized as of equal importance with the sexual instinct.

It is generally admitted that the further back into infancy one pursues the phantasies of children, the more aggression does one find, and it is impossible to believe that the whole of this aggressive potential springs from frustration at the breast. Even the most lovingly reared children generally go through a phase of rebellion at about the age of three, in which they assert their own individuality in powerfully aggressive terms. Adolescents repeat this pattern at a later stage. Indeed, adolescents need to rebel, and when there is nothing to rebel against, invent imaginary figures upon whom to vent their wrath. The notable increase in juvenile crime in the last twenty years may actually be connected with the lessening of external frustration. If parents cannot be treated as scapegoats, aggression finds other objects to attack.

Moreover, it is important to realize that no new discovery would ever be made if men were not intolerant of the old and violently assertive of the new. Aggression and creation march hand in hand and, as Bernard Berensen once put it, "Genius is the capacity for productive reaction against one's training".

In considering war and the opportunity which it provided for expressing the aggressive side of our nature, most writers have either fallen into the trap of millenniary pacifism or else exalted war with a kind of schoolboy jingoism which I find utterly repulsive. War is a great and pressing evil which we must get rid of, but it is no use supposing that we can change human nature into something pacific and gentle. For, in our efforts to realize our full potential, struggle and opposition are absolutely necessary. If enemies do not exist, we promptly invent them, as anyone who has served on a committee must know. Our mistake is to think that anything is achieved by destroying our enemies. On the contrary, we ought to struggle to preserve them.

The affluent society cushions us against hunger, against disease, and against destruction, and in doing so, deprives us of any opportunity to test ourselves to the limit, to struggle or to die. No wonder the old-fashioned kind of war was popular. For, in our humdrum passage from our well-sprung perambulators to our decorous coffins, what opportunities have we for heroism, for self-sacrifice, or for identifying ourselves with causes which transcend our petty struggles for recognition, for status, or for dominance in the human pecking order? Until nuclear weapons finally precluded heroism, war did provide a field in which men realized potentialities of courage and endurance which seldom come their way in peace time. To wait until senile decrepitude puts an end to one's protracted plush existence is not necessarily an agreeable prospect, and many natures only find fulfilment in circumstances where they are exposed to risk or at least discomfort.

The human tendency to welcome difficulties and dangers is a very interesting one. To my mind it is not fully explained by the psycho-analytic theory of masochism. Of course there are people who torture and punish themselves, consciously or unconsciously, for past offences. Of course there are people who can never allow themselves any pleasure, and who make even the smallest task into a difficult, painful examination which they never pass. On the other hand, not every task which a man may set himself falls into this category and it cannot be assumed that to welcome a certain amount of difficulty and danger is invariably pathological. To attempt the North Face of the Eiger may be a piece of masochistic folly, but not everyone who enjoys physical exertion followed by relaxation is attempting to bolster a shaky masculinity. "Strength", said Leonardo da Vinci, "is born of constraint and dies in freedom" and without a certain constraint, how is a man to discover his own strength? It seems to me that an attitude to life which takes some pleasure in overcoming obstacles or facing risks is preferable to one in which everything unpleasant is avoided and the easiest path invariably preferred, and, in the past, war has afforded men opportunities for stretching themselves to the limit, thus not only proving their strength, but evoking potentialities perhaps undreamed of in peaceful conditions.

Human potentialities which are unused tend to cause trouble to their owner. The buried talent does not lie quietly in the ground, but sets up subterranean disturbances which manifest themselves in the form of anxiety and other symptoms. Freud showed us how the repression of sexuality could lead to neurotic symptoms which have far-reaching effects on the health and happiness of the whole being. To deny expression to so important a part of human nature is to run the risk that nature will protest.

In the same way it seems probable that the denial or repression of our aggressive drives is liable to cause disharmony within ourselves, however desirable it may be that we should get rid of them. Judging from the popularity of violence on the cinema screen and television, the eagerness with which we watch boxing or wrestling, and the delight with which we read of murder, it seems certain that, automatically and inevitably, we are constantly seeking opportunities for the vicarious expression of aggressive drives. Can it be that our form of civilization is at fault? Erich Fromm has pointed out that the highest rates of suicide and alcoholism are to be found in countries which are generally considered the most democratic, peaceful and prosperous. Denmark, Switzerland, Finland, Sweden and the U.S.A. head the list. Suicide is obviously a turning in of aggression against the self, and alcoholism is equally an expression of the way in which men destroy themselves. In this country, over 5,000 people a year commit suicide, whilst 30,000 attempt it; and although we do not commit many murders, the three-fold increase in our prison population since 1938 and the racial disturbances in Notting Hill and elsewhere, surely attest the presence of violent destructive impulses which we contain within us only with difficulty. In war, these impulses used to find an acceptable channel for discharge. They can do so no longer except in wars between nations who do not possess nuclear weapons. Any satisfaction which some megalomaniac statesmen might obtain from the despatch of a nuclear missile is likely to be swiftly curtailed by his own annihilation; and although another Hitler might prefer to destroy the world and himself together, he will not thereby be providing us with an outlet for our own aggressive drives.

It seems to me just possible that we shall succeed in avoiding the holocaust; a view which has received support from the events of the Cuban crisis. But the outlawing of nuclear war, or even the abolition of war altogether, does not solve the problem of our own aggressiveness. In some ways it makes it more difficult, and that is why we ought to consider possible substitutes for war.

In a recent book, Professor Rapoport of the University of Michigan, who works at the Center for Research on Conflict Resolution, has discussed the various ways in which mathematical theory can be applied to the study of human conflict. He calls his book *Fights, Games and Debates*, and he advances the hope that our fights may be changed into games and our games into debates. His viewpoint has the signal virtue of recognizing that the alternative to war is not necessarily peace. Clausewitz defined war as "The continuation of policy by other means." Our only hope is that we can continue war by other means than the primitive one of killing each other.

This is not an impossible task, for it has been solved by other members of the animal kingdom. As Professor Eckhard Hess wrote recently of fighting between members of the same species: "Actions that injure the opponent have been removed to the end of the sequence of fighting behaviour by raising the threshold for its release to a very high level. This resulted in the development of tournaments with a very small likelihood of actual bodily harm, a development which has clear survival value to the species, since fighting behaviour will maintain its function of spacing out members of the same species without causing injury to species members."

It may seem stupidly naive to think that we can substitute ritual struggles for war; but in my view this has already happened. It is not so long ago that we found our enemies from north of the border a serious menace against whom we sent punitive expeditions. Now we defeat them at Twickenham, or are ourselves defeated. The struggle remains, but the form of it is different. I am not so naive as to imagine that serious conflicts of interest between sovereign states can be settled by football matches or gladiatorial encounters. But I do believe that serious conflicts of interest between sovereign states are not now the real problem, except, perhaps in those parts of the world which are grossly over-populated. The days when great powers might come into conflict because they were greedily competing for rich slices of Africa are surely over. What we have to deal with now are the problems of ideological conflicts—in other words, with psychology even more than with economics. Although we have not yet solved the problem of the underdeveloped countries, we do know more or less what to do about them, and it seems unlikely that nuclear war will break out as a form of conflict between the haves and have-nots, more especially if we can prevent the spread of nuclear weapons.

Ideological conflicts will go on so long as human nature remains as it is; it is the problem of resolving these without recourse to destructive violence which we have to deal with. I was recently taken to task for defending the vast sums spent on the space race on the grounds that it might be serving the function of a ritual conflict between East and West. Those who believe that conflict can be abolished would rather see the money spent on welfare, and I sympathize with their desire. But, so long as sovereign states exist there will be competitive struggles between them. The desire for prestige and power is stronger than the desire to see everyone cared for. And, in considering human nature, it must be taken into account. Do you remember how at the time of Cuba the newspapers were much concerned with Mr. Kruschev's pride? Was he going to be able to climb down without too much loss of prestige or not? It is absurd to think that millions of lives may depend upon whether a single individual's self-esteem is damaged, but such is human nature. If I am right in thinking that the space race serves a valuable function, we should not grudge expenditure on it.

I do not believe that ideological conflicts can be solved by the deliberate substitution of alternatives. But, because of man's innate aggression, these alternatives arise automatically if conflict in the form of war is impossible. There will always be plenty of ways in which countries can compete, whether it be in the space race, in education, in technology, or even in welfare. We ought to encourage competition in these fields as much as we possibly can.

But there is one vital step which has to be taken if we are to ensure that ritual and other forms of struggle are to become substitutes for war. This step is what Professor Rapoport has called the "assumption of similarity". In playing a game, we generally suppose that our opponent is a man like ourselves, who will think in roughly the same way, strike at the same objectives, and make the same sort of choices. In war, propaganda plays upon the paranoid tendency, latent in all of us, which enables us to regard other members of our species as totally different. It is not difficult to convince people that if a man has a skin of a different colour and professes a different faith, he is so utterly alien that he deserves to be destroyed. One may not kill one's neighbour, but that other man, the enemy, deserves nothing but death. For is he not a raper of women, a torturer of children, a bomber of civilians? —but I had better not go on. Thinking of the bombing of Dresden, of the Suez fiasco, of Hola, of Cyprus, and even of Aden, it is well for me to stop. For what I want to destroy in my enemy is what I cannot stomach in myself, and to kill him is to commit suicide. It is only when we can fully realize this truth that we can learn to value our enemy, and learn to fight him without destroying him. "Any man's death diminishes me, because I am involved in mankind: and therefore never send to know for whom the bell tolls; it tolls for thee."

FOOTNOTE

* Dr. Storr's opening remarks refer to Institute of Biology Symposium No. 13 (The Natural History of Aggression) held in London, October 1963. It was during this program that Storr's paper was first delivered). THE EDITORS

20   A Systematization of Gandhian Ethics of Conflict Resolution

—ARNE NAESS

I. INTRODUCTORY REMARKS

Since 1947 a great number of publications in the social sciences (taken in a broad sense) and in philosophy have had peaceful co-operation between the major power constellations of today as a main or subsidiary topic. Various surveys, such as the UNESCO publication *The Nature of Conflict*, show a reassuring richness in aspects and approaches. There has been little done, however, to utilize the vast potential of those attitudes of non-violence which have crystallized in more or less explicit ethical doctrines. Comparatively few publications attempt a synthesis of philosophical and social science approaches. In this paper an example of such a synthesis will be offered.

Any normative, systematic ethics containing a general norm against violence will be called an "ethics of non-violence." (The norm must, of course, exclude war for defensive purposes.) The term "violence" must cover not only open, physical violence but also injury and psychic terror. The term "hostility" would, perhaps, give rise to more adequate associations. In the following an instance of such an ethics of non-violence, that of Gandhi will be given in a condensed, systematized form. The task is important in part because it makes it more easy to distinguish essentials from non-essentials, and features due to particular historical situations from features of general, timeless, or at least very permanent validity or applicability.

II. SYSTEMATIZATION D OF GANDHIAN ETHICS OF CONFLICT RESOLUTION

This part of the paper has a definite, very limited purpose: to give a brief and highly condensed exposition of one part of Gandhi's ethics of conflict.

Reprinted by permission from the *Journal of Conflict Resolution*, 1958, *2*, 140-155.

In the realm of political action, Gandhi's views and precepts were usually explicit. According to his ethics, explicitness is a duty. His politically relevant actions were innumerable, and he made a running comment on them in terms of ethical appraisals. This makes it practicable to work out broad interrelated groups of sentences representing *rational reconstructions* or *models* covering Gandhi's ethically relevant verbal behavior.

The primary sources for this kind of inquiry are historical documents and other materials concerning Gandhi's activities, his own systematic writings, and his correspondence and the conversations, speeches, and so on, which were recorded or summarized by Mahadev Desai and others. Much of this material has already been printed and is easily available.

If I were to mention a publication which has particularly high value for rational reconstructions, I should choose the first volume of his *Non-Violence in Peace and War*. It was not written completely by Gandhi himself. It includes not only a collection of newspaper articles and letters but also recordings of conversations. They are all dated, and most of them refer to well-known political actions going on at the time. The concrete nature of the problems at issue does not reduce the philosophical value of the material. On the contrary. The interpretation of ethical texts of professional philosophers are usually hindered by an almost complete lack of reference to application in concrete situations. This holds good for Plato, Hobbes, Nietzsche, and others. Even constructed examples are sometimes lacking. Without abundant application to concrete and well-known situations, ethical doctrines are almost impenetrable to analysis.

In the following, one particular version, "D," of one particular rational reconstruction in the form of a normative system will be outlined. The system belongs to the class of systems which outline, reflect, or portray not all Gandhian thought but Gandhi's *ethics of group struggle between 1907 and 1934*. It does this, as far as I can judge, sufficiently closely and extensively to be considered an adequate rational reconstruction.

The version D is a condensed and therefore to some extent a rough exposition of the system. Concerning the adequacy of systematization D, the following should be added: The norms N1-N25 and most of the hypotheses are selected on the basis of a survey of norms and hypotheses in Gandhi's writings. Some of our formulations are rather close to those of Gandhi, others are only indirectly or in part derived from him. Our main concern has been to assure that all norms of group

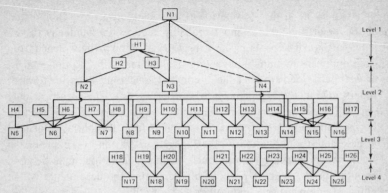

Diagram Showing Derivations in Systematization D

ethics necessary to justify and explain *satyagraha* (as described by Gandhi) are included, in a rough way, in N1-N25 and that no norm is contrary to the spirit of the formulations found in Gandhi's texts. Thus completeness or comprehensiveness has ranked high in our choice of D as a first approximation to an optimal version of the system.

The (necessary) documentation in testing the degree of material adequacy of the systematization D requires a separate article. Most of the quotations and comments required in such documentation are already published (in Norwegian) (2).

The question will not be raised as to what extent systematizations and systems may differ and still belong to the class of adequate rational reconstructions. There are presumably considerable changes in content which might be made without doing violence to the available observational material.

The ethics of group struggle is conceived in this article as a part of ethics in general, but with a certain amount of independence: the total set of its norms is derived from a small number of norms which concern group struggle and a set of (non-normative) hypotheses.

The dependence of the part upon the whole of ethics is structurally shown by the derivation of the basic norms concerning group struggle from norms of other parts of ethics. This dependence is also indicated by the fact that some of the norms of the particular version of the ethics of group struggle outlined here ("systematization D") can be derived from norms of other parts of ethics by processes of inference which do not include the basic norms of the ethics of group struggle. Thus norm N8, "Do not humiliate or provoke your opponent," in systematization D is derived from norm N4 and hypothesis H9, that is, from "If you are not able to subsume any of a group of relevant actions or attitudes as in themselves violent or constructive, then choose that action or attitude which most probably reduces the tendency to violence in the participants in the struggle" and "You invite violence from your opponent by humiliating and provoking him." But norm N8 might as well be derived from a code of conduct concerning behavior towards others, whether participants in a struggle or not. The historical data permit (of course) a number of different explications of the derivation of the top norm N1 of systematization D from top norms of general ethics. Here is one possibility, the *D-systematization. It is expressed in terms which certainly require much comment but which may be good enough for the present purpose of illustrating the dependence of the ethics of group struggle upon other parts of ethics.

*N1. Seek complete self-realization.

*H1. Complete self-realization presupposes that you seek truth.

*H2. All living beings are ultimately one.

*H3. Violence against yourself makes complete self-realization impossible.

*H4. Violence against any living being is violence against yourself (derived from *H2).

*H5. Violence against any living being makes complete self-realization impossible (derived from *H3 and *H4).

*N2. Realize non-violence and seek truth (derived from *N1, *H1, and *H5).

N1. Act in group struggle and act, moreover, in a way conducive to long-term, universal, maximal reduction of violence.

The derivation of N1 from the basic general form of self-realization permits us to picture the ethics of group struggle as an application of that norm to particular situations. It should be noted that N1 is not characteristic of consistent (or rather extreme) pacifist positions, since it may be argued, without attacking N1, that killing in group struggle may in some situations be more conducive to the long-term, universal reduction of violence than non-killing.

Systematization D

N1. Act in group struggle and act, moreover, in a way conducive to long-term universal reduction of violence.

Sentence N1 is intended to express *the* top norm of the system. All others norms are conceived to be derivable from this norm +

hypotheses. *The normative power of the system rests with N1 and N1 alone*. Instead of using the phrase "hypotheses and norms of the system," we might as well have the phrase "descriptions and prescriptions." The term "hypothesis" is used because it suggests what we wish to emphasize, the empirical, a posteriori, character of the statements. Since all norms of the system except N1 are prescribed only under the condition that certain hypotheses are true, the whole system except N1 is, in principle, open to scrutiny from a scientific point of view. That is, the validity of every single statement of the ethics of group struggle depends upon the truth and tenability of a set of empirical hypotheses, *testable only by the techniques of the social sciences*. This is asserted here in relation to systematization D only, but other systematizations would show a similar implicit dependence upon social science.

The top norm N1 is preferred to a norm simply saying "Do not use violence," because, among other things, it would be too narrow. N1 envisages a reduction of violence, not only the reduction of one's own violence. Gandhi demands not only personal abstention from violence but a conduct that does not provoke violence on the part of the opponent or anybody else affected by our conduct. Thus we should not humiliate him by certain kinds of passive resistance, because this is likely to produce hatred, which, in turn, may strengthen his disposition towards future use of violence.

There is another important aspect of N1: it requires that we act in group struggles and do not run away from the area of conflict. Here the basic attitude of the *karamayogi* reveals itself: one cannot retreat to the solitudes of the Himalayas in order to follow N1, because nonviolence by mere isolation from others is not likely to induce nonviolent behavior in others. It is by personal interaction in conflict situations that we can best reduce violence.

The use of the term "violence" in Gandhian texts is such that sometimes rather narrow and sometimes rather broad concepts can be made to fit the occurrence of the term. On the whole, "violence" is used to include much more than physical violence and injury. In this article we shall leave much of the ambiguity and vagueness untouched. For the purpose of systematization of a somewhat higher level of verbal precision than D, the following definition may suffice:

Definition 1:

"The person P is violent toward the person Q in a given situation S (or at a given time T)" shall mean the same as "The person P is injuring or coercing, or he intends to injure or coerce, or he would, if given opportunity (in that situation) injure or coerce the person Q in the situation S (or at the time T)."

The person Q may be P himself. In a next approximation to an adequate systematization, the terms "injure" and "coerce" would also either have to be carefully introduced or substituted by others.

H1. The means determine the results.

Gandhi formulated his view on this point in a most categorical way. For instance, he wrote: "Means and ends are convertible in my philosophy of life. They say 'means are, after all, means.' I would say 'means are, after all, everything.' As the means, so the end." And he has expressed his idea in this way: "The means may be likened to a seed, the end to a tree; and there is just the same inviolable connection between the means and the end as there is between the seed and the tree." H1 might also be thus formulated: The character of the means determines the character of the results.

Hypotheses H2 and H3 below are derived from H1. They may, however, be considered separately; those who hesitate to accept Gandhi's strong view of means and ends should not make their evaluation of H2 and H3 wholly dependant on H1. The latter is difficult to confirm in its extremely general form.

H2. In a group struggle you can keep the goal-directed motivation and the ability to work effectively for the realization of the goal stronger than the destructive, violent tendencies and the tendencies to passiveness and despondency only by making a constructive program part of your campaign and by giving all phases of your struggle, as far as possible, a constructive character.

A quotation from his paper *Harijan* (1946) indicates how important Gandhi found this hypothesis: "By hammering away at it through painful years," replied Gandhi, "people have begun to see that there is a potency in non-violence, but they have not seen it in all its fulness and beauty. If they had responded to all the steps that had to be taken for the effective organization of non-violence and carried out in their fulness the various items of the eighteen-fold constructive programme, our movement would have taken us to our goal. But today our minds are confused because our faith in constructive work is so weak."

We assume tacitly that the goal is acceptable from the point of view of Gandhi's ethics as a whole. This assumption is used in relation to other hypotheses and norms of systematization D. The system is conceived as a part of a general system of ethics, and directives as to how to fight for a bad cause are irrelevant. This note is important because otherwise one cannot assume, as in H2, that there is an incompatibility between goal-directed motivation and destructive, violent tendencies.

H3. Being violent counteracts long-term, universal reduction of violence.

The qualification "long-term, universal" is used in order to provide a basis for the argument that, even if the short-term result of a war or of a minor violent act may be complete suppression of a large-scale violence that at the moment is threatening, the long-term effects of the violence are likely to result in more violence than was avoided as an immediate result.

N2. Make a constructive program part of your campaign and give, as far as possible, all phases of your struggle a constructive character.

Norm N2 is conceived to be derivable from N1 and the hypothesis H2.

N3. Never resort to violence against your opponent.

This norm is conceived to be derivable from N1 and H3. Actually, as formulated above, no derivation is, of course, possible in any strictly logical sense. Such derivation would require complete formalization of the system. Here we can only offer a point of departure for explications with adequate logical relations. Remarks similar to this are called for in many other instances in the following where the terms "derive" and "derivable" are used.

N4. If you are not able to subsume any of a group of relevant actions or attitudes as in themselves violent or constructive, then choose that action or attitude which most probably reduces the tendency toward violence (i.e., violent actions or attitudes) in the participants in the struggle (partisans as well as opponents).

This norm is derived from N1 as a specification of it (it might also be conceived as derived from N1 and H1).

The next norms are derived from norms N2, N3, and N4 with the aid of further hypotheses. In order to facilitate surveying the systematization as a whole, we shall write it out in a somewhat schematic way.

A norm is said to be on level k, $k > 1$, if it is directly derived from a norm of level $k - 1$ together with certain hypotheses or as a specification of it. A hypothesis is said to be of level k, if it is used in the direct derivation of level $k + 1$.

First-level norms and hypotheses:
N1, H1, H2, H3.
Second-level norms and hypotheses:
N2. Derived from N1 and H2.
N3. Derived from N1 and H3.
N4. Derived from N1 or from N1 and H1.

H4. You can give the struggle a constructive character only if you conceive it and carry it through as a struggle *in favor of* human beings and certain values, thus eventually fighting antagonisms, but not antagonists.

H5. It will have a constructive effect on you yourself and on those for whom you struggle if you live together with them and do constructive work for them.

H6. It will create a natural basis for confidence in you among those for whom you struggle if you live together with them and do constructive work for them.

H7. All human beings have interests—at least long-term interests—in common (derivable from *H2).

H8. Co-operation on common goals reduces the chance that the actions and attitudes of the participants in the conflict will become violent.

H9. You invite violence from your opponent by humiliating or provoking him.

Thus if as part of a boycott of a university or a shop you lie down in the corridors so as to make it impossible for those opposed to the boycott to avoid stepping on you, your opponent is humiliated. He may refrain from entering the building for respectable ethical reasons or do it with resentment and anger. He is not likely to be won to your case, but, on the contrary, he will be more willing to use extreme measures in the conflict.

H10. Thorough knowledge of the relevant facts and factors increases the chance of a non-violent realization of the goal of your campaign.

Gandhi always acquired a thorough knowledge of relevant circumstances before he acted, and he warned his adherents against advocating his cause before they knew well the different aspects of the problems concerned.

H11. Secrecy and distortion or avoidance of truth reduce the chance of a non-violent realization of the goal of your campaign.

As has been indicated above, in the *D-systematization, the demand for truth was central in Gandhi's ethics.

It might be pointed out here as a subhypothesis that the intention to keep certain plans, moves, motives, and objectives secret influences our behavior so that we cannot face our opponent openly; such an intention is also more easily revealed to the opponent than we are likely to believe.

H12. You are less likely to take a violent attitude, the better you make clear to yourself what are the essential points in your cause and your struggle.

H13. Your opponent is less likely to use violent means, the better he understands your conduct and your case.

On the whole, Gandhi would insist that we inform our opponent more completely and especially by action, not mere proclamations.

H14. There is a disposition in every opponent such that wholehearted, intelligent, strong, and persistent appeal in favor of a good cause is able ultimately to convince him.

Gandhi tended to include any normal person in the intended field of validity of this hypothesis, interpreting "normal" wide enough to cover Hitler. A person's capacity to convince the opponent may be inadequate, but it can be developed.

H15. Mistrust stems from misjudgment, especially of the disposition of your opponent to answer trust with trust and mistrust with mistrust.

There are many examples in Gandhi's writings of this conception of trust and mistrust. His life likewise offers examples of the way he trusted people strongly opposed to him and the courage he thus proved. He repeatedly risked his own life by believing that he could trust his opponents when he met them personally.

H16. The tendency to misjudge our opponent and his case in an unfavorable direction increases his and our tendency to resort to violence.

H17. You can win most thoroughly with non-violent means by turning your opponent into a believer in and supporter of your case.

No effort has been made to derive some of the hypotheses from others. By suitable modifications, H15 and H17 might be derived from H14.

Third-level norms and hypotheses:

N5 (derived from N2 and H4). Conceive your struggle and carry it through as a positive struggle *in favor of* human beings and certain values, thus eventually fighting antagonisms, but not antagonists.

It may be mentioned, as an example, that Gandhi in his most famous campaign supported the people in making salt rather than instigate them against the empire salt producers and their factories. The situation desired was anticipated. One should fight the antagonism, not the antagonists.

N6 (derived from N2 and H5 or from N4 and H6). Live together with those for whom you struggle and do constructive work for them.

N7 (derived from N2 and H7 or from N4 and H7 and H8). Try to formulate the essential interests which you and your opponent have in common and try to establish a co-operation with your opponent on this basis.

N8 (derived from N3 or from N4 and H9). Do not humiliate or provoke your opponent.

N9 (derived from N4 and H10). Acquire the best possible knowledge of the facts and factors relevant to the non-violent realization of the goal of your cause.

N10 (derived from N4 and H11). Do your utmost in order to be in full accordance with the truth in your description of individuals, groups, institutions, and circumstances relevant to the struggle.

N11 (derived from N4 and H11). Do not use secret plans or moves or keep objectives secret.

N12 (derived from N4 and H12 and H13). Announce your case and the goal of your campaign explicitly and clearly, distinguishing essentials from non-essentials.

N13 (derived from N4 and H13). Seek personal contact with your opponent and be available to him.

N14 (derived from N3 or from N4 and H16). Do not judge your opponent harder than yourself.

N15 (derived from N4, H14, H15, and H16). Trust your opponent.

N16 (derived from N4, H14, and H17). Turn your opponent into a believer in and supporter of your case.

H18. You provoke your opponent if you destroy his property.

H19. Adequate understanding of your opponent presupposes personal *Einfühlung.*

H20. Avoidance of misjudgment of your opponent and his case presupposes understanding him and his case.

H21. If one keeps in mind one's own fallibility and failures, opponents are less likely to be misjudged in an unfavorable way, and their case underestimated intellectually or morally.

H22. Every political action, your own included, is likely to be based, in part, on mistaken views and to be carried out in an imperfect way.

H23. You make it difficult for your opponent to turn to support of your case if you are unwilling to compromise on non-essentials.

H24. It furthers the conversion of your opponent if he understands that you are sincere.

H25. The best way of convincing your opponent of your sincerity is to make sacrifices for your cause.

H26. During a campaign, change of its declared objective makes it difficult for opponents to trust your sincerity.

Gandhi has in mind the expansion of objectives at moments of weakness in the opponent, and contraction when it seems that the strength of the opponent has been underrated.

Fourth level of norms:

N17 (derived from N8 and H18). Do not destroy property belonging to your opponent.

N18 (derived from N14 and H19 and H20). Cultivate personal *Einfühlung* with your opponent.

N19 (derived from N10 or from N14 and H20). Do not formulate your case and the goal of your campaigns and that of your opponent in a biased way.

N20 (derived from N14 and H21). Keep in mind and admit your own mistakes and weaknesses.

N21 (derived from N14 and H21). Keep in mind and admit the possibility that you are factually or morally mistaken, even when you sincerely believe that you are not.

N22 (derived from N16 and H22 and H23). Be always willing to compromise in non-essentials.

N23 (derived from N16 and H24). Do not exploit a weakness in the position of your opponent.

N24 (derived from N16 and H24 and H25). Be willing to make sacrifices for your cause.

N25 (derived from N16 and H24 and H26). During a campaign, do not change its objective by making its goal wider or narrower.

Exemplification and Elaboration

In this section I shall illustrate how the meager outline can be taken as a starting point for a more substantial presentation. Something will be said on two of the norms of the system, N2 and N23, just in order to make them more understandable and also more subject to criticism.

"Make a constructive program part of your campaign." The importance of this norm, N2 in systematization D, stems in part from the conviction of Gandhi that, if it is ignored by some sections of the supporters of *satyagraha*, the strongest non-violent methods in fight for political freedom are rendered inapplicable. Only those who are able to take upon them the tasks of constructive community service are sufficiently mature for intense non-violent struggle. In 1930 Gandhi stressed that he could not advise civil disobedience campaigns because N2 was unlikely to be fulfilled. Insufficient constructive content of the fight for freedom would make it overwhelmingly probable that there would be violence. Gandhi was determined to stop a civil disobedience campaign in case of violence, such as happened at Chaura Chauri, where some English policemen were murdered.

Gandhi insisted on constructive definitions of goals and subgoals and demanded that Indians should work together on peaceful economic and other projects, thereby acquiring a spirit of mutual trust and a habit of sacrifice in the interest of the wider goals. In India such work was organized and planned under the name of "The Constructive Program." The norms saying that one should contribute to the implementation of the constructive program make up an integral part of the Gandhian ethics of group struggle. It is not a mere accessory.

A couple of quotations will make the point clearer. In his argumentation in January, 1930, that the atmosphere is not such that a mass civil disobedience campaign can be started, Gandhi says among other things:

> Constructive programme is not essential for local civil disobedience for specific relief as in the case of Bardoli. Tangible common grievance restricted to a particular locality is enough. But for such an indefinable thing as Swaraj [freedom], people must have previous training in doing things of All-India interest. Trust begotten in the pursuit of continuous constructive work becomes a tremendous asset at the critical moment. Constructive work therefore is for a non-violent army what drilling etc., is for an army designed for bloody warfare. Individual civil disobedience among an unprepared people and by leaders not known to or trusted by them is of no avail, a mass civil disobedience is an impossibility. The more therefore the progress of the constructive programme, the greater is the chance for civil disobedience. Granted a perfectly non-violent atmosphere and a fulfilled constructive programme, I would undertake to lead a mass civil disobedience struggle to a successful issue in the space of a few months [*Young India*, January 9, 1930].*

In the booklet *Constructive Programme*† Gandhi even says that mass civil disobedience might be dispensed with if the constructive programme were taken seriously by all concerned. He says:

> Civil disobedience is not absolutely necessary to win freedom through purely non-violent efforts, if the cooperation of the whole nation is secured in the constructive programme. . . . My handling of civil disobedience without constructive programme will be like a paralysed hand attempting to lift a spoon [quoted from Diwakar, *Satyagraha*, p. 187].‡

The constructive work is of various kinds. Here are some items of a long list: work for removal of untouchability, for spread of hand-spun and hand-woven cloth, work for other village industries, for village sanitation, for basic education through crafts, for literacy.

Gandhi also had in mind the effect upon the opponent. In the eyes of the opponent, the revolutionary seems mainly to have destruction in view. Gandhi requires methods whereby the constructive intent is made completely clear and trustworthy to the skeptical opponent.

As a demonstration against the British salt tax and salt monopoly, considered to be profoundly unjust, Gandhi and a mass of poor people marched to the sea to make salt (illegally). *While the campaign was going on*, Gandhi used much time for other tasks, such as instigating house industry and cleaning up slum quarters. The latter activity was part of the campaign and part of the struggle for *swaraj* as a whole. It was a demonstration *ad oculos* and helped the followers and opponents to fix their attention upon the positive goals rather than upon the means and the inevitable destructive components, that is, disabling British administration, etc.

One may say that the norm to partake in a constructive program is the supreme anti-antimovement norm in the system: those tendencies which are present in organization or groups favoring destruction of something (the organized anti-Semitists, anti-Communists, anti-Facists, etc.) are denounced; every action should have a clear positive, a pro-character.

I have used the norm "Give your campaign a constructive content" to illustrate the rich, scarcely surveyable material which has to be studied in order to proceed from a mere diagram toward a full presentation of Gandhi's political ethics. It should be clear from the comments and quotations that constructivity of main goals, constructivity of subgoals, and the so-called constructive program are means by which Gandhi tried to contribute to the implementation of *many* norms. And also that some norms may be viewed as occupying a lower position in relation to the norm requiring constructive work. Actually, the constructive work was a kind of partial anticipation of the condition Gandhi called *purna swaraj, real* independence, an ideal state of society. The political independence was not, as such, a constructive goal for him, since it was defined as *absence* of British domination.

Do not exploit weakness—Let us elaborate upon another norm, N23, "Do not exploit a weakness in the position of your opponent in case it is due to factors irrelevant to the struggle."

A campaign is not clearly subservient to the goal of converting the opponent, if victory in the sense of bringing the opponent to accept the conditions for terminating the *satyagraha* is caused by some misfortune he has experienced which makes it necessary for him to call off his struggle with the *satyagrahi*. In short, if by factors irrelevant to the struggle and therefore unrelated to the conversion of the opponent, the *satyagrahi* are able to get what they desire *in terms of conditions*, they should desist from asking for those conditions.

As an example, we may take what happened at the last stage of the *satyagraha* campaign in South Africa.

Gandhi fought against certain laws which he considered discriminatory against the Indians. Their repeal was the conditions of bringing the *satyagraha* campaign to a stop. The Indian leaders were planning a march as part of the *satyagraha*. When a railway strike broke out among the white employees, the government was in a dangerous position and might well have been willing to settle the conflict with the Indians in order to meet the situation created by the strike. Let me quote what Gandhi says in his narrative. Its reliable character is not contested by his adversary—and great admirer—General Smuts. Gandhi said:

> Just at this time there was a great strike of the European employees of the Union railways, which made the position of the Government extremely delicate. I was called upon to commence the Indian march at such a fortunate juncture. But I declared that the Indians could not thus assist the railway strikers, as they [the Indians] were not out to harass the Government, their struggle being entirely different and differently conceived. Even if we undertook the march, we would begin it at some other time when the railway trouble had ended. This decision of ours created a deep impression, and was cabled to England by Reuter.§

When World War II broke out, pressure was brought upon Gandhi to intensify the fight against the British. He declined taking up mass civil disobedience during the war. He said: "There is neither warrant nor atmosphere for mass action. That would be naked embarrassment and a betrayal of non-violence. . . . By causing embarrassment at this stage, the authorities must resent it bitterly, and are likely to act madly. It is worse than suicide to resort to violence that is embarrassment under the cover of non-violence."[1]

The argumentation and also the behavior of Gandhi in these two instances are in conformity with a norm such as N23. Later, during World War II, Gandhi intended to start a mass movement. This plan creates a problem for our systematization. It requires either a hypothesis that the British then, in the autumn of 1942, were no longer in a temporarily weak position, or a decision that Gandhi violated his own norms, or a decision to modify systematization D so as to make Gandhi's behavior in both 1920 and 1942 conform to the explication of his ethics. We take tentatively the view that Gandhi in 1942 violated his own norms and are consequently able to retain systematization D as adequate.

III. APPLICATION TO EFFORTS OF PEACEFUL INTER-NATIONAL CO-OPERATION

The foregoing system of norms formidably restricts the field of justifiable forms of conflict resolution. It is, however, the claim of the proponents of ethics of non-violence *that such a system does not leave out any effective form of conflict resolution*. It is presupposed that the goal is justifiable from the point of view of general ethics. It is claimed, therefore, that no effective (powerful, adequate) form is excluded for those who fight for a good cause against opponents who fight for a bad cause. The criteria of goodness offered by Gandhi and others are such that no statesmen of today would openly reject them. That is, contemporary men in power would proclaim their goals to be good in the sense required.

In view of this, the ethics of non-violence claims to give the effective means of reaching at least one of the goals of the major powers in present-day international politics, namely, that of peaceful co-operation in the minimum sense.

Grave questions arise immediately, however, when this common goal is seen in combination with other goals of more traditional kinds, such as ideological or economic domination or leadership. Here the antagonists impute to each other the most sinister designs. But the possibility that the antagonist is fighting also for bad causes, according to each of the participants, does not, according to the above, make the non-violent forms of struggle less effective. It reduces the chances that non-violent methods are wholeheartedly adopted, but not the chances of success, if adopted.

Which are now the main forms of struggle which satisfy the non-violence norms? In a general way, using a powerful slogan, they may be characterized as forms *effecting a liquidation of antagonism, not antagonists*.

In order not to lose contact with the forms that have been actually tried out with a fair degree of success, the techniques of Gandhi will be taken as representative examples. His field of action was three-fold: international politics (South Africa and India versus the British Empire), interracial, interideological conflicts (Hindu-Moslem riots, etc.), and economic conflicts (management versus labor, village industry and agriculture versus mechanization, etc.).

The more extreme forms of struggle (strikes, fasting, etc.) will here be called "Gandhian *satyagraha*" and will be considered to make up a subclass of forms satisfying the norms and hypotheses of version D. Before Gandhi resorted to *satyagraha*, his activity would go through five interrelated phases:

1. Accumulation and analysis of factual information concerning the conflict (on the spot). Unbiased exposition of the main facts relating to the conflict, with extensive use of the opponents as judges (*audi alteram partem*).

2. Clarification of essential (long-range) interests in common with the antagonists (*presupposition*: there are always common interests).

3. Tentative formulation of a limited goal for immediate action acceptable to both parties in terms of common interests and in accordance with the ultimate norm of self-realization.

4. Discussion of the tentative formulations, person to person and face to face and not merely as negotiators or representatives. Clarification of the instrumental value of the limited goal is resolving a part or aspect of the struggle.

5. In case of persistent resistance from one of the parties concerned, search for a compromise without giving up essentials. That is, search for a compromise affecting limited interests, not basic values.

If, *and only if*, these activities did not bear fruit, Gandhi would consider it justifiable (and effective) to resort to *satyagraha*.

The different forms of *satyagraha* planned and used by Gandhi were adapted to a very different situation from that confronted by a man who intends to contribute to conditions of peaceful co-operation between the major power constellations of today. There are two major differences of situations: (1) Personal contacts between the opposite groups (Indian-English, Moslem-Hindu, etc.) were very extensive and intensive in India (and South Africa). The present problem relates to groups with very small personal contact. Further, (2) in India it was mainly a question of relations between a physically mightier and a physically weaker group, whereas today there is roughly an equilibrium, both sides being eager not to let the other be physically stronger.

Gandhi's forms of *satyagraha* will therefore not be described in this paper. Gandhi himself stressed that, for him, the basic tenets of non-violence are central, not *satyagraha* as developed by him or others.

It is our contention that a renewed scrutiny of the non-violence norms, independent of Gandhian *satyagraha*, will lead to important considerations as to the attitudes and measures to be taken in international politics. It is characteristic of this situation that the five conditions necessary to justify extreme forms of struggle are not, or only in part, fulfilled.

The basic norm of the ethics of group struggle does not permit a mere personal avoidance of violence but requires us to take part in struggles and in such a way as to reduce the chances of violence in general. Applied to problems of co-operation between East and West, this means that it is the duty of all concerned to partake in solving them.

According to hypotheses H2 and H4 and to norms N2 and N5 (we refer consistently to systematization D), the struggle between Soviet-oriented and NATO-oriented powers must be given a positive content. This has, for example, the implication that one should avoid any step merely dictated *against* an institution (anti-communism, anti-imperialism) or, even worse, *against* an individual or a group of people. One's actions should have the character of being *in favor* of positive values and principles *in support* of human beings. It is, furthermore, in accordance with these hypotheses and norms if one co-operates on what are considered to be common objectives and tasks, possibly on the common tasks and objectives of mankind, both in the relation between the power groups and in areas outside them, e.g., in technically under-developed countries.

Hypothesis H3 is applicable to the idea of preventive war, preventive terror, or any other violence engaged in, in order to reduce a greater amount or a greater intensity of violence.

It is not possible here to examine each norm and hypothesis separately. N9, which stresses the importance of a thorough knowledge of the facts and factors relevant to the struggle, is of considerable importance, however.

This applies not only to statesmen but also to the people at large. Enduring efforts to create, in all countries, a first-class research and information service and education dealing with international relations are highly important.

N13 and also other norms and hypotheses stress the necessity of maximum *personal* contact between those engaged in the struggle. The evident consequence of this is a policy of opening the borders. It breaks

N13 to use a social boycott to mark moral indignation (e.g., to stop student exchanges between two countries in order to mark moral indignation concerning an action undertaken by one of these countries) or to refuse entrance into the country before certain conditions are fulfilled. There are no conditions under which one, according to Gandhi, can refuse to meet one's opponent, and, as this refers to all participants, it is not a norm affecting contact only between politicians.

The personal contact norms and hypotheses go against the assumption that real understanding of the struggle can be reached on the verbal plane. That is, the mass communication media are insufficient vehicles of information. The understanding at present between the eastern European and western European population must, according to this, be largely illusory, since personal contacts are at a near-zero level. According to H11 and other norms and hypotheses, on the other hand, the good effects of personal contacts will be reduced or impeded if they are exploited, e.g., for propaganda purposes, i.e., if the truth norm is not respected.

The above shows that even if the already developed forms of *satyagraha* will be applicable only after an increase in international personal contacts, the ethics of non-violence has considerable bearing upon the problems of coexistence.

The prescriptions resulting from application of the norms of non-violence are, on the whole, such as are put forward as recommendations from various groups of researchers in social science. The chief differences are found in relative priorities and in the ultimate justification of the actions prescribed.

IV. RESEARCH SUGGESTIONS

There is much research already done which throws light upon the tenability of the hypotheses implicit in the ethics of non-violence and also upon the chances of the ethics of non-violence being applied.

Thus the tremendous literature on the nature and consequences of prejudice, national images, black-white thinking in mass communication, patriotic history textbooks, and pressure groups against world organizations is directly relevant.

In the following some research topics are listed which concern current international conflicts:

1. Which interests do the Soviet and the NATO powers have in common, and which of these are generally acknowledged to be common interests?

We have already mentioned one common interest satisfying these criteria—the elimination of threats of annihilation. There are others, presumably, and it is of importance to know their interdependence and their status in relation to non-common interests.

Concerning research on threats see, for instance, A. I. Gladstone (29), pages 4 ff.

2. What kinds of actions could institutions or individuals from the antagonistic powers perform together in an atomosphere of co-operation in order to serve the satisfaction of the common interest? Which kinds of actions through international organizations (UN, UNESCO, etc.) fit into this context? Which conditions of work are favorable to their success?

3. How can these kinds of actions be used to give a positive content, if not to the total struggle, then at least to part of the struggle between the rival powers?

4. Especially, which role can personal contacts across the frontiers play in this context?

Which kinds of contacts—professional, tourist, student, religious, artistic, athletic, etc.—are most successful in reaching their objectives? (See G. W. Allport [10], M. B. Smith [28], C. S. Ascher [13].)

Then there are questions related to the capacity of men of good will to elicit the best in the antagonist:

5. Which factors determine to what extent a person is able to react upon an antagonist as a fellow being and to avoid reacting upon him as a symbol of an institution or a representative of a doctrine?

6. Which factors determine to what extent a person is reacted upon as a symbol or representative of an institution?

We also need research upon which of the factors operating toward minimizing the institutional or functional perception and conception can be most easily strengthened.

7. How are we to strengthen loyalties toward institutions favoring person-to-person meetings (not persons-as-symbols meetings)? (See H. Guetzkow [17].)

8. What are the factors favoring individuals acting from personal responsibility for world conflicts? (See H. C. Kelman [29], pp. 34 ff.)

The positive role of education has been shown in countless studies correlating level of education with attitudes of internationalism. There may, however, be opposite factors making the more educated more likely to let knowledge interfere with the human approach to antagonists.

9. What are the factors favoring broad conceptions of the self, conceptions that favor identification with the interests of out-groups? (See R. C. Angell [12], G. H. Mead [23], C. H. Cooley [15], and R. W. van Wagenen [30].)

10. What are the factors favoring the distinction between appreciation and friendliness, making strong disapproval consistent with consistent friendliness on the personal level? (See R. H. Blum [29], pp. 17 ff.)

11. What is the role of faith in the validity of fundamental norms of non-violence for active co-operation in a hostile environment? (See J. M. Yinger [32].)

12. What is the role of a personality-based hostility upon attitudes of co-operation and conflict toward other groups? (See Adorno et al. [9], Christiansen [14], Levison [21].)

13. Which are the factors favoring truthful, non-partisan descriptions of the political activities of the rival major powers?

14. What are the effects of secrecy upon meetings of antagonists? To what extent does it interfere with personal trust? (See H. Guetzkow [17, 18].)

There are, finally, important problems with regard to the political ethics being practiced:

15. To what extent, in different countries, are non-violence norms adhered to and practiced in political life on the local and national level? Toswhat extent is there a correlation in this respect with the attitude toward the great tension problems in world politics?

16. What have been the effects of concrete non-violence policies, compared with the probable results if other policies had been followed? To what extent and in what direction has the attitude to non-violence in general among the people concerned been influenced by these effects? (See G. Murphy [24].)

17. To what extent is the favorable attitude to non-violence in the instances found rooted in a profoundly non-violent attitude in accordance with Gandhi's ethics? To what extent is it limited to particular phenomena as military warfare? To what extent is it dependent on particular political sympathies or loyalties?

18. On which norms and on which kinds of political actions can conscious adherents of a non-violence ethics and others agree? On what tasks can they co-operate?

How can political ethics be brought into the focus of political life, research, and education?

Research upon these questions inevitably leads not only to questions of sociology, social psychology, and education but also to background questions of economics: for instance, the effect of certain economic systems upon the ability or willingness to let those attitudes of co-operation grow which favor a settlement of the most threatening conflicts in the history of mankind.

SELECTIVE BIBLIOGRAPHY

1. Dhawan, G. N. *The Political Philosophy of Mahatma Gandhi*. Bombay: Popular book depot, 1946.

2. Galtung, J., and Naess, A. *Gandhis politiske etikk* ("Political Ethics of Gandhi"). Oslo: J. G. Tanum, 1955.

3. Gandhi, M. K. *The Story of My Experiments with Truth*, Vol. I. Ahmedabad: Navajivan Press, 1927.

4. ———. *Non-Violence in Peace and War*, Vols. I, II. Ahmedabad: Navajivan Press, 1942-49.

5. Nehru, Kripalani, Bunche, Boyd Orr, Neimøller, and Others. *Gandhian Outlook and Technique*. Ministry of Education, Government of India, 1953.

6. Mühlmann, W. E. *Mahatma Gandhi: Eine Untersuchung zur Religionssoziologie und politischen Ethik*. Tubingen, 1950.

7. Shridharani, K. *War without Violence: the Sociology of Gandhi's Satyagraha*. New York, 1939.

8. Tolstoy, Leo. *The Kingdom of God Is within You*. (See Kalida Nog. *Tolstoy and Gandhi*. Patna: Pustnak Bhandar, 1950.)

Social Science Contribution to Questions
of Peaceful Co-operation: Topics
Touched on in the Present Paper

9. Adorno, T. W., Frenkel-Brunswick, Else, Levison, D. J., and Sanford, R. N. *The Authoritarian Personality*. New York: Harper & Bros., 1950.

10. Allport, G. W. "Guideline for Research in International Co-operation," *Yearbook of Social Issues*, Vol. III (1947).

11. ———. "The Role of Expectancy," in H. Cantril (ed.), *Tensions That Cause Wars*. Urbana, Ill., U. Illinois Press, 1950.

12. Angell, R. C. "Discovering Paths to Peace," in *The Nature of Conflict*. New York: UNESCO, 1957.

13. Ascher, C. S. "The Development of UNESCO's Program," *International Organization*, Vol. IV, No. 1 (1950).

14. Christiansen, B. *Attitudes towards Foreign Affairs as a Function of Personality*. Oslo, 1958.

15. Cooley, C. H. *Human Nature and the Social Order*. New York: Charles Scribner's Sons, 1922.

15a. ———. *Social Process*. New York, 1918.

16. Dunn, F. S. *War and the Minds of Men*. New York, 1950.

17. Guetzkow, H. *Multiple Loyalties*. Princeton, N. J.: Princeton Univ. Press, 1955.

18. ———. "Isolation and Collaboration: A Partial Theory of Internation Relations," *Conflict Resolution*, Vol. I, No. 1 (1957).

19. Jackson, E., *Meeting of Minds*. New York: McGraw-Hill, 1952.

20. Klineberg, O. *Tensions Affecting International Understanding: A Survey of Research*. (Social Science Research Council Bull. 62.) New York, 1950.

21. Levison, D. J. "Authoritarian Personality and Foreign Policy," *Conflict Resolution*, Vol. I, No. 1 (1957).

22. McKeon, R. (ed.). *Democracy in a World of Tensions*. UNESCO, 1951. (See answers to questions 29 and 30 concerning the value foundations of the world conflict.)

23. Mead, G. H. *Mind, Self, and Society*. Chicago: U. of Chicago Press, 1934.

24. Murphy, G. *In the Minds of Men: The Study of Human Behavior and Tensions in India*, New York: Basic Books, Inc., 1953.

25. Naess, A., et al. *Democracy, Ideology, and Objectivity*, chap. iii, "Disagreement in Controversy between 'East' and 'West' from the Point of View of Cognitive Analysis." Oslo, 1956.

26. Newcomb, T. "Autistic Hostility and Social Reality," *Human Relations*, 1947, p. 1.

27. Pool, Ithiel de Sola. *Symbols of Internationalism*. Stanford, Calif.: Stanford University Press, 1952.

28. Smith, M. B., and Casagrande, J. B. "The Cross-cultural Education Projects: A Program Report," *Social Science Research Council Items*, No. 3 (September, 1953).

29. UNESCO. SS/Coop./6

30. Wagenen, R. W. Van. *Research in the International Organization Field*. Princeton, N. J., 1952.

31. Wright, Q. *The Study of International Relations*, chapter on "International Ethics." New York: Appleton-Century-Crofts, 1955.

32. Yinger, J. M. *Religion in the Struggle for Power: A Study in the Sociology of Religion*. Durham, N. C.: Duke University Press, 1946.

FOOTNOTES

* *Young India* was a weekly newspaper, published in Ahmedabad and edited by Gandhi from 1919-1932. THE EDITORS

† *Constructive Program: Its Meaning and Place*. Ahmedabad: Navajivan Press, 1944. THE EDITORS

‡ Diwakar, *Satyagraha*. Bombay: Hind Kitales, 1940. THE EDITORS

§ See Gandhi, M. K. *Satyagraha in South Africa*. Triplicane, Madras: S. Ganesan, 1928 (1st edition); Ahmedabad: Navijivan, 1950 (2nd edition). THE EDITORS

[1] Declaration published in all Indian newspapers, October 30, 1940.

21   Three Not-So-Obvious Contributions of Psychology to Peace

—RALPH K. WHITE

There are two things in this world that don't quite fit together. One is that mushroom cloud. We try not to think about it—but it's *there*, rising, enormously, behind everything else we do. And then there's the other thing: the whole complicated spectacle of all the old causes of war going on as usual. There's the arms race, and ABM, and—much worse than ABM—that hydra-headed monster, MIRV. Most of all, there's the war in Vietnam. It stands there as a continual, glaring reminder that the United States—our own peace-loving United States—is capable of the kind of bungling that got us into that war. And then comes the thought: *if* even the peace-loving United States could bungle itself into a little war like Vietnam, what guarantee is there that we won't bungle ourselves into a big war—a nuclear war? It might be possible to exorcize the specter of that mushroom cloud if the Vietnam war did not exist. But it does exist.

From the *Journal of Social Issues*, 1969, *25* (No. 4), 23-39. Reprinted by permission of the Society for the Psychological Study of Social Issues.

The sense of bafflement is especially great perhaps among psychologists, because a good many psychologists feel that the bungling that got us into the Vietnam war, and could get us into a nuclear war, consists largely of ignoring certain fundamental *psychological* truths. Most of our American policy-makers (both Johnson and Nixon, for instance) behave as if they don't recognize certain things that we psychologists take for granted—things such as the necessity of empathy (including empathy with our own worst enemies), the dangers of black-and-white thinking, and the role of the self-fulfilling prophecy in the vicious spiral of the arms race.

Communicating with Policy-Makers

All of this strengthens the case for better communication—better communication directly between us and the policy-makers in Washington, and better communication also between us and other scholars (historians, political scientists, area specialists) who in turn influence the policy-makers a good deal more than we do.

One difficulty in communicating with these people is that from their standpoint we often sound like a little boy trying to teach Grandma to suck eggs. Many of them are experts in their own fields, people from whom we really could learn a great deal. And then we come up with these ideas that they think they have heard many times already, ideas that they often think we have dressed up in pretentious new terminology but that they regard as essentially old, familiar, and in a sense obvious ideas.

The paradox is that it is precisely these so-called obvious ideas that we often see the top policy-makers ignoring when it comes to concrete action decisions. We see that mushroom cloud coming closer because they *act* as if they couldn't see what to *us* seems obvious. So, in order to define the problem accurately, it looks as if we need three categories. First, there are the things that really are obvious, on the verbal level *and* on the action level. Second, there are the things that seem obvious on the verbal level but that are often ignored on the action level. And third, there are the things that are not obvious on either the verbal or the action level.

Difficulties in Communicating the "Obvious"

The second category, although it won't be my main focus in this paper, does seem to me the most important: namely, the things that seem obvious on the general, abstract, verbal level, but that are often ignored on the specific, concrete, action level. As examples, let's take the three ideas I've already mentioned: the necessity of empathy, the dangers of black-and-white thinking, and the role played by the self-fulfilling prophecy in the vicious spiral of the arms race.

When empathy is defined in common-sense terms like "understanding the other fellow's point of view," any policy-maker is likely to say: "Sure, I believe in that, and I try to do it all the time." The chances are he takes pride in understanding the other fellow's point of view—even when he doesn't really understand it.

Or take the black-and-white picture. Anybody who has ever seen a Western movie, and knows about the bad guys and the good guys, the black hats and the white hats, is likely to have some notion, on the verbal level, of the dangers of black-and-white thinking, even if in practice he engages in black-and-white thinking most of the time.

Or take the role of the self-fulfilling prophecy in the vicious spiral of the arms race. To some, the self-fulfilling prophecy may be a new and interesting idea—Senator Fulbright found it a new and interesting idea when he heard it from Jerome Frank*—but the vicious spiral of the arms race is an old idea that has been heard many times and might be accepted in theory even by people like Melvin Laird who in practice ignore it. What can we do then? The things that we feel are being most dangerously ignored in practice are the things most likely to make our listeners yawn.

The answer, as I see it, is *not* to stop talking about these fundamental things. It is, rather, to get right down onto the concrete action level and to talk not about these abstractions as such, but about concrete examples of them.

Bombing and Empathy

For instance, take again the notion of empathy. It seems to me that a flagrant concrete example of violation of the principle of empathy was our bombing of North Vietnam. That bombing was urged and continually supported by our most flagrant non-empathizers—the military. But its effects included a continual solidifying of opposition to us among the people in North Vietnam. It was as if we were doing our best to persuade every man, woman, and child in North Vietnam that America really *is* the devil, the wanton cruel aggressor that Communist propaganda has always said it was. Most of our military men in active service not only failed to empathize with the North Vietnamese; it looks as if they actively, though unconsciously, resisted the temptation to empathize. They shut their eyes to the best evidence available: the first-hand testimony of people like Harrison Salisbury (1967), Cameron (1966), Gerassi (1967), Gottlieb (1965), and the Quakers of the ship, *Phoenix*, who went to North Vietnam and came back saying that our bombing was solidifying opposition to us (Zietlow, 1967). They shut their eyes also to the evidence that the bombing was tending to alienate from us most of the other people in the world. And, most surprisingly, they shut their eyes to the evidence of history, represented by our own strategic bombing survey after World War II (Over-all report, 1945) which described how our bombing of Germany and Japan had had the same solidifying effect.

This kind of concrete example may jolt and antagonize some people, but it won't make them yawn. And focusing on such examples should help to make abstract concepts like empathy become more and more a part of the reality-world of the listener, on the concrete action level. It matters very little whether a policy-maker talks about empathy. It matters a great deal whether the impulse to empathize keeps coming up in his mind, at those particular moments when wisdom in action requires that he should at least try to understand the other fellow's point of view.

Communicating the "Not-So-Obvious"

Then there is that third category of psychological ideas and psychological facts that really are relatively unfamiliar to the decision-makers on both levels. I'm going to talk about three of them today: "three not-so-obvious contributions of psychology to peace." (Of course when I say "contributions" I mean potential contributions. What we have done is to learn certain things about the psychological causes of war. Whether these insights and the facts that support them ever actually contribute to peace depends on our own effort and our own skill as communicators.) Also it should be clear that these are not necessarily the most important of the not-so-obvious contributions. There are others that seem to me just as important or more so: Charles Osgood's (1962) GRIT proposal, for instance, and the experimental work Morton Deutsch has been doing (Deutsch and Krauss, 1962), and the monumental job Herbert Kelman did editing that big volume, *International Behavior* (1965). But those are pretty well known. I'm going to focus here on three that are not very well known.

THE HOVLAND PRINCIPLE IN COMMUNICATING WITH COMMUNISTS

First, there is a corollary of the Hovland principle that a two-sided presentation of an argument is more persuasive than a one-sided presentation when you are talking with people who initially disagree with you. The corollary is that *we Americans should publicly accept as much as we can honestly accept of the Communist point of view.*

To some psychologists this may seem obvious, but most of our politicians and foreign-policy makers are likely to regard it as far from obvious. To many of them it must sound like subversive doctrine—like being "soft on Communism." That is precisely why we psychologists, *if* we think the evidence supports it, ought to be saying so—clearly, and often, and with all the research evidence that we can bring to bear.

Let's look at the evidence. You are probably familiar with the impressive body of experimental data accumulated by Hovland and his colleagues (Hovland, Janis, & Kelley, 1953), on the general advantages of a "two-sided" form of persuasion—defining a two-sided argument not as a neutral position but as a genuine argument that candidly takes the stronger arguments on the other side into account. And "candidly taking them into account" means not only stating them fairly before trying to refute them, but also acknowledging any elements in them that the speaker honestly regards as elements of truth. You are probably also familiar with their more specific findings, including the finding that the two-sided approach is not always more effective. It is likely to be more effective if the audience is intelligent, or initially hostile to the viewpoint of the speaker, or both intelligent and hostile. Now comes the corollary, which is especially interesting from the standpoint of our relations with the Communist world. The Communist leaders fit exactly the Hovland prescription for the kind of people with whom one should use the two-sided approach. They are intelligent. They could hardly have maintained stability in a vast nation like the USSR if they were not at least fairly intelligent. And, to put it mildly, they are initially in disagreement with us. So it would follow that in communicating with them we should use the two-sided approach.

What Is Right in Communism?

What would it mean, concretely? It would *not* mean soft-pedaling any of the things we believe to be wrong and dangerous on the Communist side: the invasion of Czechoslovakia, for instance, or the recent regression toward Stalinism in the Soviet Union, or the anarchy and cruelty of the "great cultural revolution" in Communist China, or the assassination of village leaders in Vietnam. But it would mean coupling candor about what we think is wrong with candor about what we think is right. That raises the question: what *is* right in Communism? Is anything right? Each of us would probably have a different answer, but just to make the main point concrete I'm going to go out on a limb and mention some of the things that I personally think are right.

Most important, probably, is the depth and intensity of the Russians' desire for peace. They hate and fear war at least as much as we do. How could they not hate war, after the searing experience they went through in World War II? We can also give them credit for bearing the brunt of World War II—and winning, on that crucial Eastern Front. I know from my own experience in Moscow that nothing touches the heart of a Russian more than real appreciation, by an American, of what they suffered and what they accomplished in our common struggle against Hitler. There is real common ground here, both when we look back on World War II and when we look ahead to the future. We and the Communists, looking ahead, find ourselves on the same side in the rather desperate struggle that both they and we are waging against the danger of nuclear war.

Some other things that I personally would acknowledge include Soviet space achievements, which really are extraordinary, considering how backward Russia was in 1917; the case for Communist Chinese intervention in Korea after MacArthur crossed the 38th Parallel; the case for Communist China in the matter of Quemoy and Matsu; the Vietnamese Communist case against Diem and his American supporters; a very large part of their case against what we have been doing in Vietnam since the death of Diem. And, more basically, the proposition that the Communist countries are ahead of us in social justice. In spite of striking inequalities, my reading of the evidence is that they are definitely ahead of us in eliminating unearned income—"surplus value" —and somewhat ahead of us in diminishing the gap between rich and poor. (This and related problems are spelled out more fully in White, 1967-8.)

Research on the Need to Seek Common Ground

If all this has a subversive sound, please recall again the Hovland experiments, and also the rather large number of other experiments that bring out, in one way or another, the desirability of discovering common ground if conflict is to be resolved. For instance, there are the experiments of Blake and Mouton (see Sawyer and Guetzkow, 1965) on how each side in a controversy ordinarily underestimates the amount of common ground that actually exists between its own position and that of its adversary. There is all the research on the non-zero-sum game, and the need to keep the players on both sides from treating a *non*-zero-sum game, in which the adversaries actually share some common interests, as if it were a zero-sum game in which loss for one side always means gain for the other. There is the so-called Rapoport Debate (actually originated by Carl Rogers, apparently), in which neither side is permitted to argue for its position until it has stated, to the other side's satisfaction, what the other side is trying to establish. There is Sherif's Robbers' Cave experiment in which conflict was replaced by cooperation and friendliness when a superordinate goal—an overriding common goal—demanded cooperation (Sherif, 1958). There is Rokeach's work (1960) on the importance of common beliefs as a basis for good will. There is Kenneth Hammond's recent work on the harm done by implicit assumptions that differ on the two sides of an argument, and that are never really challenged or examined. All of these have as a common element the idea of common goals or common ground, and the desirability of common ground for conflict-resolution.

The "Modal Philosophy" and East-West Convergence

There is also my own content-analysis (White, 1949) of the values in various ideologies (American, Nazi, and Communist) using the value-analysis technique (White, 1951)—a project carried a good deal further recently by William Eckhardt (Eckhardt and White, 1967). The main upshot of that analysis was that there has apparently been a convergence of the value-systems of the Communist East and the non-Communist West. From a study of opinion and attitude surveys in a number of non-Communist countries, and of behavior data and political speeches and writings on both sides of the East-West conflict, a picture emerged of a good deal more common ground, shared by us and the Communists, than the embattled partisans on either side have ever recognized. Neither they nor we depart very far from the most commonly held political philosophy—I call it the "modal philosophy"—which with minor variations seems to characterize most of the politically conscious people in the world (White, 1957). (It is the great piling up of people in the middle zone—a very large "mode" in the statistical sense of the word "mode"—that justifies the term "modal philosophy.")

It includes three main elements. First, a preference for private ownership and free enterprise in at least the smaller economic units: the grocery store, the laundry, the repair shop, the small farm. In that respect the global majority seems to lean more toward our American way of life than toward that of the thorough-going socialists, or the Communists. A second element, though, is a strong emphasis on social welfare—helping the poor. In that respect the modal philosophy is more like Communism. And third, there is a belief in political democracy, including free speech. Most of the people in the global majority reject dictatorship, and most of them reject the word "communism" because to them it implies dictatorship, while they more or less accept the term "socialism," which to them implies democracy. In fact, the term "democratic socialism" probably comes closer than any other single term to representing what this modal philosophy is. This pattern of values and beliefs, or some not-very-wide variation from it, constitutes the great common ground that liberal Americans share, not only with millions of people who call themselves Communists but also with an actual majority of the politically conscious members of the human race.

MIRROR-IMAGE WARS AND TERRITORIAL SELF-IMAGES

A second not-so-obvious proposition is *the frequency of mirror-image wars, and the importance of overlapping territorial self-images as causes of such wars.*

There are two kinds of war. There is the mirror-image war in which each side really believes that the other side is the aggressor (Bronfenbrenner, 1961). And there is the non-mirror-image war in which one side really believes that the other side is the aggressor, while the other side, though feeling justified, doesn't really literally believe that it is the victim of aggression.

An example of a mirror-image war would be World War I. A great many Americans don't realize how well Bronfenbrenner's term, the "mirror-image," applies to what happened in 1914. A great many still picture that war as a case of outright German aggression, comparable to Hitler's aggression in 1939. The historical facts, as we know them now, do not support that belief. The Germans believed, with some factual justification, that they were the victims of aggression. They pictured Russia, France, and England as ganging up on them, and felt that unless they struck first they would be overwhelmed by enemies on two fronts. Ole Holsti and Robert North (1965) with their content-analysis of the documents of 1914, have confirmed what historians such as Fay (1928) and Gooch (1938) had already showed—that when the war actually broke out the Germans were motivated mainly by fear.

Another mirror-image war is the Vietnam war. The militants on each side clearly believe that the other side is the aggressor. The North Vietnamese see the United States as aggressing against the soil of their homeland, and, in mirror-image fashion, militant Americans see the North Vietnamese as aggressing against South Vietnam, both by a campaign of assassination in the villages and by actual troops invading the South.

There is a supreme irony in this mirror-image type of war. It seems utterly ridiculous that *both* sides should be fighting because of real fear, imagining the enemy to be a brutal, arrogant aggressor, when actually the enemy is nerving himself to fight a war that he too thinks is in self-defense. Each side is fighting, with desperate earnestness, an imagined enemy, a bogey-man, a windmill. But you can't laugh at this kind of joke. It's too bloody, too tragic. You can only stand aghast, and ask: how is it possible, psychologically, for one country, or perhaps both, to be *that* much deluded?

Then there is the other kind of war: a non-mirror-image war. Any conflict regarded by neutral onlookers as outright aggression is a case in point: Hitler's attack on Poland, for instance. He must have known, and other Germans must have known, that Poland was not attacking or threatening to attack Germany. Whatever their other justifications may have been, in this respect the German perception of the war was not a mirror-image of the perception in the minds of Germany's victims.

Since most people probably assume that the Hitler type of outright aggression is the typical way for wars to start, I did a rough check to see whether that is actually true, looking at thirty-seven wars that have occurred since 1913, and putting each of them, to the best of my ability, in one category or the other. The result was surprisingly even: 21 of the 37 wars (a little more than half) were in my judgment the mirror-image type, and 16 (a little less than half) were the non-mirror-image type. The method was rough, but it does seem clear that mirror-image wars, such as World War I and the Vietnam war, are not unusual exceptions. Their frequency is at least comparable with the frequency of non-mirror-image wars.

Overlapping and Conflict of Territorial Self-Images

Now, what can psychology contribute to an understanding of mirror-image wars, aside from applying to them Bronfenbrenner's apt and vivid term, "mirror-image"?

Actually it can contribute a number of things, several of which I've discussed in a book called *Nobody Wanted War* (White, 1968).[1] In this paper I want to focus on just one of them: the notion of the overlapping and conflict of territorial self-images.

It was a striking fact that most of the mirror-image wars in my list—16 out of 21—grew out of territorial conflicts in which there was reason to think that each side *really* believed that the disputed territory was part of itself. The surface of the world is dotted with ulcerous spots that have been the source of an enormous amount of bad blood and, often, of war: Bosnia, Alsace-Lorraine, the Sudetenland, the Polish Corridor, northern Ireland, Algeria, Israel, Kashmir, the Sino-Indian border, South Korea, Taiwan, Quemoy, South Vietnam. Every one of these ulcerous spots is a zone of overlap, where one country's territorial image of itself overlaps with another country's territorial image of *it*self.

The historians and political scientists are in general quite aware of this as a cause of war, and, under labels such as "irredentism," or simply "territorial disputes," they have given it a fair amount of emphasis. But I don't think they have given it nearly enough emphasis, and as far as I know they have never suggested an adequate psychological explanation of it. Their favorite formula, the international struggle for power, does not adequately cover it, because what needs to be explained is the special emotional intensity of the desire for power over a certain peice of territory when that territory is perceived as part of the national *self*, even though it may make little contribution to the overall power of the nation. Taiwan is a good case in point. The Chinese Communists seem fanatically intent on driving the invaders out of Taiwan—the "invaders" being us and Chiang Kai-shek—even though Taiwan would add only a little to their national power.

Identification and the Self-Image

But psychologists can offer some useful clues to an understanding of such territorial conflict. One is the notion of the self-image itself, and of how, by a process of identification, the self-image comes to include many things that were not originally part of it. We use a variety of names in referring to the self-image: many would call it simply "the self"; Kurt Lewin called it the "person." (His use of the term was broader, but I won't go into these complexities here.) But whatever we call it, I think most of us would agree that the concept of self-image plays a central role in psychology, and that the process of identification, by which other things come to be incorporated in the self-image, is also very important. Lewin, for instance, spoke of how a person's clothes come to be psychologically a part of the "person." If clothes are identified with to such an extent that they seem to be part of the person or part of the "self," then surely the territory that represents one's own nation on the map can also be part of it.

Territory in Animal Behavior

Another clue is the analogy with the territorial fighting of animals. Lorenz (1966), Ardrey (1963), Carpenter (1934), and others have described how an animal will spring to the defense of territory that it has identified with and that it seems to regard as its own. Now of course we need to be on our guard against over-hasty parallels between animal behavior and complex human behavior such as war making, but at this point the parallel seems valid, since the mechanism of identification is involved in both. In both cases, too, there is emotional disturbance when strangers—alien, unpredictable, presumably hostile strangers—are seen as impinging on land that is regarded as one's own, and therefore as part of the self.

Territorial Overlap and Intolerance of Ambiguity

Still another clue lies in the notion of intolerance of ambiguity. What calls for explanation, you remember, is the rigidity of overlapping territorial claims, usually on both sides, and the special emotional intensity of those claims. Usually each side refuses to grant for one moment that there could be a particle of validity in the other side's claim. There is a clean-cutness, a simplicity, an all-or-none quality in these territorial perceptions that is clearly a gross oversimplification of the complexity of reality. In each side's reality-world that land just *is* its own; that's all there is to it.

As an example let's take Dean Rusk, and his perception of what land belongs to whom in Vietnam. Of course Secretary Rusk didn't see South Vietnam as belonging to America, but he did apparently see it as self-evidently part of something called the "Free World," and he did assume an American responsibility to resist any Communist encroachment on the Free World. If he had not seen the problem in these simplistic terms, he would hardly have kept coming back, as he did, to the simple proposition that the Communists have to be taught to "let their neighbors alone." To him it apparently seemed self-evident that South Vietnam was a "neighbor" of North Vietnam rather than, as the Communists apparently perceive it, a part of the very body of an independent nation called "Vietnam," into which American invaders have been arrogantly intruding. To Mr. Rusk the notion that American troops might be honestly regarded by anyone as invaders was apparently an intensely dissonant thought, and therefore unthinkable.

Territorial Self-images in Vietnam

South Vietnam, I think, is almost a classical case of an area in which territorial self-images overlap and in which, therefore, each side honestly feels that it *must* expel the alien intruders. On both sides ideology is to a large extent rationalization; the chief underlying psychological factor is pride—the virile self-image—defined as having the courage to defend one's "own" land when foreigners are perceived as attacking it. In a sense you could also say that fear is a fundamental emotion in wars of this type, but it is important to recognize that the fear is mobilized by cognitive distortion—by the mistaken assumption that the land in dispute is self-evidently one's own, and that therefore anyone else who has the effrontery to exist on that land, with a gun in his hand, must be a diabolical alien "aggressor." Neither fear nor pride would be intensely mobilized—as both of them are—if it were not for this cognitive distortion. Each side feels that its manhood is at stake in whether it has the courage and the toughness to see to it that every last one of those intruders is thrown out of *its* territory. To Ho Chi Minh this proposition was apparently as self-evident and elemental as the mirror-image of it is to Dean Rusk. Neither one of them, apparently, would tolerate overlapping, and therefore ambiguous, territorial images. Frenkel-Brunswik (1949) would probably say that neither could tolerate ambiguity. We have, then, in the concept of intolerance of ambiguity, another clue to an understanding of why it is that territorial claims have such rigidity and emotional intensity. And we have the implication that *pulling apart* these overlapping images—clarifying boundaries and getting agreement on them—is one of the things that most needs to be done if we want peace. It may be, too, that deliberate withdrawal from certain hotly contested areas would on balance contribute to peace.

THE "PRO-US ILLUSION"

A third not-so-obvious proposition is that *there is a tendency to see the people in another country as more friendly to one's own side than they actually are.* Let's call this the Pro-us Illusion. It's a form of wishful thinking, obviously, but like various other forms of wishful thinking, it is seldom recognized as such by those who indulge in it.

. . . in American Perception of the USSR

One major example of it would be the long-lasting, hard-dying delusion of many Americans that most of the people in the Soviet Union are against their present rulers and on the American side in the East-West conflict. From 1917, when the Communists first came to power, until perhaps the middle 1950's this was a very widespread belief in the United States, and it contributed much to the rigidity of the militant anti-Communist policy of American policy-makers such as John Foster Dulles. The Harvard research by Bauer, Inkeles, Kluckhohn, Hanfmann, and others (1956) did a lot to put an end to this delusion, but it lingers on in some quarters. Not so very long ago a prominent United States Senator declared that the Soviet Union is "seething with discontent" and hostility to its present rulers.

. . . in Our Perception of the Bay of Pigs

Another example was the belief of many Americans, at the time of our Bay of Pigs adventure, that most of the Cuban people were intensely hostile to Castro in the same way we were, and perhaps ready to rise up against him. It is hard to tell just what was in the minds of our policy-makers at that time, but it looks as if they thought there was a good chance of some kind of uprising if we could just provide the spark to ignite it. The sad thing is that they could have known better. They had easy access to the research of Lloyd Free, a good solid piece of public-opinion survey work indicating that most of the Cuban people, less than a year earlier, were quite favorable to Castro (Cantril, 1968). But Free's evidence was ignored. According to Roger Hilsman, the policy-makers just didn't try to find out what real evidence existed on the attitudes of the Cuban people (Hilsman, 1968). They made no genuine effort to get evidence that was free from obvious bias. (The testimony of refugees in Miami, which they apparently did get, was obviously biased.) That much seems clear: their curiosity was inhibited. As to the reasons for their inhibition of curiosity, one can speculate along various lines. Perhaps it was a defense against dissonance; Festinger might say that they were embarked on an enterprise, and any doubts about the wisdom of that enterprise would have been cognitively dissonant. Or perhaps it was a defense of their black-and-white picture; they may have sensed that the information they didn't inquire into would have impaired their all-black image of Castro's diabolical tyranny over the

Cuban people, and their all-white image of themselves as liberating the Cuban people from a diabolical tyrant. Heider might say they were preserving psychological harmony or balance. In any case it looks as if they shut their eyes because they were unconsciously or half-consciously afraid of what they might see. They cherished too fondly the Pro-us Illusion—and we know the fiasco that resulted.

. . . in Our Perception of Vietnam

Now, more disastrously, there is the case of Vietnam. There, too, we more or less kidded ourselves into believing that the people were on our side. In some ways it is very much like the case of Cuba. In both cases there has been a great overestimation of the extent to which the people were pro-us, and consequently a gross overestimate of the possibility of achieving a quick military victory. In both cases, too, there has been a striking lack of interest, on the part of top policy-making officials, in the best evidence that social and political science could provide.

The irony is increased by our solemn official dedication to the great objective of enabling the people of South Vietnam to determine their own destiny. President Johnson, McNamara, Rusk, President Nixon, and others have continually talked about helping "the Viet-namese" to defend themselves against the Viet Cong and invaders from the north—as if the Viet Cong were not Vietnamese, and as if it were self-evident that most of "the Vietnamese" were gallantly resisting these attacks from within and without, and eager for our help in doing so.

Actually that was always far from self-evident. Some of you may have read my long article, "Misperception and the Vietnam War," nearly three years ago in the *Journal of Social Issues* (White, 1966). If so, you may remember the twenty-five pages of the article (pp. 19-44) that were devoted to a rather intensive effort to cover the evidence on both sides of that question and to find out how the people of South Vietnam really felt about the war. The upshot of that analysis was pessimistic; I estimated that probably there were at that time more South Vietnamese leaning in the direction of the Viet Cong—or NLF[2]—than leaning in our direction.

Since then I have revised and updated the analysis, on the basis of three more years of accumulating evidence. The new information includes all that I was able to glean during two months on the spot in Vietnam, where I had an unusual opportunity to interview well-informed Vietnamese. It includes the Columbia Broadcasting System-

Opinion Research Corporation survey, in which more than 1500 South Vietnamese respondents were interviewed (1967), the writings of Douglas Pike (1966), an outstanding authority on the Viet Cong, and a good deal of other miscellaneous evidence. None of this information is conclusive. For instance, the CBS-ORC survey obviously never solved the problem of getting peasants to speak frankly with middle-class, city-bred interviewers. But by putting together all of the various sorts of information, which is what I did in the book *Nobody Wanted War*, (pp. 29-84), we can, I think, make some fairly educated guesses.

The general upshot of the revised analysis differed from the earlier one chiefly in giving a good deal more emphasis to sheer indifference on the part of a great many of the South Vietnamese. It looks as if a large majority are now so disillusioned with both sides that their main preoccupations are simply the effort to survive, and a fervent hope that peace will come soon, regardless of which side wins. It's a plague-on-both-your-houses attitude. But the results of the earlier analysis did seem to be confirmed in that it still looks as if, among those who do care intensely about which side wins, the Viet Cong has the edge. My own very rough and tentative estimates, representing the situation in 1967, were these: something like twenty percent really dedicated on the side of the Viet Cong, something like ten percent equally dedicated on the anti-Viet-Cong side, and the remainder, something like seventy percent, relatively indifferent. Since in any political conflict the people who count are the people who care, what matters here is the estimate that, among those who *are* dedicated to one side or the other, more are against the position of the United States than for it. The upshot still seems to be that the psychological balance tips *against* the Saigon government and the intervening Americans. That is probably true even now, in 1969, and in previous years it was apparently much more true. For instance, my estimate is that in early 1965, when we first became very heavily involved, it was more like 40 to 10, not 20 to 10, in favor of the Viet Cong.

If Our Policy-Makers Had Known . . .

Suppose our policy-makers had known that most of the emotionally involved people were against us, and had known it clearly, at the time they were making those fateful commitments and staking American prestige on the outcome. Suppose that in 1961-2 when John Kennedy made his major commitment, or in 1964-5 when Johnson made his, they had said to themselves: "Of course we know that if we fight in Vietnam we will be supporting a small minority against a much larger minority." Would they have done it? Would we now have all the tragedy of the Vietnam War? All the blood, all the guilt, all the moral ignominy in the eyes of most of the rest of the world, all the sensitive intelligent young people here at home estranged from their own country? I doubt it. The American super-ego—*if* well informed—is too genuinely on the side of national self-determination, too genuinely against any clear, naked form of American domination over little countries on the other side of the world, even in the name of anti-Communism. If Kennedy and Johnson had clearly realized that the attitudes of the South Vietnamese people at that time were much more anti-us than pro-us, would this whole Vietnam mess have been avoided? I think so.

Vietnam was avoidable, just as the Bay of Pigs was avoidable. The one essential factor in avoiding both of these tragedies would have been to look hard and honestly at the best available evidence (not social-science data, in the case of Vietnam, but the testimony of the best-informed area experts, such as Joseph Buttinger). Our policy-makers in 1962 and 1965 did not look hard and honestly at the best available evidence; and the chief reason they didn't, it seems to me, was that they were clinging to an image of America as helping a beleaguered and grateful South Vietnam—not intervening in a nasty civil war in which most of those who were emotionally involved would be against us. Like the adventurers who planned the Bay of Pigs they were not really curious, because they half-knew what the answer would be if they did look honestly at the facts. They too shut their eyes and put their hands over their ears because they were cherishing too fondly the Pro-us Illusion. And we know now the disaster that resulted.

SUMMARY

The three not-so-obvious contributions (or potential contributions) of psychology to peace are:

First, a corollary of the Hovland two-sided approach: namely, that we Americans should strenuously seek common ground with the Communists, and publicly accept all we can honestly accept of the Communist point of view.

Second, the proposition that the mirror-image type of war is most likely to break out when there is overlapping and conflict of territorial self-images. It follows that reducing such overlap by clarifying boundaries, or even by deliberate withdrawal at certain points, would contribute to peace.

And third, the Pro-us Illusion, with the further proposition that if we Americans had not been indulging in it, neither the Bay of Pigs nor the Vietnam war would have occurred.

REFERENCES

Ardrey, R. *African genesis*. New York: Atheneum Publishers, 1963.

Bauer, R. A., Inkeles, A., & Kluckhohn, C. *How the Soviet system works*. Cambridge: Harvard University Press, 1956.

Bronfenbrenner, U. The mirror-image in Soviet-American relations. *Journal of Social Issues*, 1961, *17* (3), 45-56.

Cameron, J. *Here is your enemy*. New York: Holt, Rinehart, & Winston, 1966.

Cantril, H. *The human dimension*. Rutgers, 1968.

Carpenter, C. R. Behavior and social relations of the Howler monkey. *Comparative Psychological Monographs*, Johns Hopkins University, 1934.

Columbia Broadcasting System. *The people of South Vietnam: How they feel about the war*. Privately printed, March 13, 1967.

Deutsch, M., & Krauss, R. M. Studies of interpersonal bargaining. *Journal of Conflict Resolution*, 1962, *6*, 52-76.

Eckhardt, W., & White, R. K. A test of the mirror-image hypothesis: Kennedy and Khrushchev. *Journal of Conflict Resolution*, 1967, *11*, 325-332.

Fay, S. B. *The origins of the world war*. New York: Macmillan, 1928.

Frenkel-Brunswik, E. Intolerance of ambiguity as an emotional and perceptual variable. *Journal of Personality*, 1949, *18*, 108-143.

Gerassi, J. Report from North Vietnam. *New Republic*, March 4, 1967.

Gooch, G. P. *Before the war: Studies in diplomacy*. London: Longmans, Green, 1936-38. 2 vols.

Gottlieb, S. Report on talks with NLF and Hanoi. *Sane World*, September 1965, 1-6.

Hilsman, R. *To move a nation*. New York: Delta, 1968.

Holsti, O., & North, R. C. The history of human conflict. In E. B. McNeil (Ed.), *The nature of human conflict*. Englewood Cliffs: Prentice-Hall, 1965.

Hovland, C. I., Janis, I. L., & Kelley, H. H. *Communication and persuasion*. New Haven: Yale University Press, 1953.

Kelman, H. C. (Ed.). *International behavior: A social-psychological analysis*. New York: Holt, Rinehart, & Winston, 1965.

Lorenz, K. *On aggression*. New York: Harcourt, Brace, & World, 1966.

Osgood, C. *An alternative to war or surrender*. Urbana: University of Illinois Press, 1962.

Pike, D. *Viet Cong*. Cambridge: M.I.T. Press, 1966.

Rokeach, M. *The open and closed mind*. New York: Basic Books, 1960.

Salisbury, H. E. *New York Times*, January 15, 1967.

Sawyer, J., & Guetzkow, H. Bargaining and negotiation in international relations. InsH. C. Kelman (Ed.), *International behavior*, New York: Holt, Rinehart, & Winston, 1965.

Sherif, M. Superordinate goals in the reduction of intergroup conflicts. *American Journal of Sociology*, 1958, *63*, 349-356.

United States Strategic Bombing Survey. *Over-all report (European war)*. Washington, D.C., Government Printing Office, September 30, 1945.

White, R. K. Hitler, Roosevelt and the nature of war propaganda. *Journal of Abnormal and Social Psychology*, 1949.

White, R. K. *Value Analysis: The nature and use of the method*. SPSSI, 1951, p. 87.

White, R. K. The cold war and the modal philosophy. *Journal of Conflict Resolution*, 1957.

White, R. K. Misperception and the Vietnam war. *Journal of Social Issues*, 1966, *22* (3).

White, R. K. Communicating with Soviet communists. *Antioch Review*, 1967-8, *27* (4).

White, R. K. *Nobody wanted war: Misperception in Vietnam and other wars*. New York: Doubleday, 1968.

Zietlow, C. P. *Washington Post*. April 27, 1967.

FOOTNOTES

* See Jerome Frank, *Sanity and Survival: Psychological Aspects of War and Peace*. New York: Random House, 1967. Forward by Senator J. William Fulbright. THE EDITORS

[1] This book is an expanded version of "Misperception and the Vietnam War," *Journal of Social Issues*, 1966, *22* (3). *Nobody Wanted War* in a further updated edition is scheduled for paperback publication in April, 1970 (Anchor Books).

[2] The common term "Viet Cong" seems preferable to "National Liberation Front" here, since the core of the group is unquestionably Communist (which is all that "Cong" means) and the term "liberation" is question-begging Communist propaganda.

22   The Convict as Researcher

——HANS TOCH

J. Douglas Grant and I recently concluded a study on the social psychology of violence. In studying violence inside prisons we operated with a resident research staff that combined sophistication, practical experience, and the ability to inspire confidence in our informants. Our group in San Quentin prison, for instance, consisted of six men whose graduate training added up to 83 years of confinement. Their competence to study violence in prisons is obvious since five of them also qualified as subjects.

Our top researcher was an interdisciplinary social scientist for whom I cannot find enough praise. His name is Manuel Rodriguez, and his academic background consists of an eighth grade education, a term in the U. S. Army Supply School, and a short course in automobile repair.

Bur Rodriguez has other qualifications. Before the age of 18 he was arrested for malicious mischief and assault. Later he was sentenced for such offenses as armed robbery, burglary, firearms possession, narcotic addiction, and drunk driving. (I might confess that since joining us he has been arrested again, this time for driving without a license while engaged in research.)

Rodriguez has spent 15 of his 36 years behind bars, mostly in the California State Prison at San Quentin. An an inmate Rodriguez became interested in our research subject. He describes the beginning of his interest:

> I was assigned to the weight-lifting section of the gymnasium. Most of the more violence-prone inmates come here to blow off steam at one time or another. It is also sort of a refuge where an inmate can get away from the pressures of staff scrutiny and the yards. We try to keep violence nonexistent, if possible, in this section. This was part of my job, although it was not explicit. In many cases—as a peacemaker—I had to convince both would-be combatants that they could retreat without losing face or pride. Most inmates contemplating violence will usually go to a respected member of the prison community for advice on "Shall I kill this guy or not?" I and a friend of

Reprinted from *Trans-action*, September, 1967, 72-75. Copyright © September 1967, by Trans-action, Inc., New Brunswick, New Jersey.

mine were two of these persons so respected. When these guys who are straddling the fence between violence and nonviolence came to us we began to actively prescribe nonviolence. . . .

Rodriguez started out as an informed layman, with a completely pragmatic concern with violence. Today he is a sophisticated researcher, and he is an expert on the subject.

His transmutation began in early 1965 when he became a trainee in the New Careers Development Project directed by my collaborator, J. Douglas Grant. This revolutionary program is aimed at converting standard clients of professional services (such as Manuel) into dispensers of professional services—or at least into intermediaries between clients and professionals. Research work seemed one obvious target for this effort. Inmate Rodriguez was thus put to work, during his training period in prison, on the first stage of our study. His work included research design, as well as code construction, interviewing, and coding. After Rodriguez was released on parole, we were happy to hire him as a staff member.

Outside, Rodriguez has acted as our principal interviewer. He has interviewed parolees with violent records and citizens who have assaulted police officers. He is not only a sympathetic and incisive interviewer, but became unusually successful in stimulating interest among potential subjects. He is 5 feet 10 inches tall and weighs 175 pounds. He generally wears shirts that allow an unimpeded view of two arms full of tattoos. In addition, when we began the police assaulter interviews, Manuel grew a bushy moustache to make himself look—as he put it—more "subculture." This prop undergirded an invitation to participate that started with the words, "We are not a snitch outfit," but then proceeded to a thoughtful, honest exposition of our objective.

We have tried to blur the line between the observer and the observed. Each of our interviewees is invited to sit down with us to conceptualize the data obtained from him. Each one is asked to help find common denominators in the particulars obtained in the interview. Each one gets the same opportunity we do to play scientist and becomes a minor partner in our enterprise. We obtain some material of extrordinary sophistication from these nonprofessional collaborators.

RESULTS AND RAPPORT

Why do we choose to rely on these nonprofessionals? How do they serve us better than the usual research associates and assistants with the conventional technical and academic credentials?

First, and most obviously, the empirical reason: They bring us better results. They are able to establish trust where we are not, to get data that we could not get, and to obtain it in the subjects' own language. I think I can best illustrate this advantage by excerpting a brief passage out of one of our prison interviews. The respondent here is a seasoned inmate whose reputation is solidly based on a long record of violent involvements. The interviewer is one of our nonprofessional researchers—another inmate:

Q: Was it the next day that you were going through the kitchen line and that he approached you and said he was coming down and wanted his stuff, and you better be there with it?
A: He said he was coming to get me, and I better be ready. The inference was—Was I going to be ready?
Q: So you went back to the kitchen and got a shank [knife] and then went to your pad. Now this dude who was doing the talking to you now, this is the one who you were playing coon can [a card game] with? The next morning one of the dudes approached you?
A: The next morning. The same dude. When I came out of my cell in the wing this guy approached me. He lived in the wing.
Q: What is his message?
A: His message is just a play, and they were playing a pat hand. It wasn't anything different from the day before. I told him. . . .

This excerpt fits into a standardized interview schedule that was designed to tease out sequences of interpersonal events leading to violence. But it also is a snatch of conversation between two persons discussing a subject of mutual interest in the most natural and appropriate language possible. In this type of interaction, data collection occurs with no constraint, and without translations designed to please or to educate the researcher.

Another advantage to be obtained in the use of nonprofessionals is the benefit of their unique perspective in data interpretation. A well-chosen lay researcher can often be in a position to correct naive inferences by less experienced professionals. In one dramatic experience one of my research partners, inmate Hallinan of San Quentin, chided me (in writing) for drawing a hasty and incorrect conclusion from an interview we had conducted:

Your interpretation seems to be influenced by the subject's storied loquaciousness rather than the incidents themselves. Is the subject's behavior, as he claims, the result of his being an Indian leader, and having to intercede in their behalf, or is it because of his need to establish a personal reputation as a prison tough guy? I choose the latter interpretation; an interpretation based on how the subject has behaved, not how he thinks he has behaved. . . .

An Indian functions within the rigid framework of rules. "There are family codes, tribal codes, and Indian laws," is how he puts it. But there is also . . . a joint code that he is well aware of: "The cons have their own rules, and one of them is that they step on the weak." . . .

The first incident that the subject becomes involved in in is the rat-packing of an Indian child molester in order to ostracize and punish the molester, and also to solidify his position among the low-riders. So, rather than being a leader of these Indians, he is using his Indian blood to further his own ends. He wants to be a tough con, someone to be feared and respected. "The new guys that come in, no one knows about them." "Once you get a reputation you have to protect it." The above statements, and others similar in nature, were made by the subject during the course of the interview. [Their] significance is self-evident.

How does the subject go about building a reputation? As he says, fighting for home boys, and interceding for other Indians? No. Of the ten incidents—actually nine, because No. 1 and No. 9 are the same—No. 6 no violence occurred; No. 2 involved helping a friend, although the details were vague; No. 7 was a fight of his own making; No. 9 he was attacked; and No. 10 was the rat-packing incident. The remaining four involved custody. He was proudest of No. 8. In regards to this incident the following dialogue occurred:

Q: Do you think this incident helped your reputation?
A: It sure as hell did. I knocked down the Captain.
Q: How did you feel just before you knocked him down?
A: Like a big man.
Q: During?
A: I sure am doing it right this time.

The subject is also proud of the fact that at one time he had spat on the warden. . . . Obviously the subject feels that these things scare people. . . .

The word circulates that he has fought with the "bulls," implying that he will jump on a convict with little provocation. The facts are never pursued, but accepted prima facie, because those who pass on these rumors and exaggerations are the very ones who are most impressed by them. The rumor returns and the subject begins to believe his own yard reputation. . . .

Our subject has completed the building of his reputation, petty though it is, and now he and his low-rider friends can observe and honor it. Not that the cons on the yard do, but the subject feels that they do, and this is all that really matters. If anything he is tolerated, not respected and feared as he would like.

It is obvious that inmate Hallinan is not only furnishing me with a lesson in perspective, but is also demonstrating that he can compete with professionals in his methodological acumen and his ability to vividly summarize and communicate research conclusions. And although this analysis is unusually literate, because inmate Hallinan has invested much prison time in creative writing courses, much can be learned even from our most unlettered collaborators.

There is another aspect to the use of nonprofessionals which relates to a less tangible and more general advantage. Most social researchers sense some difficulty in the initial approach to subject populations of vastly different backgrounds from our own. Some of us react at this juncture with an elaborate process of ingratiation or "gaining of rapport" in which we, and the research, are presented in the (presumably) best light. This posturing is often transparently insincere and always wasteful. Worse, it usually achieves merely a wary and delicate stalemate, during which only a hit-and-run raid for data is possible before the subjects discover what has happened to them.

AVOIDING EXPLOITATION

During rare moments of honesty, we may admit that even when we induce subjects to cooperate, our dealings with them are seldom the exciting adventure we tell our students about. I say this because I suspect that the real problem is not one of communication and social distance at all—it may have nothing to do with habits or dress or the use of most vivid vocabulary. It may be that our subjects understand us only too well—that what we ask is unreasonable and unfair. After all, at best we are supplicants, and at worst, invaders demanding booty of captive audiences. In return for a vague promise or a modest remuneration we expect a fellow human being to bare his soul or to make controversial and potentially incriminating statements. The "communication" is one way—the researcher maintains his position as an "objective" recipient of non-reciprocated information.

We also make our informant aware that we are not interested in him as a person but as a "subject"—a representative of a type, or a case, or an item in a sample. He knows this because he knows who he is and who we are. He knows that he is being approached because he is the inhabitant of a ghetto or a prison, because he is a "consumer," or because he acts as an informer. And he knows that his aims are being subordinated to our own. How can he share our objectives, after all, if he cannot even see the results of the efforts in which he has participated?

I speak with considerable humility here, because I almost once again made the mistake of taking my Viennese accent and my parochial concerns into prison cells and police stations, expecting to secure frank answers to prying questions. I have done this sort of thing often in the past. This strategy strikes me now not only as naive but as offensive.

Therefore, Grant and I followed an alternative course in our violence project, such as I have briefly described. This has supplied us with linkage across cultural gaps, with highly motivated informants, with substantive expertise, with heightened analytic power, and with the feeling that we have been fair.

I shall not pretend that these benefits are automatic and free of risk. Like professional researchers, nonprofessional participants in research must be selected with care. Unintelligent or completely illiterate persons would be of limited use, as would social isolates. A cynical, exploitive, or immature outlook can create a poor prospect for programs that have the usual training resources. This is also true of rigidly held preconceptions, though to a lesser extent.

On the other hand, too close attention to selection criteria may produce a staff of quasi-professional nonprofessionals—which is also bad. They may be rejected by the subjects and even be considered a species of Uncle Tom. Not being trusted, they may have relatively limited useful knowledge or insight, contribute little, discover they are marginal members of the team, develop poor motivation.

Even careful selection will not altogether eliminate these possibilities. The nonprofessional must get training that is not only directly related to research but also can provide him with incentives, support, and a meaningful self-concept. Some of this training may be of the sort routinely encountered in graduate schools; some may be more characteristic of social movements. The nonprofessional researcher must be, in a sense, a *convert*. He must acquire a new role, a new set of values, and new models and friends while remaining in close touch with his old associates. The professional merely places others under the microscope;

but the nonprofessional must convert his own life experiences into data. While the rest of us can view research as a job, the nonprofessional must see the involvement as a mission or a crusade—or he will have trouble doing it at all.

What training then should these nonprofessionals get? First, research indoctrination, in the purest sense of the word, that aims directly at awakening curiosity and at the desire to reach latent meanings or patterns. It must try to inculcate suspicion of the unrepresentative and unique and a phobia against premature interpretation. Obviously, it must also provide tools—intensive practical instruction in the use of steps to be employed, including interviewing techniques, content analysis, survey problems, and data processing. This training must not only include general background information and acquaintance with the content of the research but also self-corrective and social skill training of the kind necessary to work with sensitive groups.

But the most critical challenge is ours. Will we treat the trained nonprofessional as a partner and colleague and respect his integrity and abilities? We must preserve the nature of our own contribution; but we must also be prepared to become members of our own team.

Nonprofessionals, if given the opportunity, can help us shape ideas, formulate designs, and analyze results. We should continue to provide intellectual discipline and a sense of perspective. For the rest, we may find ourselves in the unaccustomed position of being students to spirited and able teachers—and the benefits will be reflected in the quality of research, as well as in the resolution of ethical dilemmas that currently often leave social researchers with a bitter taste in the mouth.

Adler, A., 309, 310
Adorno, T. W., 335
Aemilius, P., 259
Akert, K., 159
Ajax, 259
Alexander, 259
Angell, R. C., 335
Anshen, R. N., 102
Appley, M. H., 11
Ardrey, R., 11, 14, 348
Arnold, M., 259
Ayllon, T., 237, 246
Azrin, N., 5, 153, 155-161, 163-168, 170, 237, 246

Baer, D. M., 237
Bauer, R. A., 350
Bandura, A., 61
Baroff, G. S., 246
Baxter, R., 27
Beach, F., 163
Beeman, E. A., 161-163
Berensen, B., 308
Berkowitz, L., 61, 132, 137, 142
Blum, R. H., 335
Brierton, G. R., 170
Bronfenbrenner, U., 345, 346
Broom, R., 29
Brown, G. D., 236, 246
Bruno, G., 55
Brutus, 259
Burlingham, D., 249
Buss, A., 2, 3
Buttinger, J., 353

Caesar, J., 262
Calhoun, J. B., 159
Cameron, J., 341
Campbell, A., 198-207
Cantril, H., 349
Carlyle, T., 32
Carrel, A., 284
Carson, R., 284
Carthy, J. D., 2
Castro, F., 350
Catania, A. C., 166, 168
Cavanaugh, J., 205
Chapman, J. J., 264
Chiang Kai-shek, 347
Christiansen, B., 335
Clausewitz, 312
Cofer, C. N., 11
Cooley, C. H., 335
Coules, J., 132

Craine, W. H., 174
Crew, F. A. E., 159
Croce, B., 214
Cullen, E., 293, 299
Curran, F. J., 248

Daniel, W. J., 153
Dart, R., 14
Darwin, C., 1, 2, 33, 34, 56
Davis, F. C., 152
Delgado, J., 170
De Monchaux, C., 3
Denton, F., 117, 126
Desai, M., 315
Deutsch, M., 340
de Vore, I., 54
Dewey, J., 106, 214, 215, 221
Dickinson, L., 264
Diomedes, 259
Dollard, J., 101
Dugdale, R., 59
Dulles, J. F., 350

Easterbrook, J. A., 142
Ebling, F. J., 2
Eckhardt, W., 344
Egger, M. K., 159
Eibl-Eibesfeldt, I., 5, 50
Einstein, A., 16, 269, 275, 276, 279
Elbert, J., 177

Ferster, C. B., 158
Feshbach, S., 132, 134, 136
Festinger, L., 350
Flory, R. K., 163
Flynn, J. P., 159
Frank, J., 340
Frederickson, E., 161
Free, L., 350
Frenkel-Brunswick, E., 349
Freud, A., 249, 255
Freud, S., 12, 54-56, 83, 110, 248, 269, 275, 276, 280, 309
Fromm, E., 312
Frosch, J., 255
Fulbright, J. W., 340

Gandhi, M., 3, 214, 215, 315-317, 319-321, 323-331, 333, 335
Geen, R. G., 142
Gellerman, S., 134, 138
Genghis Khan, 91, 281
Gerassi, J., 341

Goddard, H., 60
Gottlieb, S., 341
Glover, E., 101
Gooch, F., 346
Grant, E. C., 5
Grant, J. D., 356, 357
Guetzkow, H., 334, 336, 344

Hake, D., 158, 164, 166
Hall, C. S., 152
Hamilton, J., 246
Hammond, K., 344
Hartman, R., 32
Hector, 259
Herodotus, 32
Hess, E., 313
Hess, W. R., 159
Hilsman, R., 350
Hinde, R. A., 289, 290
Hitler, A., 91, 277, 312, 324, 343, 346
Hoar, W. S., 290
Hobbes, T., 315
Hobhouse, L. T., 214
Ho Chi Minh, 348
Hocking, W. E., 72
Hoesti, O., 345
Holz, W. C., 237, 246
Hoover, J. E., 143
Hosking, E., 166
Hovland, C. I., 342
Hull, C. L., 101
Hutchinson, R. R., 157, 158, 163, 164,
    166-168, 170
Huxley, A., 211
Huxley, J., 54

Inkeles, 350

James, W., 214
Janis, I. L., 342
Johnson, A., 101
Johnson, L., 337, 350
Johnston, J., 166, 169
Jung, C. G., 309

Kallikak, M., 60
Kelley, H. H., 341
Kelman, H. C., 334, 340
Keiser, S., 248
Keith, A., 31, 35, 101
Kennedy, J., 353
Kenny, D. T., 132
Khruschev, N., 313-314

Kislak, J. W., 163
Kitzler, G., 49
Klein, M., 248, 251, 309
Klein, S. J., 152
Kluckhohn, C., 350
Knoll, M. D., 299
Kramer, G., 45, 54
Krasner, L., 236
Krauss, R. M., 341

Laird, M., 340
Lea, H., 262
Lesser, A., 3
Levi, W. M., 165
Levison, D. J., 335
Lewin, K., 347
Lichtenberg, G. C., 22
Linton, R., 101
Loew, 142
Lorenz, K., 13, 285-286, 289-290,
    298-304
Lovaas, O. I., 245

Mabry, J. H., 166, 171, 236
MacArthur, D., 343
Marais, E., 39
Marco Polo, 32
Mason, W. A., 60
Masserman, J., 159
May, M., 101
McDougall, W., 11, 68
McKinley, W., 260
McLaughlin, R., 170
McNamara, J., 351
Mead, G. H., 335
Mees, H., 236
Metraux, A., 101
Meyr-Holzapfel, M., 52
Miller, N., 61, 101, 153, 168
Mirskaia, L., 159
Montagu, A., 5, 59
Moore, G. E., 214
Morris, D., 5, 48, 285-286, 294
Moyal, J. E., 115
Murdock, G. P., 101
Murphy, G., 335

Naess, A., 214-215, 235
Nansen, F., 16
Napoleon Bonaparte, 91, 281
Nietzsche, F., 316
Nixon, R., 339, 351
Norman, D. G., 157

North, R., 345

Oehlert, B., 52
O'Kelly, L. W., 153, 168
Osgood, C., 341

Parke, R. D., 142
Patton, S., 263
Pavlov, I., 167
Pepitone, A., 132
Pike, D., 352
Plato, 316
Potter, H. B., 38

Ranson, 159
Rapoport, A., 312, 314
Reed, W., 101
Reichling, G., 132
Reynolds, E., 33
Reynolds, G. S., 166, 168
Richardson, J., 166, 168
Richardson, L., 115-118, 128, 130
Richter, C. P., 153
Risley, R. T., 236, 237
Rodriguez, M., 356, 357
Rogers, C., 344
Rokeach, M., 344
Romney, G., 205
Roosevelt, T., 262
Rosecrance, R., 115-117
Rusk, D., 348, 349, 351
Royce, J., 214
Russell, B., 214, 309

Salisbury, H., 341
Sawyer, J., 344
Schaeffer, B., 245
Schepers, G. W. H., 29
Schilder, P., 248, 249
Schneirla, T. C., 298
Schuman, H., 197-206
Sclater, W. L., 37, 40
Scott, J. P., 9, 153, 161
Seligman, C. G. and B. Z., 101
Seward, J. P., 165
Shortridge, G. C., 37
Shotwell, J., 101
Simmons, J. Q., 246
Skinner, B. F., 109, 111, 152-154,
    166, 168, 169, 211, 300
Smith, E., 27-29
Smith, S., 166
Sorokin, P., 115

Speck, R., 60
Stachnik, T. J., 166, 171, 236
Steckel, L. C., 153, 168
Steinmetz, S. R., 101, 263
Stevenson-Hamilton, J., 37
Stone, L., 2
Strabo, 32

Tate, B. G., 245
Taylor, A. E., 214
Tedeschi, R. E., 156, 164
Thibault, J. W., 132
Thompson, T., 163
Thorpe, W. H., 299
Thucydides, 259
Tinbergen, N., 13, 51
Tolstoy, L., 264
Touch, J., 245
Toynbee, A., 115
Troland, L. T., 11
Tyler, V. O., 236, 246

Ullman, L. P., 236
Ulrich, R., 5, 110, 153-161, 163-170,
    174, 175, 236
Utsurikawa, N., 152

van Wagenen, R. W., 335
Vernon, W., 166, 168
Von Haartman, 293
von Holst, E., 159
von Saint-Paul, U., 159

Waller, W., 102
Walters, R. H., 61, 142
Warner, L., 102
Washburn, S., 54
Wasman, M., 159
Watson, J., 250
Wayne, J., 62
Wechsler, D., 248, 249
Wells, H. G., 268
Wells, L. H., 35
Whaley, D., 245
Wheeler, G. C., 101
Wheeler, W. M., 71
White, A. C., 37
White, R., 343, 347, 351
Williams, J. H., 45
Wolf, M. M., 236, 237
Wolff, P. C., 163, 164, 166, 169
Wolfson, W. Q., 255
Wood, W., 37

Wright, Q., 101, 115, 118-122, 130

Yerkes, R. M., 101, 152
Yinger, J. M., 335

Zietlow, C. P., 341
Zuckerman, S., 152

SUBJECT INDEX

ABM, 338
Action-specific energy, 61
Africa, 28, 83, 88, 91, 92, 313
  South, 329, 331
Aggression
  anxiety, 132
  arousal of, 138
  biology of, 289
  displaced, 229
  eliciting stimuli, 143, 149
  German, 345
  impulses of, 80
  innate, 314
  institutionalized, 228, 231
  intergroup, 303
  interspecies, 177
  interspecific, 3
  intraspecific, 3, 44, 45, 49
  operant, 167-168, 169-170
  personal, 229, 231
  positive aspect of, 308
  prepared, 97
  respondent, 167-168, 169-170
  ritualized, intraspecific, 54
    redirected, 53
  socially sanctioned, 228
  spontaneous, 289
  stimulus elicited, 141-142
  suppressed, 309
  suppression of, 194-196
  "true," 53
  waning effect of, 304
Aggressive:
  activities, of a nation state, 3
    vicarious, 132, 133, 136, 138, 139
  associations, 136, 138
  behavior, redirected, 52
    ritualized, 53
    physiology of, 291
  coloration, 51
  cue properties, 142, 149
  drive, 52, 133, 142, 311-312
  fantasies, 138
  impulses, 86, 280
  motivation, 139
  patriotism, 86
  responses, 142

  urge, 304
  words, 134, 142
Aggressiveness
  as instinctual behavior, 78
  "natural," 83
  physiological, 83
Agonistic behavior, 1
Alcohol, 276
Algeria, 347
Amount of violence index, 117
Amount of war index, 124
Anger, 1, 3, 67, 72, 81-83, 85, 86, 132,
      139, 142, 148, 292, 322
  arousal, 136, 142, 145
    instigators, 142, 148
  impulse, 74, 78
  psychology of, 83
Animal sociology, 52
Animism, 78
Animus, 231
Antelope, 29, 38
Anthropophagy, 32
Anti-Communists, 328, 332, 349, 353
Anti-Fascists, 328
Anti-Imperialism, 332
Anti-militarist party, 265
Anti-poverty program, 205
Anti-Semitists, 327
Anti-social:
  action, 213
  behavioral patterns, 195, 196
  habits, 182
  practices, 221
Anxiety, 217, 224, 251, 255, 256, 311
Appeasement
  ceremonies, 53
  signals, 292
Appetitive behavior, 52
Armed:
  attack, 97
  conflicts, 3
  contest, 93
  force, 88
  raids, 93
  robbery, 356
Arms race, 338-340
Army, 256, 260, 268

discipline, 266
life, 263
Arsonists, 203
Assassination, 342, 346
Assault, 86, 356
Assertiveness, 303
Assyria, 90
Attack, 46, 51, 53, 287, 291, 295
  redirected, 303, 309
  species-specific, 45
  urge to, 289
Attack-avoidance system, 291
Attitude, 2
  changes, 227
  emotional, 302
  motivational, 302
Australia, 32, 40, 83
Australopithicine stage, 31, 36
Australopithicus africanus, 9, 28, 40
Austria, 40
Austro-Hungarian Empire, 100, 102, 273, 279
Authoritarian personality structure, 213
Authority, 24, 84
Aversions, 68, 138
Aversive stimuli, 152, 158, 170
  cold, 158
  heat, 158
  noise, 158
Avoidance, 291
  responses, 173

Baboons, 28, 30, 37-40, 54, 81
  skulls, 29
Babylonia, 90
Bay of Pigs, 350, 353, 354
Beetles, 37
Bestiality, 264
Birds, 28, 36-41, 44, 66, 68, 70, 72, 74, 75, 81, 184-195, 293, 294, 300
  bank swallow, 69
  chickens, 159
  European cuckoo, 71
  kingfisher, 69
  robins, 69
Birth control, 304, 309
Black Americans, 197, 199-200
Black and white thinking, 339, 340
Blitzkrieg, 97
Blood
  bad, 59
  tax, 265-267

Bloodshed, 288, 291, 303
Bond behavior, 53
Bosnia, 347
Boxing, 35, 312
Boycott, 202, 321
  social, 332
British Empire, 281, 282, 331

Cain, mark of, 33
Campus violence, 4
Cannibal feasts, 90
Cannibalism, 32, 33, 77, 93
Captive animals, 45
Carnivores, 30, 53, 67, 294
Carnivorous:
  diet, 36
  habits, 31
  propensity, 36
Castration, 163
Cat, 67, 159, 165, 182, 185-195
Catharsis, 62, 105, 137
Children, 4, 62, 81, 177, 297, 301, 302, 310
  postencephalic, 251
Chimpanzees, 5, 29-31, 36
China, 32, 262
  Communist, 343
Christianity, 98
Cities, 19, 196-198, 204, 234, 259, 277, 279, 280
Civic disorder, 78
Civil:
  disobedience, 326-328, 330
  disorder, 200
Coexistence, 333
College students, 4
Combat, 120
  ceremonial, 72
  soldiers, 224
Constructive program, 320, 326-327
Communism, soft on, 342
Communist leaders, 342
Communists, Russian, 23
Compass of Motives, 22
Competition, 184, 187, 195, 210
Concentration camps, 82
Conflict
  civil, 126, 129
  east-west, 344, 349
  emotional, 254
  ideological, 313
  industrial, 220
  inter-group, 219
  motivational, 291, 292

movements, 292
territorial, 346
world, 335
Congo, 269
Conquest, 18, 91, 93
Conscription, 265, 267
Control group, 6
Cooperation, 69, 71, 83, 86, 314, 335
  peaceful, 330
Cooperative behavior, 176
Counteraggression, 223, 224, 229, 231
Counterdisplay, 72
Cow, 40
Cowardice, 297
Crabs, 37
Crime, 83
  genetic causes of, 60
  juvenile, 309
Criminal conduct, 262
Crowd, violent, 210
Crowding, 68
Cruelty, 342
Cuban crisis, 312
Cursing, 241
Czechoslovakia, 40, 273, 342

Darwinism, 196
Death, 45, 72-74, 82, 296, 304, 308,
    314
  drive, 55
  violent, 261
  wishes, 248
  see also Instinct, death
Defending the fatherland, 127
Defense, 9, 97
  of civilization, 119
  of family, 44
  self, 270
  territorial, 52, 195
Defensiveness, 138
Defiance, 253
Demonstrations, 200
  nonviolent, 202
Denmark, 97, 312
Deprivation
  in opportunities, 254
  maternal, 60
  of love, 249, 251
  peer, 60
Desegregation, 220
Destructiveness, innate, 273, 276,
    280-281
Detroit, 197-207
Dictatorship, 345

of reason, 24
Diem, 343
Discrimination, 206
Displacement, 106
Display, 47
Disorder, "spontaneous," 203
Distress, 68
  signals, 297
  see also Appeasement
Dog, 29, 67, 81, 182, 185-195
Dominance, 169, 188, 311
  orders, 182
Dominance-submission relationship,
    195
Domination, 187
  economic, 330
  social, 198
Domestication, 54, 270, 278
  human, 277
Dreams, 256
Drugs, in suppressing aggressive
    behavior, 244
Ducks, 53

Ego defenses, 141
Egypt, 90
Emotion, 210
Elephant, 29, 45
Empathy, 215
Enemy, wickedness of, 128
England, 40, 262, 273, 346
Escape, 9, 46, 156-158
Esperanto, 309
Ethology, 13, 55, 286, 287, 294
Europe, 275
  post-war, 273
Evolution, 289, 292
  adaptive divergent, 293
  cultural, 295, 296
  phylogenetic, 11
  social, 261
Evolutionary mechanisms, 1
Exploratory behavior, 301
Extinct animals (various species), 29
Extinction, 105, 106, 158

Facial expressions, 295
Famine, 303, 304
Fantasy, 224, 251
  sexual, 252, 269
Fascism, 277
Fear, 1, 12, 47, 82, 138, 199, 256,
    263, 269, 291, 295, 296, 343,
    345, 349

Feminism, 261, 262
Fighting, 1, 9, 155-158, 161, 163, 195, 196, 284
  angry, 93
  antecedents of, 86
  biological need for, 76
  ceremonial forms, 72
  collective, 78, 93
  defensive, 74, 288, 295
  destructive forms, 72, 76
  evolution of, 49
  infra-human, 67
  inter-species, 164
  intra-species, 288
  mass, 45
  men, 288
  mouth, 47
  movements, 154
  organized, 90
  posture, 173
  rate of, 160
  reflexive, 170, 176
  ritualized, 50
  rival, 44
  stereotyped, 174, 176
  territorial, 293, 348
Finland, 312
Firearms possession, 356
Fish, 40, 44, 46-48, 51, 54, 163, 293
  stickleback, 290, 295
France, 20, 40, 99, 345
Frederick the Great, 265
Free speech, 345
Frustration, 103-105, 159, 203, 217, 227, 251, 254, 288, 309, 310

Gang, organized, 97
Garden of Eden, 40
Geese, 53
General systems, 114
Genetics, 66
Germany, 40, 89, 97, 99, 102, 255, 260, 273, 341, 345
Ghetto, 198-201, 204-207, 361
Gibbon, 27, 28, 36, 37, 40
Gladiatorial encounters, 313
Glory, love of, 258
Goats, 29
Goose, 69
Gorilla, 29, 36
Great apes, 35
Great Britain, 99
Greece, 259

GRIT proposal, 341
Guilt, 1, 137, 138, 224, 226, 352
  arousal, 138
Guinea pigs, 164, 192

Habituation, 51, 196
Hamsters, 164
Hard-hats, 4
Hardihood, 261, 266
Hate, 80, 81, 83, 85, 86, 102, 343
Hatred, 1, 3, 22, 24, 28, 231, 256, 319
Hawaiian Islands, 262
Head banging, 250
Head butting, 254
Headhunting, 33, 77, 90, 95
Heroism, 128, 260, 263, 311
Hippopotamus, 29
Holland, 40
Homocide, 54, 143, 276
Homocidal threats, 254
Homologous motor pattern, 48
Hostile:
  acts, 149
  environment, 335
  impulses, 234
Hostility, 3, 80, 81, 134, 184, 190, 194, 213, 256, 314, 349
  infantile, 248
  latent, 229, 230, 232
  repressed, 229
Hovland principle, 342
Humanism, 278, 283
Human, man-hunting, 90
Human sacrifices, 93
  sacrificial practices, 33
Human suffering, 127
Hunting, 182, 190, 258
Hyaenas, 29, 294, 303
Hysteria, mass, 304

Imitation, 13
  of an aggressive model, 62
Imperialism, 86, 88, 91, 93, 94, 259, 283
Impulses, destructive, 269, 312
Impulse to destroy, 73
India, 331
Indifference to suffering, 33
Indoctrination, 86
Infanticide, 33
Inhibition, 132, 138, 143, 149, 170
Innate releasing mechanism, 13
Innate teaching mechanism, 300

Insanity, 83
Insectivores, 28
Insects, 37, 38, 71
Instinct, 21, 35, 72, 270, 297
    death, 2, 23, 248, 280, 307, 309
    destructive, 21-24, 59
    erotic, 248
    human, 55, 248
    military, 260
    miscarrying of, 55
    sex, 310
    sex recognition, 73
Instinctive:
    behavior, 13
    drive, spontaneous, 55
    tendencies, 12
Intention, 3
    movement, 6, 47
Ireland, 274
    Northern, 347
Isolate monkey, 60
Isolates, social, 362
Isolation, 13, 22, 99, 168, 184, 193,
        195, 299, 319
Israel, 347
Italy, 89, 99, 273, 277

Jackal, 29
Japan, 99, 255, 260, 262, 341
Jealousy, 12, 81
    sexual, 83
Jingoism, 259

Kashmir, 347
Kerner commission, 198, 201
Killing, 73, 288, 295, 303
    in group struggles, 317
    men, 284
    of the less fit, 66
    the enemy, 224
    tools, 297
Kinship system, 87
Korea, 343, 347

Lambs, 37, 38
Latin America, 117
Law and order, 294
League of Nations, 16, 20, 100, 275,
        282
Lemurs, 37
Leopard, 37
Lion, 29, 45, 67, 210, 294
Lizards, 37, 39, 44, 45, 49, 50

Looting, 204

Malaysia, 83
Mammals, 54, 66, 68, 71, 72, 74, 75
Manly virtues, 267
Manliness, 266, 267
Maori, of New Zealand, 89
Marines, 256
Marital tabus, 278
Masochism, 311
Masochistic personalities, 217
Matsu, 343
Megalomaniac statesmen, 312
Melanesia, 32, 88
Mental hospital, 82
Mexico, 89
Mice, 4, 36, 152, 164
Middle Ages, 280
Militancy, 199
Militarism, 65, 75, 265
Militarist, 66, 76
    authors, 262
    theory, 67, 71
Military, 339
    affairs, 268
    apparatus, 268
    aptitudes, 264
    career men, 4
    effort, 303
    feelings, 258
    force, 78
    honor, 266
    machine, 89
    mind, 266
    operations, 78
    organization, 88, 93, 220
    party, 264
    professional, 261
Minorities, 100
    status of, 88
    "subject . . . ," 89
MIRV, 338
Mobbing, 51
Modification
    behavior, 211
    of deviant behavior, 236
Monkey, 28, 37, 44, 164, 166, 170,
        286
    see also Isolate monkey
Murder, 312
    see also Homocide
Murderer, mass, 288, 291
Muscular strength, 17

Mythology, 87

Naked ape, 293
Napoleonic wars, 89, 102
Narcissism, 278
  clan, 278
  national, 278
National:
  consciousness, 277
  egoism, 97
  hatreds, 93
  Rifle Association, 8
  solidarity, 278
Nationalism, 86, 88, 89, 91, 93, 94,
    99, 100, 277
  European, 277
  militant, 273
  wars of, 90
Nationalist movements, 219
Nations, 19, 218, 262, 270, 271, 275,
    277
  as armies, 265
NATO, 332, 333
Natural selection, 66, 75
Naziism, 98
Nazis, 230, 232
Navy, 260
Neuroses, 249
Neurotic, 269
  symptoms, 311
New Guinea, 32, 83
Newark, 203
New Careers, 357
New Zealand, 32, 88
Nietzscheism, 65
Non-resistance, 74
Nonviolence, 3, 315, 316, 318, 320,
    329, 335, 357
  ethics of, 315, 334
Nonviolent:
  action, 225
  army, 327
  means, 217
  methods, 326
  political movement, 214
  problem solving, 8
  struggle, 326
Non-zero-sum game, 344
North America, 32
Norway, 97
Noxious stimuli, 2

Oedipus:

complex, 278
  conflict, 253
  situation, 252
Operant behavior, concept of, 149
Oppression, 100
Orangutan, 36
Out-group, 228
Oysters, 37

Pacification, 74
Pacific civilization, 267
Pacifism, 261, 264, 270, 310
Pacifist:
  movements, 219
  values, 233
Pacifists, 26, 65, 66, 318
Pain, 110, 152, 253
  and fear economy, 264
Painful stimulation, 153
Palestine, 273
Passive resistance, 319
Pathology, cultural, 86, 95
Patients
  hyperaggressive, 226
  mental, 4
  retarded, 236
  wheel chair, 237
Patriotic pride, 266
Patriotism, 261, 277
Pax Britannica, 281, 282
Pax Romana, 19, 281, 282
Peace, 65, 70
  advocates, 264
  causes of, 272
  economy, 264, 265
  party, 261
  perpetual, 77
  world, 75
Pearl Harbor, 256
Peck order, 292, 294
  human, 311
Persona, 51
Persia, 90
Peru, 89
Phillipines, 262
Pig, 29, 30, 40, 41
Pigeons, 5, 65, 67, 69-71, 73, 74, 158,
    165
Pithecanthropus-Sinathropus, 32
Play, 68, 188, 189, 195
Plunder, 258, 259
Poland, 89, 97, 273, 346
Police, 97, 234, 361

force, 275, 276, 303
Policeman, 54, 203
Polynesia, 32
Population density, 159, 160, 291, 303
Pornographic literature, 234
Poverty, 264
Power, 206
   lust for, 93
   urge to, 310
Predation, 3
Predator, 45, 51, 288
Prehistoric past, 278
Prey-predator relationship, 44
Primates, 5, 36, 40, 73
Prisons, 82, 234, 356, 361
Propaganda, 86, 276, 314
   communist, 340
Propagandism, 268, 333
Protest, 203-205, 251, 254
   peaceful, 200
Pro-us illusion, 349, 354
Prussia, 277
Prussianism, 65
Psychic:
   energy, 2
   terror, 315
   traumas, 249
Psychoanalysis, 21, 55
Psychoanalytic:
   movement, 309
   theory, 12
Psychopathic personalities, 251
Psychosis, mass, 270
Psychotherapy, 224, 306
Pugnacity, 12, 73, 78-80, 83, 85, 86,
      93, 258, 260
   impulse of, 81
   instinct of, 68
Punishment, 104, 105, 158, 170, 176,
      211
   fear of, 105

Quakers, 341
Quemoy, 343, 347
Quick temper, 61

Rabbits, 184, 188, 192, 194
Races, 19
Racial relations, 100
Racism, 206
Radicalizing experience, 206
Radical movements, 219
Rage, 61, 250, 251
Rank, privileges of, 81

"Rape of the earth," 284
Rapoport Debate, 344
Rats, 4-6, 44, 45, 159, 168, 184, 187
   see also Rodents
Rebellion, 205, 208, 310
Redirection, 51, 52, 54, 303
   see also Aggression, redirection of
Reflex fatigue, 160
Reflexive behavior, 154
Reign of Terror, 33
Releasers, 49
Releasing:
   mechanism, 47
   stimulus, 13
Reptiles, 49
Revenge, 17
Revolution(s), 96, 205, 211, 280
   Communist, 129
   cultural, 98, 343
   European, 129
   French, 89, 118
Revolutionary, 328
Rewards, vicarious, 62
Rhinoceros, 29
Riots, 197-203, 205
   Hindu-Moslem, 331
Ritualization, 46-48, 52
   phyletic and cultural, 49, 54
Robbers' Cave experiment, 344
Robbery, organized, 91
Rodents, 28, 39, 45, 153
Roman:
   gladiatorial shows, 33
   Empire, 281
Romans, 19
Rome, 259
Russia, 99, 262, 345

Sadism, 230-232, 281
Sarpedon, 259
Satyagraha, 316, 326, 329, 331-333
Savagery of civilization, 95
Scalping, 33
Scandinavia, 40
Screaming, 237, 241, 244, 250, 254
Scorpions, 37, 39
Secret society, 84
Segregation, 206
Selection pressure, 49
Self-:
   assertion, 310
   biting, 60, 250
   destructive behavior, 236

determination, 353
fulfilling prophecy, 339, 340
image, virile, 349
preservation, 270, 281
realization, 317
sacrifice, 310
Sensitive periods, 301, 302
Sex behavior, 189
Sexuality, repression of, 311
Sheep, 29, 41
Sheep's paradise, 261
Shock, electric, 110, 142, 144, 153,
    154, 168, 192, 245
Sierra Leone, 37
Signal function, 292
Size of violence, 121
Size of war indicator, 127
Slave:
  raiding, 93
  wars, 90
Slavery, 33, 90, 92
Slender Loris, 37
Snipers, 203
Social:
  ape, 294
  contagion effects, 225, 227
  hierarchy, 196
  learning theory, 61
  life, interspecies, 188
  rank order, 44
  welfare, 345
Socialistic equilibrium, 265
Sollwert, 300, 301
South America, 32
Soldiers, 268, 297
Spain, 260
Sparta, 260
Species-specific behavior, 195
Space:
  program, 303
  race, 313
Spacing out, 69
"Spirit of unity," 98
Stalinism, 335
State sovereignty, 77, 100
Stress, of peace, 307
Strike, labor, 231, 329, 331
Struggle
  class, 206
  ethics of group, 317
  for existence, 66
  group, 316, 317, 319
  racial, 206
  ritual, 313

social and political, 219, 220
to the death, 66
Sublimation, 248, 303
Sudetenland, 347
Suicidal, 218
  threats, 254
Suicide, 312, 314
Sumatra, 32
Superego, 228, 276, 352
  controls, 224, 225, 230
Sweden, 97, 312
Switzerland, 99

Tantrums, 236, 251, 252
Tarantulas, 37
Tarsier, 36
Taungs discovery, 29
Taiwan, 347
"Technocratic" instrument of violence,
    96
Territorial:
  behavior, 291
  disputes, 347
  species, 291
  rights, 81
Territorialism, group, 292, 294
Territory, 51, 52, 54, 67, 68, 72
Terror, preventive, 333
  see also Psychic terror
Third world, 8
Threat, 47, 49, 53, 74, 254, 292, 307
  behavior, 46, 51
  gestures, 295
Threats of annihilation, 334
Tibet, 32
Tiger, 33
Time-out, 244
"Token economy," 211
Torture, 33
Totalitarian:
  status, 78, 283
  systems, 95
Totalitarianism, 86, 95-100, 277
Tournament fights, 85
Transvaal, 39
Treaty of Trianon, 273
Treaties of Versailles, 273
Trial and error learning, 106
Triumph ceremony, in ducks and
    geese, 53
Trobriand Islands, 88
Truth, 65, 318, 323, 333, 342
Turkeys, 73
Typical intensity, 48, 49

Unconscious motivation, 149
UNESCO, 314, 334
United States of Europe, 277
Upright posture, 31
Utopia, 261, 264
Utopianism, 265

Vacuum response, 61
Vegetarianism, 41
Vengeance, 204
Viet Cong, 351, 352
Vietnam, 343, 347, 348, 351, 354
    North, 341
Vindictiveness, 81
Violence
    abstention from, 319
    against yourself, 318
    frontier community, 219
    in TV, 62
    organized, 86
    reduction of, 318, 321
    sanctioned, 225, 226, 232
    social psychology of, 356
    spontaneous, 78
    systemic, 125
    use of, 319
    warlike, 275
    witnessed, 142
Violent:
    means, 223
    tendencies, 319, 321

War, 8, 16-26, 66, 75-78, 91, 92,
        98-100, 109, 116-121, 211, 219,
        230, 249, 255, 274-275, 280,
        294, 296, 306-307, 310, 321
    abolition of, 312
    against, 258
    antecedents of, 87
    anthropological theory of, 77
    balance of power, 118-121
    biological argument for, 65
    biological determinants of, 78
    Boer, 260
    civil, 21, 118-121, 126-127, 272,
        273, 275
    comradeship of, 307
    constitutional intolerance of, 26
    cyclic patterns in, 115
    declaration of, 118
    function, 265, 266, 308
    future, 127
    generation without, 128

grim humor of, 127
historical, 95, 96
imperial, 118-121
international, 276
in Vietnam, 338, 346
machine, 96
Mexican and Spanish, 272
mirror-image, 345, 353
modern, 96, 258
need for, 307
nuclear, 312, 313, 338, 343
of conquest, 19
. . . ordeal instituted by God, 263
of 1812, 272
party, 261, 266
Peloponnesian, 259
popularity of, 307
preventive, 332
regime, 261
retaliatory and revenge, 128
Revolutionary, 27
Second, 33, 329, 341, 343
symposium on, 77
taxes, 259
total, 95, 96, 100
world without, 309
Warfare, 1, 65, 66, 77, 78, 83, 87, 90,
        91, 93, 95, 109, 209, 327, 335
    against nature, 267
Warlike, 260
Warmongers, 21
Watts, 200
Weapons, 17, 45, 50, 87, 143-146,
        147, 287, 297, 303
    battleship, 268
    bombs, 297, 303
    fangs, 45
    firearms, 143
    fists, 35, 297
    gun, 109, 146, 150, 256, 349
    knife, 256
    nuclear, 307, 311, 312
    rifle, 268
    spear, 297
Whales, 293
Witchcraft, 102
Withdrawal, 291, 292
Wolf, 54, 294
World government, 309

XYY syndrome, 60

Zero-sum game, 344
Zululand, 38